Bunyaviridae

Molecular and Cellular Biology

Edited by

Alexander Plyusnin

Department of Virology
Haartman Institute
University of Helsinki
Helsinki
Finland

and

Richard M. Elliott

Centre for Biomolecular Sciences
School of Biology
University of St Andrews
St Andrews
UK

Caister Academic Press

Caister Academic Press
Norfolk, UK

www.caister.com

British Library Cataloguing-in-Publication Data
A catalogue record for this book is available from the British Library

ISBN: 978-1-904455-90-5

Description or mention of instrumentation, software, or other products in this book does not imply endorsement by the author or publisher. The author and publisher do not assume responsibility for the validity of any products or procedures mentioned or described in this book or for the consequences of their use.

Cover design adapted from Figure 5.1

Printed and bound in Great Britain

Contents

Other Books of Interest

Extremophiles	2012
Bacillus	2012
Microbial Biofilms	2012
Bacterial Glycomics	2012
Non-coding RNAs and Epigenetic Regulation of Gene Expression	2012
Brucella	2012
Molecular Virology and Control of Flaviviruses	2012
Bacterial Pathogenesis	2012
Bunyaviridae	2011
Emerging Trends in Antibacterial Discovery	2011
Epigenetics	2011
Metagenomics	2011
Nitrogen Cycling in Bacteria	2011
Helicobacter pylori	2011
Microbial Bioremediation of Non-metals	2011
Lactic Acid Bacteria and Bifidobacteria	2011
Essentials of Veterinary Parasitology	2011
Hepatitis C: Antiviral Drug Discovery and Development	2011
Streptomyces: Molecular Biology and Biotechnology	2011
Alphaherpesviruses: Molecular Virology	2011
Recent Advances in Plant Virology	2011
Vaccine Design: Innovative Approaches and Novel Strategies	2011
Salmonella: From Genome to Function	2011
Insect Virology	2010
Environmental Microbiology: Current Technology and Water Applications	2010
Sensory Mechanisms in Bacteria: Molecular Aspects of Signal Recognition	2010
Bifidobacteria: Genomics and Molecular Aspects	2010
Molecular Phylogeny of Microorganisms	2010
Nanotechnology in Water Treatment Applications	2010
Iron Uptake and Homeostasis in Microorganisms	2010
Caliciviruses: Molecular and Cellular Virology	2010
Epstein-Barr Virus: Latency and Transformation	2010
Lentiviruses and Macrophages: Molecular and Cellular Interactions	2010
RNA Interference and Viruses: Current Innovations and Future Trends	2010
Retroviruses: Molecular Biology, Genomics and Pathogenesis	2010
Frontiers in Dengue Virus Research	2010
Metagenomics: Theory, Methods and Applications	2010
Microbial Population Genetics	2010
Borrelia: Molecular Biology, Host Interaction and Pathogenesis	2010
Aspergillus: Molecular Biology and Genomics	2010

Contributors

Gjon Blakqori
Centre for Biomolecular Sciences
University of St Andrews
St Andrews
UK

gb41@st-and.ac.uk

Michèle Bouloy
Institut Pasteur
Unité de Génétique Moléculaire des Bunyavirus
Paris
France

mbouloy@pasteur.fr

Anne Bridgen
Freelance Virologist
Croft Dhn
Newtonmore
UK

abridgen1@gmail.com

Richard M. Elliott
Centre for Biomolecular Sciences
School of Biology
University of St Andrews
St Andrews
UK

rme1@st-andrews.ac.uk

Natalia Frias-Staheli
Laboratory of Virology and Infectious Disease
Center for the Study of Hepatitis C
The Rockefeller University
New York, NY
USA

nfrias@rockefeller.edu

Richard Kormelink
Laboratory of Virology
Wageningen University
Wageningen
Netherlands

richard.kormelink@wur.nl

Rafael A. Medina
Department of Microbiology
Mount Sinai School of Medicine
New York, NY
USA

rafael.medina@mssm.edu

Alexander Plyusnin
Department of Virology
Haartman Institute
University of Helsinki
Helsinki
Finland

alexander.plyusnin@helsinki.fi

Tarja Sironen
Department of Virology
Haartman Institute
University of Helsinki
Helsinki
Finland

tarja.sironen@helsinki.fi

Christina F. Spiropoulou
Viral Special Pathogens Branch
Division of High-Consequence Pathogens and Pathology
National Center for Emerging and Zoonotic Infectious Diseases
Centers for Disease Control and Prevention
Atlanta, GA
USA

ccs8@cdc.gov

Antti Vaheri
Department of Virology
Haartman Institute
University of Helsinki
Helsinki
Finland

antti.vaheri@helsinki.fi

Olli Vapalahti
Department of Virology
Haartman Institute
University of Helsinki
Helsinki
Finland

olli.vapalahti@helsinki.fi

Preface

The previous book devoted to bunyaviruses was published more than a decade ago (*The Bunyaviridae*, edited by R.M. Elliott, Plenum Press, New York, 1996), and while it has been, and still is, widely used and cited, considerable data have accumulated to warrant a new treatment of the family. The content of the current monograph illustrates the impressive progress achieved in this field in recent years, not only with the recognition of new viral species and generation of huge amounts of sequence data, but also with mechanistic studies, in particular with regard to counteracting host innate immune responses, through the application of 'reverse genetics' as was forecast in 1996. Hence, we asked contributing authors to concentrate on reviewing newly acquired information.

Two nomenclature changes have been accepted by the scientific community since the 1996 *Bunyaviridae* book. Firstly, the virion glycoproteins are now referred to as Gn and Gc, successfully replacing the former (and potentially misleading) designations G1 and G2. Secondly, the *Bunyavirus* genus has been renamed as *Orthobunyavirus,* allowing the term 'bunyavirus' to be used as a descriptor when considering the family as a whole, and avoiding possible confusions between familial and genus-specific traits. Thus, the family presently comprises five genera: *Orthobunyavirus, Hantavirus, Nairovirus, Phlebovirus* and *Tospovirus.* The first four genera include animal-infecting viruses while viruses in the *Tospovirus* genus infect plants. Further taxonomic changes since 1996 include attempts to provide workable definitions of the species within each genus; like all aspects of virus taxonomy these definitions remain fluid and on occasion controversial.

Descriptions of the properties and replication cycles of viruses in each of the five genera are described in individual chapters (Chapters 1 to 6), while aspects of diagnostics for the animal-infecting bunyaviruses are described in Chapter 7. Chapter 3 is dedicated to genetics and evolution of hantaviruses. The final chapter includes our thoughts on the current challenges and future prospects.

We are grateful to all authors of this monograph who contributed with their excellent expertise in various aspects of bunyavirus research. Sadly, we have to acknowledge the untimely death of one of the pioneers of tospovirus research, Rob Goldbach, who had originally agreed to write that chapter, and thank Richard Kormelink for providing the tospovirus chapter in unhappy circumstances.

Lastly, we would like to thank Annette Griffin of Horizon Scientific Press for her invaluable help.

Alexander Plyusnin
Richard Elliott

Molecular Biology of Orthobunyaviruses

1

Richard M. Elliott and Gjon Blakqori

Abstract

The *Orthobunyavirus* genus comprises over 170 named viruses, including the prototype of the *Bunyaviridae* family, Bunyamwera virus, as well medically significant viruses such as La Crosse and Oropouche viruses, and pathogens of veterinary importance like Akabane and Cache Valley viruses. In this chapter we review the molecular characteristics of orthobunyavirus genomes and encoded proteins, and details of the orthobunyavirus replication cycle. We describe how the development of reverse genetics has paved the way to understand interactions of orthobunyaviruses with the host cell, and in particular the evasion of the host innate immune response. Finally we review the differences in virus replication in vertebrate and invertebrate cells.

Introduction

The *Orthobunyavirus* genus is the largest of the five genera in the *Bunyaviridae* family, containing over 170 named viruses, and includes Bunyamwera virus, the prototype bunyavirus. Bunyamwera virus was isolated by Smithburn and colleagues from *Aedes* mosquitoes in the Semliki Forest in Uganda (Smithburn *et al.*, 1946), and gave its name sequentially to the 'Bunyamwera group' of arboviruses (Casals and Whitman, 1960), the 'Bunyamwera supergroup' (Casals, 1963), the *Bunyaviridae* family and then the *Bunyavirus* genus (Bishop *et al.*, 1980). It was the first bunyavirus whose complete genome sequence was determined (Lees *et al.*, 1986; Elliott, 1989a,b), and it was the first segmented negative-strand virus from which infectious virus was recovered by transfecting cells with cloned cDNAs (Bridgen and Elliott, 1996). As more viruses were classified into the *Bunyaviridae* family and differences between viruses placed in the various genera became more apparent, the frequently used term 'bunyavirus' when referring to viruses in the family as a whole became confusing, and in 2005 the *Bunyavirus* genus was renamed *Orthobunyavirus* (Nichol *et al.*, 2005). Thus nowadays viruses in the *Orthobunyavirus* genus are referred to as 'orthobunyaviruses'. In this chapter we review the molecular biological properties of viruses in the *Orthobunyavirus* genus, acknowledging that molecular dissection is limited to a handful of the viruses classified therein. We concentrate on the more recent research findings, building on material discussed in previous reviews (Pattnaik and Abraham, 1983; Elliott, 1990a–c, 1996, 2008; Bouloy, 1991; Schmaljohn and Hooper, 2001).

Characteristics and classification

The distinguishing characteristics of orthobunyaviruses are (1) the pattern of sizes of the genomic RNA segments (Table 1.1); (2) the pattern of sizes of the viral structural proteins (Table 1.2); and (3) the consensus nucleotide sequences at the 3′ and 5′ termini of the viral RNA segments (Table 1.3) (Nichol *et al.*, 2005). However, very few viruses have been characterized at this level, and like all aspects of viral taxonomy, nomenclature and generic characteristics must be regarded as fluid until more data become available. Extensive serological studies, using complement fixation (CF), neutralization (NT) and haemagglutination inhibition (HI) assays, had grouped the viruses into 18 serogroups (Anopheles A, Anopheles B, Bakau, Bunyamwera, Bwamba, Group C, Capim, California, Gamboa, Guama, Koongol, Minatitlan, Nyando, Olifanstlei, Patois, Simbu, Tete and Turlock), each serogroup containing 'viruses' or 'types' that were antigenically distinct but related. Other isolates of these viruses have been classified as 'subtypes' or 'varieties' using alternative assays or monoclonal antibodies (Bishop, 1990; Calisher, 1996). While this serological classification of viruses proved convenient, it did produce anomalies owing to the propensity for genome segment reassortment (see below), where different serological tests reflected antigenic relationships of proteins encoded by different genome segments e.g. CF is mediated by the S segment-encoded nucleocapsid protein whereas NT and HI antibodies are directed against the M segment encoded glycoproteins (Calisher, 1996).

The International Committee on Taxonomy of Viruses (ICTV) formalized the definition of a viral species in its seventh report as 'a polythetic class of viruses that constitute a replicating lineage and occupy a particular ecological niche' (van Regenmortel *et al.*, 2000). Thus, for the first time the ICTV study groups were obliged to define and create lists of species within each genus. In these authors' view, for many virus families, including the *Bunyaviridae*, there were insufficient data to undertake this task with sufficient rigour. Nevertheless, such a list was devised and thus the *Orthobunyavirus* genus currently comprises 48 species plus three tentative species (Table 1.4) (Nichol *et al.*, 2005). Species are primarily defined by cross-NT and cross-HI relationships; where data are available, they indicate that one species cannot reassort with any other species, and that the amino acid sequences of the N proteins differ by more than 10% (Nichol *et al.*, 2005). A coordinated effort to determine the complete genome nucleotide sequences of all 170+ recognized orthobunyaviruses would help in generating a rational classification scheme.

At least 30 orthobunyaviruses have been associated with human disease (Table 1.4), causing a range of symptoms such as febrile illness, encephalitis, or haemorrhagic fever, while other orthobunyaviruses cause abortion or teratogenic effects in pregnant livestock. Many human infections by orthobunyaviruses result in self-limiting, but sometimes severe, fever. A prime example is Oropouche fever, caused by Oropouche virus (OROV), which

Table 1.1 Pattern of sizes of viral RNA segments in *Bunyaviridae* genera (sizes given in kb)

RNA segment	Genus				
	Orthobunyavirus	*Hantavirus*	*Nairovirus*	*Phlebovirus*	*Tospovirus*
L	6.9	6.5	12.2	6.4	8.9
M	4.5	3.6	4.9	3.5	4.8
S	1.0	1.7	1.7	1.7	2.9
Total	12.4	11.8	18.8	11.6	16.6

Table 1.2 Pattern of sizes of viral structural proteins in *Bunyaviridae* genera (sizes given in kDa)

Protein	Genus				
	Orthobunyavirus	*Hantavirus*	*Nairovirus*	*Phlebovirus*	*Tospovirus*
L	250	250	450	250	330
Gc	110	55	75	65	75
Gn	35	70	35	55–70	45
N	25	50	50	30	30

Table 1.3 Consensus 3′ and 5′ terminal nucleotide sequences of genome RNAs

Orthobunyavirus	3′ UCAUCACAUGA..............................UCGUGUGAUGA 5′
Hantavirus	3′ AUCAUCAUCUG ....AUGAUGAU 5′
Nairovirus	3′ AGAGUUUCU ....AGAAACUCU 5′
Phlebovirus	3′ UGUGUUUC ....GAAACACA 5′
Tospovirus	3′ UCUCGUUA...CUAACGAGA 5′

is the second most common arboviral disease after Dengue fever in Brazil (Pinheiro *et al.*, 1981). Recurrent epidemics have involved tens of thousands of patients, and are economically significant owing to man-hours lost from work. In the past 40 years, Oropouche fever has emerged as a major public health problem in tropical areas of Central and South America (Azevedo *et al.*, 2007). OROV is transmitted by culicoid midges and unusually for an orthobunyavirus infection, humans develop significant viraemia such that uninfected midges can acquire the virus after biting. In the Midwestern USA, La Crosse virus (LACV) causes severe encephalitis and aseptic meningitis in children and young adults (McJunkin *et al.*, 2001). Around 75–100 cases of La Crosse encephalitis requiring hospitalization are reported annually (Calisher, 1994), and up to 57% of these patients require intensive care. More than 10% of the hospitalized patients will have long-lasting neurological deficits (McJunkin *et al.*, 2001), with severe economic and social consequences (Utz *et al.*, 2003). However, since less than 1.5% of LACV infections are clinically apparent, it is estimated that more than 300,000 infections occur annually in the Midwestern USA alone (Calisher, 1994). Related viruses including California encephalitis, Jamestown Canyon, and snowshoe hare have also been associated with human illness in North America while Inkoo and Tahyna viruses cause human disease in Europe. An orthobunyavirus, originally designated Garissa virus, was isolated from human haemorrhagic fever cases during a large disease outbreak in East Africa in 1997 and 1998 (Bowen *et al.*, 2001). The virus was subsequently recognized as being the same as the previously identified Ngari virus, and characterized as a naturally occurring reassortant (see later). Cache Valley virus in the Americas and Akabane (AKAV) virus in Australasia and the Middle East are responsible for fetal death and congenital abnormalities among ruminants. More extensive descriptions of orthobunyavirus disease are provided in reviews by (Gonzales-Scarano *et al.*, 1996; Elliott, 1997; Zeller and Bouloy, 2000; Nichol, 2001; Service, 2001).

Genome structure

All members of the *Bunyaviridae* share a common negative-sense coding strategy of the four structural proteins: the L RNA segment encodes the L protein, the RNA-dependent RNA

Table 1.4 List of species in the *Orthobunyavirus* genus

Species	Virus	Abbreviation	Vector	Distribution	Disease
Acara virus	Acara virus	ACAV	Mosquitoes	SA/NA	–
	Moriche virus	MORV	Mosquitoes	SA	–
Akabane virus	Akabane virus	AKAV	Mosquitoes Culicoid flies	Africa, Asia, Australia	Cattle
	Sabo virus	SABOV	Culicoid flies	Africa	–
	Tinaroo virus	TINV	Culicoid flies	Australia	–
	Yaba-7 virus	Y7V	ND	Africa	–
Alajuela virus	Alajuela virus	ALJV	Mosquitoes	NA	–
	San Juan virus	SJV	Mosquitoes	SA	–
Anopheles A virus	Anopheles A virus	ANAV	Mosquitoes	SA	–
	Las Maloyas virus	LMV	Mosquitoes	SA	–
	Lukuni virus	LUKV	Mosquitoes	SA	–
	Trombetas virus	TRMV	Mosquitoes	SA	–
Anopheles B virus	Anopheles B virus	ANBV	Mosquitoes	SA	–
	Boraceia virus	BORV	Mosquitoes	SA	–
Bakau virus	Bakau virus	KETV	Mosquitoes	Asia	–
	Ketapang virus	BAKV	Mosquitoes	Asia	–
	Nola virus	NOLAV	Mosquitoes	Africa	–
	Tanjong Rabok virus	TRV	ND	Asia	
	Telok Forest virus	TFV	ND	Asia	–
Batama virus	Batama virus	BMAV	ND	Africa	–
Benevides virus	Benevides virus	BVSV	Mosquitoes	SA	–
Bertioga virus	Bertioga virus	BERV	ND	SA	–
	Cananeia virus	CNAV	Mosquitoes	SA	–
	Guaratuba virus	GTBV	Mosquitoes	SA	–
	Itimirim virus	ITIV	ND	SA	–
	Mirim virus	MIRV	Mosquitoes	SA	–
Bimiti virus	Bimiti virus	BIMV	Mosquitoes	SA	–
Botambi virus	Botambi virus	BOTV	Mosquitoes	Africa	–
Bunyamwera virus	Bunyamwera virus	BUNV	Mosquitoes	Africa	Human
	Batai virus	BATV	Mosquitoes	Asia	–
	Birao virus	BIRV	Mosquitoes	Africa	–
	Bozo virus	BOZOV	Mosquitoes	Africa	–
	Cache Valley virus	CVV	Mosquitoes	NA	Sheep, cattle
	Fort Sherman virus	FSV	Mosquitoes	SA	Human
	Germiston virus	GERV	Mosquitoes	Africa	Human
	Iaco virus	IACOV	Mosquitoes	SA	–
	Ilesha virus	ILEV	Mosquitoes	Africa	Human
	Lokern virus	LOKV	Mosquitoes Culicoid flies	NA	–
	Maguari virus	MAGV	Mosquitoes	SA	–
	Mboke virus	MBOV	Mosquitoes	Africa	–
	Ngari virus	NRIV	Mosquitoes	Africa	Human
	Northway virus	NORV	Mosquitoes	NA	–
	Playas virus	PLAV	Mosquitoes	SA	–

Table 1.4 continued

Species	Virus	Abbreviation	Vector	Distribution	Disease
Bunyamwera virus	Potosi virus	POTV	Mosquitoes	NA	–
	Santa Rosa virus	SARV	Mosquitoes	NA	–
	Shokwe virus	SHOV	Mosquitoes	Africa	Human
	Tensaw virus	TENV	Mosquitoes	NA	–
	Tlacotalpan virus	TLAV	Mosquitoes	NA	–
	Tucunduba virus	TUCV	Mosquitoes	SA	–
	Xingu virus	XINV	Mosquitoes	SA	Human
Bushbush virus	Bushbush virus	BSBV	Mosquitoes	SA	–
	Benfica virus	BENV	Mosquitoes	SA	–
	Juan Diaz virus	JDV	ND	NA	–
Bwamba virus	Bwamba virus	BWAV	Mosquitoes	Africa	Human
	Pongola virus	PGAV	Mosquitoes	Africa	Human
California encephalitis virus	California encephalitis virus	CEV	Mosquitoes	NA	Human
	Inkoo virus	INKV	Mosquitoes	Europe	Human
	Jamestown Canyon virus	JCV	Mosquitoes	NA	Human
	Keystone virus	KEYV	Mosquitoes	NA	–
	La Crosse virus	LACV	Mosquitoes	NA	Human
	Lumbo virus	LUMV	Mosquitoes	Africa	Human
	Melao virus	MELV	Mosquitoes	SA	–
	San Angelo virus	SAV	Mosquitoes	NA	–
	Serra do Navio virus	SDNV	Mosquitoes	SA	–
	Snowshoe hare virus	SSHV	Mosquitoes	NA	Human
	South River virus	SORV	Mosquitoes	NA	–
	Tahyna virus	TAHV	Mosquitoes	Europe	Human
	Trivittatus virus	TVTV	Mosquitoes	NA	–
Capim virus	Capim virus	CAPV	Mosquitoes	SA	–
Caraparu virus	Caraparu virus	CARV	Mosquitoes	SA, NA	Human
	Apeu virus	APEUV	Mosquitoes	SA	Human
	Bruconha virus	BRUV	Mosquitoes	SA	–
	Ossa virus	OSSAV	Mosquitoes	NA	Human
	Vinces virus	VINV	Mosquitoes	SA	–
Catu virus	Catu virus	CATUV	Mosquitoes	SA	Human
Estero Real virus	Estero Real virus	ERV	Ticks	NA	–
Gamboa virus	Gamboa virus	GAMV	Mosquitoes	NA	–
	Pueblo Viejo virus	PVV	Mosquitoes	SA	–
Guajara virus	Guajara virus	GJAV	Mosquitoes	SA, NA	–
Guama virus	Guama virus	GMAV	Mosquitoes	SA, NA	Human
	Ananindeua virus	ANUV	Mosquitoes	SA	–
	Moju virus	MOJUV	Mosquitoes	SA	–
	Mahogany Hammock virus	MHV	NND	NA	–
Guaroa virus	Guaroa virus	GROV	Mosquitoes	SA, NA	Human
Kairi virus	Kairi virus	KRIV	Mosquitoes	SA	Horse

Table 1.4 continued

Species	Virus	Abbreviation	Vector	Distribution	Disease
Kaeng Khoi virus	Kaeng Khoi virus	KKV	Nest bugs	Asia	–
Koongol virus	Koongol virus	KOOV	Mosquitoes	Australia	–
	Wongal virus	WONV	Mosquitoes	Australia	–
Madrid virus	Madrid virus	MADV	Mosquitoes	NA	Human
Main Drain virus	Main Drain virus	MDV	Mosquitoes Culicoid flies	NA	Horse
Manzanilla virus	Manzanilla virus	MANV	ND	SA	–
	Buttonwillow virus	BUTV	Culicoid flies	NA	–
	Ingwavuma virus	INGV	Mosquitoes	Africa, Asia	Pig
	Inini virus	INIV	ND	SA	–
	Mermet virus	MERV	Mosquitoes	NA	–
Marituba virus	Marituba virus	MTBV	Mosquitoes	SA	Human
	Gumbo Limbo virus	GLV	Mosquitoes	NA	–
	Murutucu virus	MURV	Mosquitoes	SA	Human
	Nepuyo virus	NEPV	Mosquitoes	SA, NA	Human
	Restan virus	RESV	Mosquitoes	SA	Human
Minatitlan virus	Minatitlan virus	MNTV	ND	NA	–
	Palestina virus	PLSV	Mosquitoes	SA	–
M'Poko virus	M'Poko virus	MPOV	Mosquitoes	Africa	–
	Yaba-1 virus	Y1V	Mosquitoes	Africa	–
Nyando virus	Nyando virus	NDV	Mosquitoes	Africa	Human
	Eret 147 virus	ERETV	Mosquitoes	Africa	–
Olifantsvlei virus	Olifantsvlei virus	OLIV	Mosquitoes	Africa	–
	Bobia virus	BIAV	Mosquitoes	Africa	–
	Dabakala virus	DABV	Mosquitoes	Africa	–
	Oubi virus	OUBIV	Mosquitoes	Africa	–
Oriboca virus	Oriboca virus	ORIV	Mosquitoes	SA	Human
	Itaqui virus	ITQV	Mosquitoes	SA	Human
Oropouche virus	Oropouche virus	OROV	Mosquitoes Culicoid flies	SA	Human
	Facey's Paddock virus	FPV	ND	Australia	–
	Utinga virus	UTIV	ND	SA	–
	Utive virus	UVV	ND	SA	–
Patois virus	Patois virus	PATV	Mosquitoes	NA	–
	Abras virus	ABRV	Mosquitoes	SA	–
	Babahoya virus	BABV	Mosquitoes	SA	–
	Pahayokee virus	PAHV	Mosquitoes	NA	–
	Shark River virus	SRV	Mosquitoes	NA	–
Sathuperi virus	Sathuperi virus	SATV	Mosquitoes Culicoid flies	Africa, Asia	–
	Douglas virus	DOUV	Culicoid flies	Australia	–
Simbu virus	Simbu virus	SIMV	Mosquitoes Culicoid flies	Africa	–

Table 1.4 continued

Species	Virus	Abbreviation	Vector	Distribution	Disease
Shamonda virus	Shamonda virus	SHAV	Culicoid flies	Africa	–
	Peaton virus	PEAV	Culicoid flies	Australia	–
	Sango virus	SANV	Mosquitoes Culicoid flies	Africa	–
Shuni virus	Shuni virus	SHUV	Mosquitoes Culicoid flies	Africa	–
	Aino virus	AINOV	Mosquitoes Culicoid flies	Asia, Australia	–
	Kaikalur virus	KAIV	Mosquitoes	Asia	–
Tacaiuma virus	Tacaiuma virus	TCMV	Mosquitoes	SA	–
	CoAr 1071 virus	CA1071V	Mosquitoes	SA	–
	CoAr 3627 virus	CA3627V	Mosquitoes	SA	–
	Virgin River virus	VRV	Mosquitoes	NA	–
Tete virus	Tete virus	TETEV	ND	Africa	–
	Bahig virus	BAHV	Ticks	Asia, Europe	–
	Matruh virus	MTRV	Ticks	Africa, Europe	–
	Tsuruse virus	TSUV	ND	Asia	–
	Weldona virus	WELV	Culicoid flies	NA	–
Thimiri virus	Thimiri virus	THIV	ND	Africa, Asia	–
Timboteua virus	Timboteua virus	TBTV	Mosquitoes	SA	–
Turlock virus	Turlock virus	TURV	Mosquitoes	NA, SA	–
	Lednice virus	LEDV	Mosquitoes	Europe	–
	Umbre virus	UMBV	Mosquitoes	Asia	–
Wyeomyia virus	Wyeomyia virus	WYOV	Mosquitoes	SA	Human
	Anhembi virus	AMBV	Mosquitoes	SA	–
	BeAr 328208 virus	BAV	Mosquitoes	SA	–
	Macaua virus	MCAV	Mosquitoes	SA	–
	Sororoca virus	SORV	Mosquitoes	SA	–
	Taiassui virus	TAIAV	Mosquitoes	SA	–
Zegla virus	Zegla virus	ZEGV	ND	NA	–

SA, South America; NA, North America; ND, not described.

polymerase; the M segment encodes the two virion glycoproteins Gn and Gc; and the S segment encodes the nucleocapsid protein. Many bunyaviruses also encode non-structural proteins, and there is variation between the genera in the coding strategies for non-structural proteins, both in whether non-structural proteins are present, their position relative to structural proteins and indeed their coding-polarity on the genome segments (Nichol *et al.*, 2005). So far all orthobunyaviruses analysed encode a non-structural protein on the M segment termed NSm that is sandwiched between the coding regions for Gn and Gc to form a polyprotein precursor. On the other hand, although it was thought a non-structural protein on the S segment termed NSs was characteristically encoded in an internal overlapping reading frame to the N coding region, recent nucleotide sequence analysis has shown that several orthobunyaviruses have lost the NSs ORF (Mohamed *et al.*, 2009).

The three genomic RNA segments are only found encapsidated by the N protein in the form of ribonucleoprotein complexes (RNP) and not as naked RNA molecules, indicating that encapsidation occurs concomitant with RNA synthesis. Similarly the antigenomic RNA (also known as viral complementary RNA or replicative-intermediate RNA) is also only found as RNP, and indeed it is RNP, as for all negative-sense viruses, that is the functional template for all viral RNA synthetic events (Elliott, 2005). The encapsidation signal that mediates interaction with the N protein is located within the 5′ of genomic and antigenomic RNA based on studies with BUNV and Jamestown Canyon virus (Osborne and Elliott, 2000; Ogg and Patterson, 2007). N protein binds single-stranded rather than double stranded RNA (as might be found in the panhandle, see below), and evidence suggests that N binds in a structure-, rather than sequence-, specific manner.

Each genome segment can be divided into three regions: 3′ non-coding region (3′NCR), coding region, and 5′ non-coding region (5′NCR). The terminal 11 nt of each genome segment are apparently highly conserved for all orthobunyaviruses, and the 3′ and 5′ ends are invertedly complementary to allow base pairing to form panhandle structures that are characteristic of segmented negative-sense RNA viruses. However, in many cases nucleotide sequences have been determined using oligonucleotide primers based on the consensus sequence derived from the first genomes sequenced, and so that the extent of variation within the very termini could be more extensive than currently appreciated. Immediately after the conserved terminal sequence, short segment-specific sequences are found (only 3 or 4 nt) that are conserved between viruses in the same serogroup (Elliott *et al.*, 1991), and thereafter the sequences are highly variable between the segments, both within one virus and between analogous segments of different viruses. The lengths of the NCR are also highly variable, even within viruses in the same serogroup, whereas the coding regions are absolutely conserved in length, e.g. in Bunyamera serogroup viruses, the N protein ORF comprises 699 nt whereas the S segment 3′NCR varies between 26 and 87 nt, and the 5′NCR between 169 and 226 nt (Dunn *et al.*, 1994). Between viruses in different serogroups the extent of variation is even greater, with some viruses having extremely short NCRs e.g. Oropouche virus (OROV) S segment has a 3′NCR of only 14 nt and a 5′NCR of 44 nt (Table 1.5).

In addition to the N protein encapsidation signal, the NCRs contain the viral promoters, mRNA transcription termination signals and probably the RNP-packaging signals (Dunn *et al.*, 1995; Osborne and Elliott, 2000; Flick and Pettersson, 2001; Weber *et al.*, 2001), and thus it is remarkable the degree of variation, both in length and sequence, observed in the NCRs for these common functions. Experiments using minigenome constructs (genome analogues containing a reporter gene such as luciferase in place of the viral coding sequence; see below), have been used to compare promoter efficiency, and showed that in mammalian cells the M segment promoter was the most active, the L promoter showed intermediate activity and the S promoter had the lowest activity (Barr *et al.*, 2003; Kohl *et al.*, 2004a). When the assay was performed in mosquito cells, the M promoter was again the most active but the L and S segment promoters showed similar activities (Kohl *et al.*, 2004b). Both complementarity and defined nucleotide sequence are required for a functional promoter (Barr and Wertz, 2004; Kohl *et al.*, 2004a), and single point mutations within the promoter region can have marked effects, either abolishing or enhancing promoter activity (Kohl *et al.*, 2003a, 2004a). Creation of recombinant BUN viruses carrying deletions in the S segment NCRs showed that the genomic 5′NCR was more sensitive to mutation than the

Table 1.5 Sequence characteristics of representative orthobunyavirus S RNA segments

		S segment nucleotides			Amino acids		
Serogroup	Virus	Total	5′NCR	3′NCR	N	NSs	Reference
Anopheles A	ANAV	946	52	156	245	No NSs	Mohamed *et al.* (2009)
Anopheles B	ANBV	991	35	213	247	No NSs	Mohamed *et al.* (2009)
Bakau	NOLAV	961	85	174	233	101	Yandoko *et al.* (2007)
Bunyamwera	BUNV	961	85	174	233	101	Elliott (1989b)
Bwamba	BWAV	1096	81	310	234	92	Lambert and Lanciotti (2008)
California	LACV	978	78	192	235	97	Bowen *et al.* (1995)
Group C	ITQV	920	76	139	234	98	Nunes *et al.* (2005)
Nyando	NDOV	959	79	178	233	101	Yandoko *et al.* (2007)
Simbu	OROV	754	44	14	231	91	Saeed *et al.* (2000)
Tete	TETEV	1022	73	172	258	No NSs	Mohamed *et al.* (2009)
Turlock	MPOV	955	85	147	240	95	Yandoko *et al.* (2007)

3′NCR, and that internal 3′NCR sequences (i.e. between the conserved terminal sequence and the coding sequence) played a role in regulating transcription and regulation. Deletion of internal 5′NCR sequences resulted in attenuation of virus replication but did not affect levels of S segment RNAs (Lowen and Elliott, 2005). Attenuation of replication was also achieved by rearranging coding and non-coding sequences: a recombinant virus was created containing wild type M and S segments but a chimeric L segment where the L ORF was flanked by the M segment NCRs (Lowen *et al.*, 2005). This virus was attenuated both in cultured cells and in mice, and suggests a strategy of generating live attenuated viruses with vaccine potential.

Reverse genetics

Reverse genetics systems employ recombinant DNA technology to introduce targeted genetic modifications that can be subsequently assessed *in vitro* or *in vivo*, thus suggesting a causative link between a resulting phenotype and the genetic modification (Kawaoka, 2004). With regard to bunyaviruses, three types of reverse genetics systems have been established that allow reconstitution of the complete virus life cycle or particular aspects thereof. In minireplicon systems, virus-like genome analogues, consisting of a reporter gene that is flanked by viral regulatory sequences, are transcribed and replicated by plasmid-expressed viral proteins in tissue-cultured cells. Overexpression of the viral glycoproteins Gn and Gc leads to the production of virus-like particles (VLPs) that permit packaging and transfer of minireplicons or authentic genomes. Finally, rescue systems are used to generate infectious recombinant viruses from cloned full-length genomes. In practice, reverse genetics systems are used to address a broad range of basic and applied questions in virology, such as probing the function of viral genes and non-coding regulatory sequences, the expression of reporter genes as markers, the use of VLPs or rationally designed attenuated viruses as vaccines and the retargeting of tissue tropism for gene therapy purposes. Furthermore, minireplicon and VLP systems allow the study of certain aspects of high containment viruses in a low containment facility.

Bunyavirus reverse genetics systems are technically demanding, which is reflected by the still limited number of systems that have been established so far. Major obstacles that concern all negative-strand RNA viruses (NSVs) are the mandatory need for the error-prone reverse transcription of the RNA genome into DNA prior to cloning and the expression of at least viral nucleoprotein and polymerase to ensure transcription and replication. Minireplicons and genomes are usually transcribed intracellularly from transfected cDNA plasmids. Most reverse genetics systems for NSVs rely on the bacteriophage T7 DNA-dependent RNA polymerase (T7RNAP) for this purpose (Kawaoka, 2004). Advantages of using T7RNAP are the functionality in a broad spectrum of cell types and its cytoplasmic localization, the cell compartment where most NSVs, including bunyaviruses, replicate. However, transcription initiation by T7RNAP is most efficient with a guanoside (G) as the starting nucleotide and transcription termination is sloppy (Milligan *et al.*, 1987; Kuzmine *et al.*, 2003). Since bunyavirus genomes start with an adenosine (A) or uracil (U), transcripts usually contain one or more non-templated G nucleotides at the 5′ end in order to insure high transcriptional levels. Although the extra nucleotides are tolerated in some rescue systems, they might decrease recognition by the viral polymerase or nucleoprotein and possibly be detrimental when trying to rescue other bunyaviruses. This problem can be overcome by employing posttranscriptional trimming by self-splicing ribozymes to yield authentic 5′ and 3′ ends.

T7RNAP is a non-cellular enzyme and hence has to be delivered in trans. Common ways to provide T7RNAP are infection with recombinant vaccinia viruses, co-transfection of expression plasmids or the usage of transgenic cell lines. Transcription of precise minireplicons and genomes can also be achieved by the cellular, nuclear localized DNA-dependent RNA polymerase I (RNAPI). Originally developed for reverse genetics systems of orthomyxoviruses, the RNAPI-approach has been successfully adopted in bunyavirus minireplicon and rescue systems (Neumann *et al.*, 1999). However, promoter sequences of RNAPI are species-specific, thus restricting the choice of cell lines in which a system can be used. In addition, transcripts of RNAPI driven reverse genetics systems may undergo splicing in the nucleus and minireplicons tend to have lower signal-to-noise ratios.

Minireplicon systems have been established for the orthobunyaviruses BUNV, LACV, and AKAV (Dunn *et al.*, 1995; Blakqori *et al.*, 2003b; Ogawa *et al.*, 2007a). The BUNV system was driven by T7RNAP, whereas LACV and AKAV minireplicons were transcribed by RNAPI. All systems required the co-expression of N and L protein for robust induction of the minireplicon-based reporter, confirming that encapsidated RNA in form of a ribonucleoprotein particle (RNP), but not naked RNA, is the template for transcription and replication by the L polymerase. After reaching optimal levels, increasing amounts of both N and L led to down-regulation of minireplicon activity, particularly in the LACV system (Blakqori *et al.*, 2003b). No such effect was observed in the AKAV system, where reporter activity increased in a dose-dependent manner (Ogawa *et al.*, 2007a). In all orthobunyavirus minireplicon systems reported so far, NSs was found to be inhibitory.

A VLP system for BUNV has been established and assisted to explore the roles of Gn and Gc in particle production (Shi *et al.*, 2007a). Immunization with purified VLPs of Rift Valley fever phlebovirus raised protective antibody titres in mice (Naslund *et al.*, 2009). The immunogenic potential of VLPs of BUNV or other orthobunyaviruses has not yet been tested.

The recovery of recombinant BUNV from transfected cDNA-containing plasmids by Bridgen and Elliott in 1996 was the first rescue of a segmented NSV. The generation of NSVs from transfected plasmids is in general more efficient when the genome is provided in antigenomic (i.e. positive) sense, a phenomenon that is probably associated with the illegitimate assembly of seeding RNPs, which seems to be more difficult with negative sense transcripts (Conzelmann, 1998). In addition, hybridization between unencapsidated genome-sense transcripts and mRNAs derived from helper (expression) plasmids could occur, effectively inactivating both types of RNA. The original BUNV rescue protocol included the transfection of six T7RNAP-driven plasmids, three full-length antigenomic constructs and three correspondent expression plasmids that would provide all six viral proteins (Bridgen and Elliott, 1996). Cells were superinfected with a T7RNAP-expressing vaccinia helper virus. Although it was possible to generate recombinant BUNV with this approach, it was time-consuming and efficiency was low, most likely due to competition by the replicating helper virus, particularly as both viruses acquire envelopes from the Golgi complex. It was later discovered that the full-length antigenomic constructs permit low-level synthesis of viral proteins that was sufficient to support rescue, making the co-transfection of protein-expressing support plasmids obsolete (Lowen *et al.*, 2004). There is obviously no need for high initial protein levels in the BUNV rescue system since additional expression from support plasmids did not increase efficiency over the three-plasmid approach. The need for helper virus infection was circumvented by utilization of a cell line that constitutively expressed T7RNAP (Buchholz *et al.*, 1999). It is now possible to recover severely attenuated viruses that possibly would have not been able to establish with the original approach. The improved three-plasmid based protocol has since been adopted to rescue LACV and the phlebovirus RVFV (Blakqori and Weber, 2005; Ikegami *et al.*, 2006). In case of LACV, rescue was only possible with the three-plasmid approach but not when additional expression-plasmids were co-transfected. This is most likely due to the oppressive effects of L and N overexpression observed in the minireplicon system. Another orthobunyavirus, AKAV, has been recovered by means of RNAPI transcription, as has the phlebovirus RVFV (Ogawa *et al.*, 2007b; Billecocq *et al.*, 2008). In contrast to T7RNAP-driven rescues, RNAPI-dependent systems require transfection of at least five plasmids, three antigenomic full-length constructs plus expression plasmids for N and L, for successful virus recovery. The restriction of the three-plasmid approach to T7RNAP-driven rescue systems is currently not understood but might be explained by the recently discovered cytoplasmic capping activity of eukaryotic cells (Otsuka *et al.*, 2009), by differential transcript levels or inherent differences between T7RNAP and RNAPI transcripts.

Reverse genetics systems are valuable tools that help to improve our understanding of the molecular biology of viruses and provide new ways of vaccine design. Since reverse genetics systems are still only available for a small number of viruses, a major focus is to overcome the apparent technical difficulties in order to establish rescue systems for other important members of the *Orthobunyavirus* genus and the *Bunyaviridae* family.

Viral proteins

L protein

The bunyavirus L protein, the RNA-dependent RNA polymerase (RdRp), so named because of its large size (>200 kDa), is the catalytic enzyme for both viral RNA synthetic events, transcription and replication. Comparison of the available orthobunyavirus L protein sequences with those of other RNA polymerases, enabled definition of four amino acid motifs, called A to D, that comprise the 'polymerase module' conserved in all RNA polymerase (Poch *et al.*, 1989; Delarue *et al.*, 1990). These are located in the middle of the bunyavirus polymerase. Subsequently, a fifth motif upstream of motif A, termed pre-motif A, was identified and shown to be a feature of all RdRp, and a motif downstream of motif D, termed E, was found to be conserved in all segmented virus RdRp (Muller *et al.*, 1994). The critical role for polymerase activity was confirmed by mutagenesis of several of these conserved residues in the BUNV L protein (Jin and Elliott, 1992). Conserved motif C contains the actual catalytic domain for nucleotide polymerization and for all segmented negative-sense RNA viruses contains the conserved tripeptide SDD; interestingly, mutation of SDD to GDN, the corresponding motif in non-segmented negative sense viruses abolished polymerase activity (Jin and Elliott, 1992). The N-terminal domain of orthobunyavirus L proteins contains a conserved PD-(D/E)xK nuclease motif and the isolated domain of LACV L indeed exhibited nuclease activity *in vitro* (Reguera *et al.*, 2010). The nuclease domain of LACV bears structural similarities to that of the PA subunit of the influenza A virus polymerase, suggesting a common origin of the cap-snatching process of segmented NSVs (Reguera *et al.*, 2010). A further conserved region downstream of motif E has been described that is specific for bunyavirus RdRp (Aquino *et al.*, 2003). No functional analyses of these residues has been reported.

Using reverse genetics, Shi and Elliott (2009) were able to insert the V5 epitope tag (Southern *et al.*, 1991) into the BUNV L protein at two positions towards the C terminus and recover viable recombinant viruses. By immunofluorescent staining of infected cells with antibodies to the V5 tag, the L protein was shown to have a cytoplasmic distribution, displaying a punctate to reticular pattern with a concentration in the perinuclear area of the cell. Cell fractionation studies showed that a portion of L was associated with intracellular membranes (in agreement with microscopic investigations reported by Fontana *et al.* (2008; see later). Interaction between L and N proteins was shown by co-immunoprecipitation experiments, and immunostaining with specific antibodies indicated that the majority of L protein colocalized with N in infected cells. The epitope tagged L protein provides a valuable reagent with which to investigate the association of host proteins with L that could modulate its different modes of RNA synthesis.

Gn and Gc proteins

The orthobunyavirus glycoproteins Gn and Gc are type I integral membrane proteins that are modified by N-linked glycosylation (Schmaljohn and Hooper, 2001; Nichol *et al.*, 2005; Shi *et al.*, 2005). They are synthesized from a polyprotein precursor molecule that also includes the non-structural NSm protein sandwiched between them (Fazakerley *et al.*, 1988). The marked size difference between Gn (approx 35 kDa) and Gc (approx 110 kDa) is characteristic for orthobunyaviruses (Table 1.2). The glycoproteins are rich in cysteine residues (about 5% of all residues), most of which are conserved in position between different viruses, and thus enable the proteins to make significant intra- and inter-molecular disulphide bonds. Indeed it is known that Gn and Gc form a heterodimer in the ER (Shi *et al.*, 2004a, 2005). The Golgi targeting-retention signal has been mapped to the transmembrane (TM) domain of Gn, and Gn will locate to the Golgi if expressed on its own. If Gc is expressed alone it remains in the ER, and requires the chaperone-like role of Gn (which can be expressed as a separate gene as well as in polyprotein form) for correct folding and trafficking to the Golgi (Lappin *et al.*, 1994a; Shi *et al.*, 2004a).

Bioinformatic analyses indicated that the Gc protein of La Crosse orthobunyavirus (LACV) shares several sequence and structure motif similarities with the E1 fusion protein I of Sindbis and Semliki Forest alphaviruses, therefore suggesting that Gc functions as a Class II fusion protein (Garry and Garry, 2004). Mutational analyses of a hydrophobic region (residues 1066–1087) of LACV Gc that is well conserved among different orthobunyaviruses (Cortez *et al.*, 2002), supported this notion (Plassmeyer *et al.*, 2005a, 2007). However, as mutations in the CT of BUNV Gn severely affect membrane fusion (Shi *et al.*, 2007a), Gn must also play an important role in the fusion process. In agreement with experimental data for LACV Gc, the N-terminal region of the BUNV Gc ectodomain (up to residue 880) was shown not to be required for membrane fusion though the residues flanking the fusion peptide (residues 1058–1079 in BUNV Gc) are structurally critical for the conformational change in Gc that occurs during the fusion process (Shi and Elliott, 2009). Recombinant LACV carrying mutations in the Gc fusion peptide are impaired for growth in both mammalian and insect cells but are still neurotoxic to neuronal cells (Soldan *et al.*, 2010), implying that the fusion peptide is a determinate of neuroinvasiveness but not necessarily neurovirulence.

BUNV glycoproteins possess three potential N-linked glycosylation sites, one site on Gn (at N residue 60) and two sites on Gc (N624 and 1169). Alignment of the glycoprotein gene sequences of 20 orthobunyaviruses indicates that site 1 (N60 for BUNV) on Gn and site 3 (N1169 for BUNV) on Gc are conserved for all viruses while site 2 (N624 for BUNV) on Gc is conserved among viruses in the Bunyamwera serogroup (Briese *et al.*, 2004b). The strict conservation of N-glycosylation sites suggests that the presence of N-glycans plays an important role in protein function. This was evaluated by generating N-glycosylation-site deficient viruses using reverse genetics, and the results showed that the glycan on Gn (N60) was essential for BUNV replication, whereas the two glycans on Gc were dispensable for virus replication but contributed to efficient virus infection, perhaps in protein folding, intracellular protein trafficking or virus infection. Elimination of the glycosylation site on Gn resulted in severe misfolding of both Gn and Gc proteins, which were then retained in the ER. It is also true for Hantaan virus (Shi and Elliott, 2004), as well as other viruses (Doms *et al.*, 1985; Hebert *et al.*, 1997; Helenius and Aebi, 2001), that N-glycosylation sites near the N terminus of the glycoprotein are the most sensitive to elimination, presumably because these sites first engage the protein folding machinery in the ER to initiate the correct folding process.

Although both N-glycosylation sites on BUNV Gc protein are modified by carbohydrate addition, the fact that mutant Gc proteins lacking either or both sites were able to target to Golgi complex suggests that the N glycans on Gc are more likely involved in some aspect of viral replication rather than promoting protein folding per se (Shi *et al.*, 2005). The mutant viruses lacking either or both N-glycans were able to infect cultured mammalian cells, indicating that N-glycans on Gc appear not to be essential for binding of the virus to the (unknown) host cell receptor. However, the mutant viruses lacking both N-glycans on Gc exhibited a smaller plaque size and generated lower titres than wt BUNV, perhaps indicative that the major effect of N-glycans on the Gc protein is on a late stage of viral assembly or release.

The generation of recombinant BUNV with deletions at the N-terminus of Gc (Shi *et al.*, 2009) and the previous characterization of mutant Maguari orthobunyaviruses (MAGV) that also had deletions at the N-terminus of Gc (Pollitt *et al.*, 2006b) indicate that the

N-terminal domain of the larger glycoprotein is not required for infection of cultured cells. Furthermore, analysis of the BUNV deletion mutants suggested a model for domain structure of Gc (Shi *et al.*, 2009), and opened the possibility for inserting foreign sequences into Gc. To this end recombinant BUNV have been created which express chimeric Gc proteins with either enhanced green fluorescent protein (eGFP) or mCherry fluorescent protein in place of the deleted Gc sequences (Shi *et al.*, 2010). The recombinant viruses are moderately attenuated for growth on mammalian cells but grow to sufficiently high titres to be valuable reagents to explore at the gross level the BUNV replication cycle. Isolated recombinant viruses fluoresce brightly when activated by UV light as seen by conventional microscopy, and have allowed the processes of virus assembly at the Golgi, trafficking of progeny virus through the cytoplasm and egress at the cell membrane to be followed. Interestingly, the eGFP-Gc protein was not stable after repeated passage, and the eGFP sequence as well as amino acids extending into NSm were excised by an in-frame deletion of the viral M segment. In contrast the mCherry-Gc protein was stable for at least ten passages. In the future, advanced microscopical techniques, such as real-time observations and single-particle tracking, will permit detailed dissection of the BUNV replication cycle in cultured cells.

NSm protein

The orthobunyavirus NSm protein is around 16–18 kDa in size and is expressed between Gn and Gc in the M segment encoded polyprotein precursor. The exact termini of NSm have only been determined experimentally for snowshoe hare orthobunyavirus (Fazakerley *et al.*, 1988), and the coordinates for NSm of other viruses inferred by sequence similarity (Elliott, 1990a; Briese *et al.*, 2004a). The C-termini of Gn proteins have the consensus sequence R-X-X-R (Fazakerley *et al.*, 1988; Elliott, 1990a; Nakitare and Elliott, 1993a; Campbell and Huang, 1999; Briese *et al.*, 2004a) so that cleavage is most likely mediated by a furin-like enzyme. The BUNV NSm protein was found to localize to the Golgi in virus-infected cells and when expressed alone from a transfected cDNA clone (Nakitare and Elliott, 1993b; Lappin *et al.*, 1994b). The hydropathy profile and computer-aided domain prediction analysis of NSm revealed that the protein has features common to transmembrane proteins (Elliott, 1990b), and BUNV NSm was shown to be an integral membrane protein by its resistance to sodium carbonate extraction (Fujiki *et al.*, 1982), and its partitioning with the viral glycoproteins, in membranes isolated from infected cells. Furthermore, NSm was shown to interact with the C-terminus of Gn. NSm did not affect the co-translational cleavage of the BUNV glycoprotein precursor, but seemed to be required for the efficient maturation of the two glycoproteins (Shi *et al.*, 2004b). The fact that the Golgi complex houses the bunyavirus factory where virus particles mature and bud (Kuismanen *et al.*, 1982; Salanueva *et al.*, 2003; Novoa *et al.*, 2005) suggested that NSm may be involved in the process of virus assembly and morphogenesis. However, characterization of a mutant of the related Maguari virus indicated that the C-terminal two-thirds of NSm are not essential (Pollitt *et al.*, 2006a). The absolute requirement for the N-terminus of NSm for virus assembly was demonstrated using the VLP assay and the ability or not to recover viable viruses by reverse genetics. It is thought that the N-terminus of the NSm protein acts as the internal signal sequence for NSm itself, but in viruses with extensive deletions of the C-terminus of NSm and the N-terminus of Gc (as in MAGV mutants and recombinant BUNV) this region can also act as the signal sequence for Gc. A viable virus expressing a hybrid NSm-eGFP protein, was created by reverse genetics demonstrating that incorporation of a foreign sequence did not interfere with NSm function or indeed processing of the M segment precursor protein. This correlated with data from Bupp *et al.* (1996), who showed that insertion of a portion of the beta-galactosidae gene into the La Crosse virus NSm coding region did not affect processing or intracellular targeting of the glycoproteins when expressed via vaccinia virus vectors. Other roles for NSm in mosquito cells or in animals remain to be elucidated.

N protein

The bunyavirus nucleocapsid protein, N, is a multifunctional protein that in addition to encapsidating each of the three genome and antigenome segments, interacts with itself to form oligomers, with the viral L (RNA polymerase) protein, with the carboxy terminal regions of either or both of the virion glycoproteins to stabilize the virion structure, and also with host cell proteins like the poly(A)-binding protein (PABP). The N protein is also suggested to play a part in regulating the switch from transcription to replication activities of the viral polymerase (Shi *et al.*, 2005). Data from the available sequenced isolates, representing 11 serogroups, show that the N proteins range from 231 to 258 amino acid residues in length (Table 1.5), though viruses within a serogroup, in general, have the same length of N protein. Phylogenetic studies of N protein sequences, where available, are congruent with antigenic relationships (Bowen *et al.*, 1995; Nunes *et al.*, 2005; Watts *et al.*, 2009).

In infected cells the BUNV N protein forms multimers (Leonard *et al.*, 2005; Mohl and Barr, 2009), and chemical cross-linking studies of deletion mutants indicated that both N- and C-terminal amino acids are involved in self-interaction, suggesting head-to-head and head-to-tail interactions occur (Leonard *et al.*, 2005). Extraction of the RNA contained in synthetic BUNV RNP showed the stoichiometry to be 12 nt per N protein monomer (Mohl and Barr, 2009). A conserved arginine at position 94 was found to be essential for RNA binding, with an arginine at position 40 and lysine at 50 contributing to binding (Walter *et al.*, 2010). From analyses of an extensive series of single amino acid substitution mutants of the BUNV N protein, the central region of the protein was also shown to participate in N–N interaction, and this region overlapped an internal portion of N, slightly more towards the C-terminus, associated with RNP packaging (Eifan and Elliott, 2009).

Eifan and Elliott (2009) also characterized four temperature-sensitive (ts) mutants of BUNV, generated by reverse genetics, with single amino acid changes in the N protein. Previously only a single orthobunyavirus ts mutant mapping to the S genome segment, from over 210 ts mutants reported (Pringle, 1996), had been described, namely Maguari virus ts23 (Pringle and Iroegbu, 1982; Pringle, 1996). The paucity of ts mutants mapping to the S segment has been ascribed to the overlapping coding strategy for N and NSs, and the generation of four mutants in BUNV N protein testifies to the power of reverse genetics. The BUNV mutants could be divided into two groups, those that were defective in antigenome synthesis but not mRNA transcription and those that were replication-defective but transcription competent, and suggests that different domains within N are associated with different RNA synthesis activities. MAG ts23 is also defective in replication at non-permissive temperature (unpublished results of Pritlove and Elliott, quoted in Pringle, 1996). The mechanism by which residues in N modulate template activity requires elucidation; one possibility is that transcription and replication require different cellular cofactors that bind to distinct regions on the N protein.

The generation of crystals of BUNV N protein that diffract to 3.3 angstroms has been reported (Rodgers *et al.*, 2006), and in the near future a 3-D structure of the protein should be available. This will greatly help our understanding of the many functions of the orthobunyavirus N protein.

The non-structural protein NSs

The small genome segment of orthobunyaviruses is transcribed into a single mRNA. While the full-length mRNA encodes the nucleocapsid protein N, most viruses in the genus additionally express a smaller non-structural protein, designated NSs. Translation of NSs is initiated from an alternative overlapping reading frame that starts towards the 5′ end of the mRNA. Orthobunyavirus NSs proteins range from 91 to 109 amino acids in length and are largely hydrophobic. Immunofluorescence analysis with plasmid-expressed FLAG-tagged protein showed that BUNV NSs localizes to the nucleus as well as to yet-to-be-specified granular structures in the cytoplasm (Thomas *et al.*, 2004). Although there is no striking overall resemblance to other known viral or cellular proteins, the carboxy-terminal half of NSs of California serogroup viruses shares some similarity in sequence and function with the proapoptotic *Drosophila* protein Reaper (Colon-Ramos *et al.*, 2003). Intriguingly, a recent genetic analysis revealed that viruses within the Tete and Anopheles A and B serogroups do not encode a NSs protein but in turn have slightly larger N proteins (Mohamed *et al.*, 2009). Whether these enlarged N proteins can functionally substitute NSs is currently unknown.

Studies conducted with mainly BUNV and LACV revealed that NSs has multiple functions in virus replication and host cell interaction. Preventing translation of NSs protein from an S segment expression plasmid enhanced reporter activity in both BUNV and LACV minireplicon systems, whereas overexpression of NSs had the opposite effect, implying that NSs is a negative regulator of replication by the viral polymerase (Weber *et al.*, 2001; Blakqori *et al.*, 2003b). This seems to be a preserved function as NSs proteins of the related orthobunyaviruses Guaroa virus and Lumbo virus also proved inhibitory in the BUNV minireplicon system (Weber *et al.*, 2001). However, NSs was found to be non-essential for virus viability as demonstrated by the rescue of recombinant NSs-deficient BUNV, LACV, and AKAV (Bridgen *et al.*, 2001; Blakqori and Weber, 2005; Ogawa *et al.*, 2007b). The relevance of the inhibitory effect of NSs seen in the minireplicon system is still unclear since lack of NSs only slightly affected BUNV titres in BHK cells and was non-detrimental for LACV in Vero cells. Nonetheless, NSs-deficient BUNV and LACV appeared to be strongly attenuated in type I interferon (IFN)-competent mice and mammalian cell lines, identifying NSs as a virulence factor and antagonist of the interferon system (Bridgen *et al.*, 2001; Blakqori *et al.*, 2007). As discussed in detail later, NSs globally inhibits mRNA transcription by the cellular DNA-dependent RNA polymerase II (RNAPII). This has a strong inhibitory effect on the IFN-beta promoter, thereby preventing the activation of innate immune responses. Besides down-regulating mRNA transcription, NSs of BUNV and California serogroup viruses also interfered with protein synthesis in *in vitro* translation systems (Colon-Ramos *et al.*, 2003; Hart *et al.*, 2008). Thus, NSs contributes to the prominent host cell shut-off that is observed in many mammalian cell lines infected with NSs-expressing orthobunyaviruses by inhibiting both transcription and translation. In spite of the many similar properties shared by BUNV and LACV NSs, they distinctly affect apoptosis. While lack of NSs turned BUNV into an inducer of apoptosis, LACV NSs itself is proapoptotic and, accordingly, a recombinant virus that lacks NSs was less apoptogenic than wild-type virus (Kohl *et al.*, 2003b; Blakqori and Weber, 2005). LACV NSs has also been reported to counteract RNA interference (RNAi) in mammalian cells (Soldan *et al.*, 2005).

Only little is known about the role of NSs in the insect host. Like in IFN-deficient mammalian cells, NSs was non-essential for viability and growth efficiency of NSs-deficient LACV in the *Aedes albopictus*-derived cell line C6/36 (Blakqori *et al.*, 2007). The same study found that NSs was also dispensable for mediating resistance to superinfection with homologous virus and, in contrast to observation in mammalian cells, unable to inhibit RNAi. Furthermore, no degradation of RNAPII or influence on the phosphorylation status of its C-terminal domain was observed (Verbruggen *et al.*, 2010). Based on these findings, NSs has no obvious function in the infection cycle in invertebrate cells. However, NSs proteins of orthobunyaviruses differ considerably in size as well as in amino acid composition and studies to assess a possible role of NSs in the context of infection of whole mosquitoes are still missing. Such studies are highly desirable, particularly in light of recent findings that C6/36 cells have an unusual RNAi response, which is due to a defect in Dicer-2 activity or expression (Scott *et al.*, 2010).

Taken together, orthobunyavirus NSs proteins are involved in multiple virus–host cell interactions in mammalian cells, whereas a functional role in the insect host is still elusive. The predominant biological role of NSs appears to be the inhibition of the type I IFN system of mammalian hosts, a measure that ensures high viraemia that in turn increases the probability of infection of blood-feeding mosquitoes.

Viral replication cycle

The primary interaction with target cells is mediated by one or both of the virion glycoproteins binding to the cellular receptor(s). Receptors have not been identified for any orthobunyavirus, but their wide host range *in vitro* suggests a ubiquitous cell surface molecule(s) is probably involved, at least for initial virus binding. From studies on LACV treated with protease, which preferentially cleaves the larger Gc glycoprotein, it has been suggested that Gc is the major attachment protein for vertebrate cells, whereas Gn may be more important for attachment to mosquito cells (Ludwig *et al.*, 1989, 1991). However, while this is an attractive model, other studies indicate a role for Gc in attachment to both types of cells (Kingsford *et al.*, 1983; Sundin *et al.*, 1987; Hacker *et al.*, 1995; Hacker and Hardy, 1997), and this issue needs to be reinvestigated. The role of Gc, particularly its N-terminal ectodomain, in virus replication is complicated by the observation that two non-ts revertants (R1 and R2) of a ts mutant of Maguari virus (MAGV, also an orthobunyavirus) contain considerable deletions in the N-terminus of Gc (R1 has 239 residues deleted and R2 has 431 amino acids missing; (Elliott *et al.*, 1984; Murphy and Pringle, 1987; Pollitt *et al.*, 2006b). Experiments with OROV have shown that virus entry is via endocytosis in clathrin-coated vesicles, and that blocking acidification of endosomes inhibits virus replication (Santos *et al.*, 2008). In common with many other enveloped viruses bunyaviruses can fuse cells at acidic pH, and for orthobunyaviruses this is accompanied by a conformational change in Gc (Jacoby *et al.*, 1993; Plassmeyer *et al.*, 2005b; Shi *et al.*, 2007b).

Transcription and replication

Like other negative-strand RNA viruses, after uncoating of the virion in the cytoplasm the bunyavirus genome segments are transcribed into viral mRNAs (primary transcription) by the L protein. Transcription is a primer dependent process and short oligonucleotides corresponding to the capped 5′ ends of host mRNAs are used as primers. These are generated by an endonuclease activity that maps to the L protein (Jin and Elliott, 1991, 1993). Transcriptase activity can be demonstrated in detergent-disrupted virion preparations and is stimulated by methylated cap analogues such as mGpppAm, or naturally capped mRNAs such as alfalfa mosaic virus RNA 4 (Bouloy and Hannoun, 1976; Patterson *et al.*, 1984). Thus bunyaviruses, like influenza viruses, employ 'cap-snatching' as a mechanism to prime transcription. In contrast to influenza viruses, which use newly produced mRNAs in the nucleus as primer-donors, bunyaviruses target mature mRNAs in the cytoplasm, and thus their replication is not sensitive to actinomycin-D or alpha-amanitin, again in contrast to influenza viruses (Vezza *et al.*, 1979; Bishop *et al.*, 1983; Patterson and Kolakofsky, 1984; Eshita *et al.*, 1985; Rossier *et al.*, 1986; Kolakofsky *et al.*, 1987; Jin and Elliott, 1993). A consequence of this method of transcription is that bunyaviruses do not need to encode a capping enzyme like the non-segmented negative-strand RNA viruses.

Bunyavirus transcription is unique among negative-sense RNA viruses in that functional viral mRNA synthesis requires on-going protein synthesis (Abraham and Pattnaik, 1983b; Pattnaik and Abraham, 1983; Patterson and Kolakofsky, 1984; Vialat and Bouloy, 1992), a finding that at face value appears incompatible with the presence of a virion transcriptase. In the absence of protein synthesis only short transcripts are produced *in vivo* and *in vitro*. If the *in vitro* reaction is supplemented with rabbit reticulocyte lysate, however, full-length RNAs are synthesized. The translational requirement is not at the level of mRNA initiation, but rather during elongation or, more precisely, to prevent the transcriptase from terminating prematurely. A model to account for these observations proposed that in the absence of ribosome binding and protein translation the nascent mRNA chain and its template can base-pair, thereby preventing progression of the transcriptase enzyme (Raju and Kolakofsky, 1987). Recently this model has been tested using BUNV model templates containing translational stop codons (Barr, 2007), and the results showed that translation of nascent mRNAs prevented transcription termination. Thus orthobunyavirses couple transcription and translation, a feature commonly found in prokaryotes, but rare in eukaryotic cells.

Synthesis of orthobunyavirus mRNAs terminates about 40 nucleotides (M and L segments) or 100 nucleotides (S segment) before the end of the genome RNA template (Patterson and Kolakofsky, 1984; Eshita *et al.*, 1985; Cunningham and Szilagyi, 1987; Bouloy *et al.*, 1990; Jin and Elliott, 1993). Analysis of transcription-termination within the BUNV S segment using the minigenome system identified a 33 nt region within the 5′NCR that signalled mRNA termination (Barr *et al.*, 2006). Within this region a 6 nt sequence (3′-GUCGAC-5′) played a major role. Interestingly, mutation of the 33 nt region to prevent termination revealed a second functional termination sequence downstream. This second sequence contained a pentanucleotide 3′-UGUCG-5′ that overlaps the major upstream signal. The pentanucleotide, or slight variation of it, is conserved in several other orthobunyavirus S segments, suggesting a common mode of transcription-termination on S segments. However, while the BUNV L segment also contains the pentanucleotide sequence in its 5′NCR, the motif could not be identified in the M segment, indicating transcription termination signals are not universally conserved. Orthobunyavirus mRNAs are not demonstrably 3′ polyadenylated (Abraham and Pattnaik, 1983a; Pattnaik and Abraham, 1983; Bouloy *et al.*, 1984), but many have the potential to form stem–loop structures that are probably involved in enhancing translation of viral mRNAs (Blakqori *et al.*, 2009).

In order to replicate the negative-sense genome RNA, a full-length complementary, positive-sense RNA, the antigenome, must be synthesized. This molecule differs from the positive-sense mRNA in that it does not have the 5′ primer extension and the 3′ end extends to the 5′ terminus of the genomic RNA template. Only the viral L and N proteins are required for replication (Dunn *et al.*, 1995), and experiments with minigenome systems show that the NSs protein negatively affects polymerase activity (Weber *et al.*, 2001; Blakqori *et al.*, 2003a). It is not known what controls the switch from transcription to replication, but by analogy with other negative-sense RNA viruses the level of N protein synthesized in the cell probably plays a role. In addition, two events differ between transcription and replication: the requirement of a primer for initiation and recognition of the mRNA termination signal. There may be different forms of the polymerase protein, perhaps modified by association with cellular proteins, that undertake these different modes of RNA production. In the infected cell antigenomes are only found assembled into nucleocapsids, and therefore encapsidation of the nascent antigenome may prevent its interaction with the template, thereby overcoming the mRNA termination signal. These possibilities await detailed investigation.

Virion assembly

Maturation of bunyaviruses characteristically occurs at the smooth membranes in the Golgi apparatus, and hence is inhibited by monensin, a monovalent ionophore, that disrupts transport through the Golgi (Cash, 1982). The viral glycoproteins accumulate in the Golgi complex and cause a progressive vacuolization. Using recombinant expression systems it has been shown that the targeting of the bunyavirus glycoproteins to the Golgi is a property of the glycoproteins alone, and does not require other viral proteins or virus assembly (Nakitare and Elliott, 1993a; Bupp *et al.*, 1996). The Golgi targeting and retention signal resides in Gn, and localization of Gc to the Golgi requires it to heterodimerize in the ER with Gn (Lappin *et al.*, 1994b; Shi *et al.*, 2004a). NSm localizes independently to the Golgi and does not appear to influence localization of the glycoproteins (Nakitare and Elliott, 1993a; Lappin *et al.*, 1994b). Substantial deletions can be made in NSm while retaining virus viability (Pollitt *et al.*, 2006b; Shi *et al.*, 2006), but the N-terminus of NSm is critically required for virus assembly.

Detailed electron microscopic studies of BUNV-infected cells have shown that assembly occurs in a 'viral factory' that is built around the Golgi complex and comprises repetitive units of Golgi stacks, mitochondria, components of the rough ER and virus-derived tubular structures with a globular head (Salanueva *et al.*, 2003). Inside the Golgi stack annular particles (immature precursors) transform into dense intermediate viral particles, and this first maturation step depends on a functional trans-Golgi. These viral forms are hence found within vesicles and their glycoproteins differ in their endo-β-*N*-acetylglucosaminidase H (endo H) sensitivity: the first form is completely endo H sensitive whereas the second immature form has glycoproteins partially resistant to endo H digestion (Novoa *et al.*, 2005). A second maturation occurs when the virus leaves the cell and is visualized as changes to the exterior of the particle resulting in a defined glycoprotein spikes. The glycoproteins on extracellular particles are also partially endo H resistant (Madoff and Lenard, 1982; Novoa *et al.*, 2005).

Analysis of the unique tubular structures in the Golgi stacks of infected cells showed they comprised a tubular domain and globular domain at one end, with the tubes open to the cytoplasm. The viral polymerase and N protein were found mainly in the globular head, and viral RNPs were released from disrupted, purified tubes (Fontana *et al.*, 2008). The NSm protein was also detected in the tubes and in intracellular intermediate viral forms but not in extracellular virions, suggesting NSm possibly plays a role as a scaffold during virion assembly. Viral tubes contain cellular proteins such as actin and myosin I, and both tube assembly and viral morphogenesis were sensitive to drugs that affect actin. Advanced imaging and 3-D reconstructions of infected cells showed that the tubes link cellular organelles e.g. mitochondria, to the Golgi and interact with intracellular viral forms, thus providing a route for cellular factors required for genome replication. Based on these observations a model has been proposed (Fig. 1.1) whereby viral RNPs are transported from sites of replication in the globular domain to the cytoplasm where they condense on Golgi membranes modified by the insertion of Gn and Gc, and promote budding of immature particles into Golgi-derived vesicles. The absence of a matrix-like protein in bunyaviruses, which for other viruses probably functions as a bridge between the nucleocapsid and the glycoproteins, suggests that direct interactions between the bunyavirus nucleocapsid and the carboxy terminal (CT) domains of the glycoproteins may be a prerequisite for budding. Indeed direct evidence for the requirement of the CT of both glycoproteins has been obtained using a virus-like particle assay (see above) (Shi *et al.*, 2007a). After budding into the Golgi cisternae, vesicles

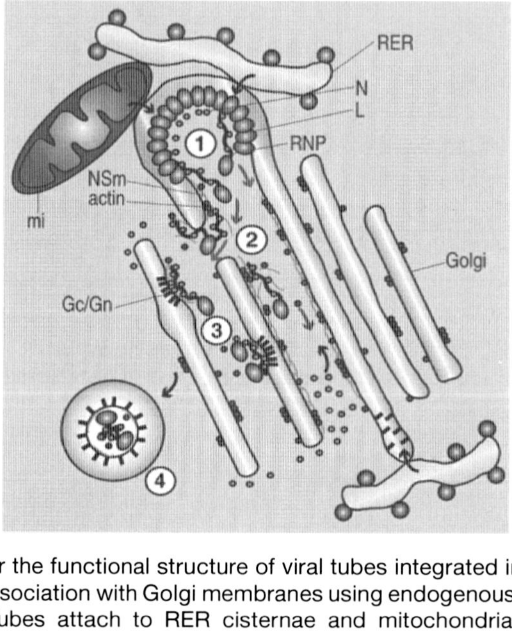

Figure 1.1 Model for the functional structure of viral tubes integrated in the viral factory. Viral tubes assemble in association with Golgi membranes using endogenous actin and polymerized NSm as scaffold. Tubes attach to RER cisternae and mitochondria that provide the cell factors needed for viral replication and assembly. Viral polymerase (L) concentrates in the globular domain. Replicated viral RNA (step 1) gets protected by interaction with N molecules, abundant around Golgi membranes and inside tubes. N molecules enter the tubes through their openings to the cytoplasm. With the help of actomyosin-based motors assembled RNPs then travel towards the cytoplasm (step 2), where they bind to the cytoplasmic domains of viral glycoproteins (Gc/Gn) concentrated in the nearby Golgi membranes. Budding events (step 3) lead to new viral particles that incorporate some amounts of the scaffold NSm protein and actin inside. Assembled immature viral particles are then ready for maturation and secretion (step 4). From Fontana *et al.* (2008).

containing viral particles are transported to the cell surface via the exocytic pathway, eventually releasing their contents to the exterior. Intriguingly, while it is believed that similar assembly and maturation processes occur in both mammalian and insect cells, infected mosquito cells are able to tolerate this gross perturbation of cellular architecture and maintain viability (eventually leading to a persistently infected state) whereas mammalian cells show extreme cytopathic effects and are killed by the infection (Borucki *et al.*, 2002b).

Genome packaging

A key problem faced by all segmented genome viruses is how they package the full genetic complement into a single infectious virion. While this topic has been extensively studied in influenza virus (reviewed by Hutchinson *et al.*, 2009), it is less well understood for orthobunyaviruses. Conserved sequence elements within 20–33 nt of the 5′ and 3′NCRs of the BUNV M segment were identified as important in packaging of a minigenome into virus like particles, and that minigenomes containing the NCRs from the three segments had different packaging efficiencies. Minigenomes were packaged without the requirement for viral coding sequences, and the packaging of a BUNV S segment-derived minigenome was not enhanced by inclusion of N protein coding sequence (Kohl *et al.*, 2006); this apparently contrasts the situation with influenza viruses where packaging signals encompass both non-coding and coding regions. However, much further work is required to elucidate the mechanism of bunyavirus RNP packaging.

Evolutionary potential of bunyaviruses and genome plasticity

Viral RNA-dependent RNA polymerases are known to be error-prone because, unlike DNA polymerases, they lack a proofreading ability. This is a significant factor in the accumulation of mutations leading to genetic heterogeneity ('drift') found in viral RNA populations (Drake and Holland, 1999; Moya *et al.*, 2000). Relatively few wide-scale analyses of genetic drift of orthobunyaviruses have been reported. RNAse T1-generated oligonucleotide fingerprinting studies on field isolates of La Crosse virus indicated that no two isolates were identical (Klimas *et al.*, 1981). On the other hand, direct nucleotide sequencing of the M genome segments of three La Crosse virus isolates obtained from the brains of children who died in 1960, 1978, and 1993 revealed remarkable genetic stability: only 51 nucleotide changes were detected out of 4526 bases over this 33-year period. This suggests that only some genetic variants of the virus are virulent for humans (Huang *et al.*, 1997).

Genome segment reassortment results in more dramatic antigenic changes ('shift'), as is well known in the generation of pandemic influenza viruses (Nelson and Holmes, 2007). Following a mixed infection of a cell with two parental viruses, progeny may be produced containing segments derived from the different parents. For a 3-segmented genome virus, six possible reassortment genotypes are possible (Pringle, 1996). Reassortment is restricted to closely related bunyaviruses, and even then certain combinations of viruses appear genetically incompatible. In a large scale analysis of reassortment between Bunyamwera serogroup viruses, the L and S segments appeared to co-segregate but this co-segregation was lost when reassortant viruses were used as parents (Iroegbu and Pringle, 1981; Pringle and Iroegbu, 1982; Pringle, 1996). However, in other studies on California serogroup virus such genetic linkage was not observed (Gentsch and Bishop, 1976; Gentsch *et al.*, 1980;

Rozhon *et al.*, 1981; Urquidi and Bishop, 1992; Reese *et al.*, 2008). The molecular basis for restriction in reassortment potential remains to be elucidated but presumably involves compatibility of viral protein–protein and protein–RNA interactions.

Orthobunyavirus reassortment also occurs in mosquitoes following ingestion of two viruses in a single blood meal or in separate feedings (mimicking interrupted feeding due to the host's physical reaction to the mosquito) provided the two events are temporally close. If the time between the infections is greater than 2 or 3 days, there is resistance to infection with the second virus (Beaty *et al.*, 1985; Chandler *et al.*, 1990, 1991; Borucki *et al.*, 1999). Reassortment has been documented in salivary glands and ovaries, and reassortant viruses can be transmitted both transovarially and by feeding. Such experimental studies support genomic analysis of bunyaviruses isolated from mosquitoes caught in the field, where evidence for naturally occurring reassortment has been obtained, e.g. among LACV isolates (Reese *et al.*, 2008) or Group C viruses (Nunes *et al.*, 2005). The genetic potential for reassortment of the vast number of known bunyaviruses should not be underestimated when considering the possibility of new and emerging pathogenic viruses. This is strikingly illustrated by the characterization of Ngari virus that has been associated with haemorrhagic fever in East Africa. Genome analysis showed that the L and S segment were derived from Bunyamera virus whereas the M segment was closely related to the M RNA segment of Batai virus, originally isolated in Malaysia (Gerrard *et al.*, 2004; Briese *et al.*, 2006; Yanase *et al.*, 2006). Interestingly, haemorrhagic disease has not been reported for either of the parental viruses.

Analysis of non-ts revertants of MAGVts8, which had ts lesion(s) mapping to the M segment (Murphy and Pringle, 1987; Pollitt *et al.*, 2006b) revealed the remarkable plasticity of the orthobunyavirus genome. Sequence analysis of the ts8 showed three amino acid substitutions towards the M terminus of Gc. Two non-ts revertants were isolated following passage at non-permissive temperature and expressed Gc proteins that were smaller than the parental Gc. Nucleotide sequence determination showed that in these viruses internal in-frame deletions on the M segment had occurred in order to excise the altered amino acids (Pollitt *et al.*, 2006b). As mentioned above, a similar phenomenon occurred upon passage of viruses expressing a chimeric Gc-EGFP gene (Shi *et al.*, 2010). Exploiting reverse genetics it has been possible to rearrange the NCR and coding sequences of the viral genome, while the viruses seem tolerant to gross deletions in the NCR sequences (Lowen *et al.*, 2005; Lowen and Elliott, 2005).

Evasion of the innate immune system

The innate immune system of vertebrates rapidly recognizes and counteracts microbial infections at the cellular level in an attempt to reduce spread whilst a specific immune response can mount. Instead of specific antigens, common building blocks of viruses and bacteria such as nucleic acids and cell wall components, collectively known as pathogen-associated molecular patterns (PAMPs), are perceived as non-self by so-called pattern recognition receptors (PRRs). Based on their subcellular localization, these sensory proteins can be divided into two groups, membrane-bound and cytoplasmic, both of which contain members that recognize virus-derived PAMPs. All PRRs identified to date that are involved in the induction of antiviral responses bind to nucleic acids, which could represent either a viral genome or replication intermediates. Immune cells express Toll-like receptors (TLRs), transmembrane proteins that are located on the cell surface or in internal vesicular structures like endosomes, lysosomes, and phagosomes (Iwasaki and Medzhitov, 2004). They enable the cell to detect pathogens that are either present in the extracellular environment or have been taken up by endocytosis or phagocytosis.

Virus-sensing TLRs are activated by double-stranded RNA (TLR3), single-stranded RNA (TLRs 7 and 8) and unmethylated DNA (TLR 9), respectively. Their importance in inducing antiviral responses has been demonstrated for viruses such as mouse cytomegalovirus (TLR3), vesicular stomatitis virus and influenza virus (TLR7 and 8), and herpes simplex virus 2 (TLR9) (Lund *et al.*, 2003, 2004; Tabeta *et al.*, 2004). Fibroblasts, on the other hand, recognize invading viruses through either cytoplasmic PRRs like the RNA helicases RIG-I (retinoic acid inducible gene I) and MDA5 (melanoma differentiation-associated protein 5), as well as the DNA sensor STING (stimulator of interferon genes), which is localized in the ER membrane (Andrejeva *et al.*, 2004; Yoneyama *et al.*, 2004; Ishikawa *et al.*, 2009). Although similar in structure, consisting of two N-terminal CARD (caspase-recruitment domains) domains and a central RNA-binding helicase domain, RIG-I and MDA5 have evolved to detect distinct RNA species. Recent findings suggest that full activation of RIG-I depends on the simultaneous binding of two separate features of single-stranded viral RNAs: a triphosphate group at the 5′ end and double-stranded regions (Schmidt *et al.*, 2009). By contrast, activation of MDA5 is triggered by large web-like RNA structures (Pichlmair *et al.*, 2009). Loss of RIG-I or MDA5 in knock-out mice showed that most RNA viruses are exclusively recognized by one or the other sensor. While RIG-I was mainly important for recognition of paramyxoviruses, vesicular stomatitis virus and influenza virus, MDA5 was involved in detection of picornaviruses (Kato *et al.*, 2006). However, for some viruses both RIG-I and MDA5 seem to be important for recognition, whereas for others both sensors are redundant for detection (Takeuchi and Akira, 2009). Induction of innate immunity in response to cytoplasmic non-CpG DNA is mediated by STING, an ER-located transmembrane protein that is essential for limiting herpes simplex virus 1 infection in mice (Ishikawa *et al.*, 2009).

Binding of PAMPs by their cognate PRRs activate different signalling cascades that eventually converge to induce the expression of type I (alpha/beta) interferons (IFNs) in order to establish an antiviral state (Randall and Goodbourn, 2008). Interferons are secreted cytokines that act in an autocrine and paracrine manner. Upon binding to IFN receptors that are located on the cell surface, they activate the so-called JAK-STAT signalling cascade that results in the expression of IFN-stimulated genes, many of which have antiviral function. Some ISGs like the protein kinase R (PKR), endoribonclease L (RNase L), and Mx proteins have been studied in detail and their mode of action is well understood, but for the majority of ISGs knowledge about their antiviral activity is scarce. Basal levels of PKR are elevated in response to IFNs and it is activated through binding to dsRNA, a replication by-product that occurs in many virus-infected cells (Randall and Goodbourn, 2008). Activated PKR phosphorylates the eukaryotic translation initiation factor eIF2α, which leads to a general shut-off of protein synthesis (Farrell *et al.*, 1977). Besides PKR dsRNA also activates the 2′–5′ oligoadenylate synthetase (2–5 OAS) which produces short 2′–5′ oligoadenylate molecules that in turn serve as potent activators of RNase L. Once activated, RNase L will indiscriminately cleave viral and cellular single stranded RNAs (ssRNAs), thereby inhibiting both virus replication as well as host cell protein synthesis (Randall and Goodbourn, 2008).

The 5′ triphosphorylated RNA genomes of LACV induce IFN-beta production via recognition by RIG-I (Verbruggen *et al.*, 2010), thereby triggering an antiviral state that can effectively inhibit virus replication. The cytoplasmic MxA protein, a large GTPase of the dynamin superfamily whose expression is strictly IFN-dependent, greatly contributes to the observed antiviral activity (Haller and Kochs, 2002). Lack of MxA enhances LACV virulence in knockout mice, whereas constitutively expressed MxA suppresses LACV replication in transgenic mammalian and insect cells, the latter of which naturally lack this protein (Frese *et al.*, 1996; Miura *et al.*, 2001). Exactly how MxA interferes with LACV infection is not completely understood, however both GTP binding as well as oligomerization through the stalk region are required for its antiviral activity (Pitossi *et al.*, 1993; Gao *et al.*, 2010). MxA sequesters LACV nucleoprotein to the perinuclear area, which presumably makes it unavailable for virus replication (Kochs *et al.*, 2002). The depletion of N particularly affects replication of the longer M and L genome segments while synthesis the S segments remains stable (Frese *et al.*, 1996). Other bunyaviruses that have been shown to be sensitive to MxA include RVFV, Dugbe virus, Hantaan virus, Puumula virus and Crimean–Congo haemorrhagic fever virus (Frese *et al.*, 1996; Bridgen *et al.*, 2004). MxA also inhibits a broad range of other RNA and DNA viruses, but the sensitivity-conferring PAMP they share has yet to be identified.

In order to successfully establish infection, animal viruses have to find ways to evade the IFN system. Over the past decade a great variety of mechanisms have been uncovered about how viruses target the IFN-system at different levels: they prevent production of type I IFNs, interfere with the signalling cascade that leads to expression of IFN-stimulated genes (ISGs) or block the actions of individual antiviral ISGs (Haller *et al.*, 2006). In cell culture, replication of BUNV was more affected when IFN was applied prior to infection and the virus blocked induction of IFN-beta transcription in infected cells, indicating that BUNV is sensitive to type I IFNs but has measures to foreclose their synthesis in the first place (Weber *et al.*, 2002; Streitenfeld *et al.*, 2003). A recombinant BUNV that lacks expression of NSs was more sensitive to the action of type I IFNs and turned out to be an inducer of IFN-beta transcription in cell lines and mice. Complementary, deletion of the type I IFN system restored replication and virulence to levels comparable to wild-type BUNV (Weber *et al.*, 2002). Similar results have been found with LACV (Blakqori *et al.*, 2007). Thus, NSs appears to be the main IFN-antagonist of orthobunyaviruses. Rather than targeting a single step or protein of the IFN-system, NSs inhibits the innate immune response by exerting a general block of mRNA transcription by RNAPII, which has a profound inhibitory effect on the IFN-beta promoter (Thomas *et al.*, 2004). Interestingly, OROV, a member of the Simbu serogroup, was markedly less prone to IFN-pre-treatment than BUNV and other orthobunyaviruses, indicating that it might possess additional inhibitory capacity towards IFN-signalling or one or more ISGs (Livonesi *et al.*, 2007). Besides viral factors that actively inhibit the IFN response, viruses also seem to be passively protected from mosquito saliva *in vivo*, which contains factors that modulate the production of cytokines (Zeidner *et al.*, 1999). Experimental mosquito-mediated infection with LACV was more efficient than syringe-mediated inoculation in mice and chipmunks (Borucki *et al.*, 2002a). These findings may also explain how NSs-deficient orthobunyaviruses that are transmitted to mammals could still propagate in face of an innate immune system.

Innate immunity of insects differs from that of mammals and has only recently become a focus of intense research. Most obviously, insects are devoid of an interferon system and mainly rely on RNAi and Toll-mediated pathways to fight off viruses and other microbial infections (Fragkoudis *et al.*, 2009). Overexpressed LACV NSs protein was able to inhibit RNAi responses in mammalian cells (Soldan *et al.*, 2005). However, a recombinant LACV that lacks NSs expression was still able to establish persistent infection of *Aedes albopictus*-derived C6/36 cells that produced siRNAs directed against the LACV S segment (Blakqori *et al*, 2007). Recent studies found that C6/36 cells are defective in Dicer-2 activity and produce small LACV-specific RNAs that are markedly longer than siRNAs normally found in other insect cell lines (Brackney *et al.*, 2010; Scott *et al.*, 2010). Thus, C6/36 cells apparently do not possess a complete set of small RNA-mediated antiviral pathways. Additional experimental data from other mosquito cell systems is needed to assess whether small RNAs have antiviral function directed against orthobunyaviruses and how those might escape from such activity. Sequence analysis of several orthobunyaviruses of the Tete and Anopheles A and B serogroups revealed that they lack a functional NSs reading frame (Mohamed *et al.*, 2009). Current evidence suggests that, at least with regard to LACV and viruses that lack NSs, even if it were the case that orthobunyaviruses could evade an RNAi response in insects, this is unlikely accomplished by the NSs protein. Further investigation is hampered by the coding strategy of NSs as it is difficult to mutate NSs without affecting the overlapping N protein.

Apoptosis

In multicellular organisms, the deliberate disposal of cells that are damaged, proliferate or no longer needed during development has an important regulatory function in growth and infection control. Various extrinsic and intrinsic stimuli such as oxidative stress, UV radiation, and microbial infection can trigger the programmed cell death cascade known as apoptosis. Once activated, apoptotic signalling-pathways that originate at either the cell surface (e.g. via the Fas receptor), the nucleus (e.g. via p53) or mitochondria (e.g. via Bcl-2 family proteins), will unfold series of events that ultimately lead to the self-destruction of the cell. The main part of the damage is accomplished by proteases of the caspase family and activated nucleases, which are found at the tail end of the signalling pathways (Widlak and Garrard, 2005; Best, 2008). Major signs of apoptosis are the formation of inclusion bodies, fragmentation of chromatin and subsequently the whole nucleus, loss of mitochondrial integrity and changes in cell membrane polarization.

Viruses can be activators or inhibitors of apoptosis. In general, the ability to inhibit apoptosis is seen in viruses that persistently infect cells like human papilloma virus type 16, whereas viruses with a lytic infection cycle often induce cell death, for example influenza A virus (FLUAV) (Hinshaw *et al.*, 1994; Filippova *et al.*, 2007). As mentioned before, orthobunyaviruses can establish either a lytic or persistent infection, depending on the host cell. Insect cells are persistently infected and cell viability is not affected. Consequently, there are no reports about apoptosis induction by any orthobunyavirus in insect cells. By contrast, in mammalian cells orthobunyaviruses generally display a lytic replication cycle that eventually results in cell death. The orthobunyaviruses LACV, OROV, AKAV, and Aino virus have been shown to induce apoptosis (Pekosz *et al.*, 1996; Lim *et al.*, 2005; Acrani *et al.*, 2010). For LACV, this could be partially prevented by overexpression of the anti-apoptotic protein Bcl-2 (Pekosz *et al.*, 1996). A recombinant virus that lacked NSs expression was less apoptogenic than NSs-expressing wt virus and the NSs of LACV and other California serogroup viruses triggered apoptotic events such as mitochondrial cytochrome c release and caspase 3 activation in *Xenopus* oocyte extracts (Colon-Ramos *et al.*, 2003; Blakqori and Weber, 2005). These NSs proteins share sequence as well as functional similarity with the pro-apoptotic *Drosophila* protein Reaper and both Reaper and NSs bind to and inhibit the anti-apoptotic regulator Scythe, which was proposed to be the basis for their pro-apoptotic function (Colon-Ramos *et al.*, 2003).

In contrast to the situation with LACV, deletion of NSs turned an otherwise non-apoptogenic BUNV into an apoptosis inducer (Kohl *et al.*, 2003b). The anti-apoptotic function of BUNV NSs is thought to be due to its ability to block signalling of activated IFN regulatory factor 3 (IRF3), a pathway that not only induces transcription of IFN genes but also activates programmed cell death. The dissimilar capabilities of orthobunyaviruses to either induce or block apoptosis and the different functions of NSs proteins in these processes shows that overall apoptosis does not seem to be an important factor in the life cycle of these viruses. This is supported by the findings that virus yields were not greatly influenced by apoptosis, whether blocked or triggered by a particular virus or being prevented by overexpression of Bcl-2 (Pekosz *et al.*, 1996; Bridgen *et al.*, 2001; Blakqori and Weber, 2005). It rather looks like induction or inhibition of apoptosis is a side effect of LACV and BUNV ability to block protein translation and IFN induction and that any possible detrimental effects are negated by a fast replication cycle.

Lytic versus persistent infection in mammalian and insect hosts

Orthobunyaviruses are arthropod-borne viruses (arboviruses) that are maintained in nature through complex intra- and interspecies infection cycles that involve invertebrate and vertebrate hosts. In both systems, cells are turned into molecular factories that are reprogrammed to produce and release progeny virus. However, infection can result in fundamentally different outcomes for the cell. Mosquitoes and derived cell lines are persistently infected and shed virus particles without signs of compromised viability. Most mammalian cells, on the other hand, succumb rapidly and infection of animals and humans is often associated with mild or severe forms of disease. Defining the molecular basis for the differences displayed in the two host systems would greatly help to understand cellular functions and might identify new targets for antiviral strategies.

In mammalian cells, infection with orthobunyaviruses often leads to damage and death. Transcription and translation, as well as the innate immune response of the host cell are often compromised and apoptosis can be induced. An exception is the differentiated neuronal cell line NT2N, in which LACV establishes a productive infection in the absence of cytopathic effect. By contrast, the undifferentiated progenitor cell line uNT2 goes into apoptosis upon LACV infection, indicating that the differential state of the cell contributes to how the infection affects the host (Pekosz *et al.*, 1996).

In sharp contrast to their aggressive replication strategy in mammalian systems, orthobunyaviruses establish a non-damaging, lifelong persistent infection in their mosquito vectors. Primary infection occurs in the mid-gut (through blood-feeding on small mammals), reproductive tract (by venereal transmission) or egg (by transovarial transmission) but can subsequently disseminate to many tissues like neural ganglia, fat body and salivary glands (Borucki *et al.*, 2002a). The mild nature of infection is of important function for the maintenance of orthobunyaviruses in the insect vector. For instance, it allows the efficient transovarial transmission of LACV from infected female *Aedes triseriatus* mosquitoes to their offspring. Although LACV is actively replicating in the developing egg and larvae, no detrimental effects are observed and females that have been infected by this route even seem to have mating advantages over non-infected mosquitoes (Reese *et al.*, 2009). Much of the knowledge about the persistent infection of insect cells is derived from experiments in cell culture. The cells are derived from mosquito larvae and it is not known from which particular tissue they originated. One of the commonly used mosquito cell line is the *Aedes albopictus*-derived line C6/36, which was originally used to grow flaviviruses (Igarashi, 1978). Persistent infection of orthobunyaviruses in insect cell culture is characterized by oscillating titres (Elliott and Wilkie, 1986). This reflects the situation in nature, where, after a primary infection that is followed by high viraemia, virion production is drastically reduced but will rise again upon blood feeding (Borucki *et al.*, 2002a). This response to nutrition is probably mimicked in cell culture by cell passaging and provision of fresh growth medium. Thus, particle production apparently correlates with increased metabolism and an active cell cycle. Self limitation by encapsidation, and hence silencing, of the nucleocapsid protein-expressing S segment mRNA by LACV as well as the occurrence of L segment-derived defective interfering (DI) RNAs in BUNV-infected C6/36 cells have been discussed to contribute to persistent infection (Hacker *et al.*, 1989; Scallan and Elliott, 1992).

Another phenomenon that is observed in persistently infected insect cells and mosquitoes is the interference with superinfection by homologous and closely related heterologous orthobunyaviruses (Newton *et al.*, 1981; Beaty *et al.*, 1983). There is no protection towards more distinctly related viruses from the same or other genera within the *Bunyaviridae* or other virus families, pointing towards a sequence-specific mechanism as the basis for interference. Sindbis virus-expressed partial and full-length genome sequences of the LACV S segment, but not the M segment, are sufficient to mediate resistance to homologous virus *in vitro* and *in vivo* (Powers *et al.*, 1996). Interestingly, LACV S segment-specific small RNAs were detected in infected cells, suggesting that the protection to superinfection is indeed conferred by RNAi (Blakqori *et al.*, 2007). Small RNA profiling by deep sequencing revealed that the majority of LACV-specific small RNAs produced in C6/36 cells are 24 to 28 nucleotides long, thus longer than the 21 nucleotide-long RNAs normally found in other mosquito and fly cell lines (Brackney *et al.*, 2010; Scott *et al.*, 2010). These longer RNAs found in C6/36 cells result from a lack in Dicer-2 expression or function (Scott *et al.*, 2010). The size distribution of LACV-specific siRNAs indicates that resistance to superinfection in this cell line is likely mediated by a micro RNA or Piwi-interacting RNA pathway. It is currently not known how the persisting virus escapes such activity. However, a recombinant LACV that lacks expression of NSs is equally able to impose interference, arguing against a NSs-dependent mechanism (Blakqori *et al.*, 2007).

Shut-off of host cell protein synthesis and translation of viral proteins

In order to favour their own replication, many fast-growing viruses resort to the drastic measure of partial or complete shut-off of host cell protein production. Interference with expression of host genes can be achieved on the transcriptional as well as the translational level. Recent results suggest that orthobunyaviruses, like viruses of the *Picornaviridae, Flaviviridae* and *Caliciviridae* families, inhibit both, transcription of mRNAs and translation of host cell proteins. While many viruses of the above mentioned families achieve this by encoding proteases that cleave and thereby disable transcription- and/or translation factors, none of the proteins encoded by orthobunyaviruses possess any known proteolytic activity. Consequently, orthobunyaviruses must have evolved a distinct strategy to cause shut-off.

Transcription of eukaryotic mRNAs is conducted by the DNA-dependent RNA polymerase II (RNAPII). An initiation phase that comprises assembly of the transcription complex on the promoter in order to start polymerization of a short transcript that is subjected to 5′ capping and a subsequent elongation phase that is followed by 3′ polyadenylation can be distinguished. Both phases are regulated by phosphorylation events at the C-terminal domain (CTD) of RNAPII, which consists of multiple copies of a hepta-repeat that contains several serine residues (Orphanides and Reinberg, 2002). Phosphorylation of the serine at position five, which is facilitated by the cyclin-dependent kinase 7 (CDK 7), leads to initiation of transcription (Komarnitsky *et al.*, 2000). Subsequent phosphorylation of the serine at position two, catalysed by a complex that consists of CDK 9 and cyclin T, is required for transcription elongation and 3′ end processing (Price, 2000; Ahn *et al.*, 2004). Dephosphorylation of the CTD by various phosphatases enables RNAPII to engage in a new round of transcription. In BUNV-infected cells, phosphorylation at serine two was strongly inhibited, leading to a block of host cell mRNA synthesis (Thomas *et al.*, 2004). Infection with a NSs-deficient BUNV and overexpression studies demonstrated that the NSs protein is sufficient and necessary to block RNAPII activity. BUNV NSs was found to interact with the protein MED8, which is part of the Mediator complex that assists RNAPII in transcription (Leonard *et al.*, 2006). This interaction might lead to the observed block in phosphorylation of the CTD. Furthermore, MED8 is part of a complex that forms an ubiquitin ligase (Brower *et al.*, 2002). It is conceivable that NSs not only inhibits RNAPII activity but also influences its stability in the long run (Leonard *et al.*, 2006). Such instability, specifically of the elongating IIo form of RNAPII, is induced by the NSs of LACV (Verbruggen *et al.*, 2010). NSs triggers the degradation of RNAPII via a proteasome-dependent pathway that bears similarities to the DNA-damage response. The inhibition of RNAPII by NSs of BUNV and LACV affects transcription from both constitutively active promoters like the SV40 immediate early promoter as well as inducible ones like the IFN beta promoter (Thomas *et al.*, 2004; Verbruggen *et al.*, 2010). Thus, NSs contributes to both, host cell protein shut-off and inhibition of the type I-IFN system.

Besides blocking transcription, there is mounting evidence that orthobunyaviruses also downregulate host cell protein synthesis. Efficient translation of most eukaryotic mRNAs depends on the presence of a 5′ 7-methylguanosine cap and a 3′ polyadenosine (poly(A)) tail. In the closed-loop model of translation, the ends of an mRNA molecule are brought into close proximity through binding of a number of proteins. These include, among others, the cap-binding eukaryotic translation initiation factor 4E (eIF4E) at the 5′ end, the poly(A)-binding protein (PABP) at the 3′ end and eukaryotic translation initiation factor 4G (eIF4G) which tethers eIF4E to PABP (Hentze, 1997). The circularisation of mRNAs is thought to enhance translation by increasing the chance of ribosomal subunits that are released from the 3′ end to quickly re-engage at the 5′ end to initiate another round of protein synthesis.

The translation of polyadenylated mRNAs is inhibited in BUNV-infected mammalian cells and the NSs protein of BUNV and California serogroup viruses, as well as nucleocapsid protein N and L polymerase of BUNV, were found to be inhibitory in *Xenopus* oocyte and rabbit reticulocyte *in vitro* translation systems (Colon-Ramos *et al.*, 2003; Hart *et al.*, 2008). Furthermore, infection with BUNV leads to the nuclear redistribution of PABP, which normally shuttles between the cytoplasm and the nucleus but predominantly resides in the former (Blakqori *et al.*, 2009). PABP is crucial for the translation of polyadenylated mRNAs in that it facilitates nuclear-to-cytoplasm transport, confers protection from nucleases and mediates efficient translation by taking part in establishing the closed-loop formation. Nuclear confinement of PABP by BUNV is likely to interfere with the normal translation process of host cell proteins. The nuclear redistribution of PABP was found to be less pronounced with a virus that lacked NSs-expression, indicating that this accessory protein participates in the process (Blakqori *et al.*, 2009). The mechanism by which NSs contributes to the translational block has not yet determined. However, chemical inhibitors of RNAPII can also trigger nuclear confinement of PABP and it is therefore possible that NSs indirectly causes nuclear redistribution by blocking RNAPII activity (Afonina *et al.*, 1998). The fact that a NSs-deficient virus was still able to mediate nuclear confinement of PABP strongly suggests that there are other contributing factors. Pull-down experiments revealed that BUNV N binds to PABP and both proteins co-localize in the cytoplasm prior to nuclear redistribution of PABP (Blakqori *et al.*, 2009). It is currently not known whether this interaction has an inhibitory effect on translation and if N is somehow involved in the nuclear redistribution of PABP. Interestingly, binding of N to PABP was enhanced in the presence of RNA. Messenger RNAs bound by PABP might be the source of the 5′ cap structures needed by the L polymerase to initiate transcription of viral mRNAs. The effects of BUNV on PABP have only recently been discovered and many aspects of this interaction need to be elucidated.

The prominent shut-off exerted by BUNV and other orthobunyaviruses in mammalian cells presents them with a dilemma: although it seems that they do not require the steady support of host cell factors for their own mRNA transcription, they do depend on the translational machinery for protein production and also for replication (Raju and Kolakofsky, 1987). By blocking the usual translation process, orthobunyaviruses have to employ an alternative approach to ensure synthesis of their own proteins. Recent findings with the BUNV S segment suggest that the key around the self-imposed translational inhibition lies within the 3′ end-located untranslated region (UTR). When fused to a reporter, the 3′ UTR was able to stimulate translation almost to the same extend as a poly(A) tail (Blakqori *et al.*, 2009). However, translation efficiency was unaffected by depletion of PABP, indicating that the 3′ UTR does not simply act as a surrogate for a poly(A)-tail. Furthermore, unlike rotavirus, the BUNV S segment 3′ UTR-mediated translation in the absence of vial proteins. It rather looks like the 3′ UTR either binds directly to eIF4G or attracts other, yet unknown cellular proteins to achieve formation of a closed-loop for efficient translation.

Orthobunyaviruses, exemplified by BUNV and LACV, successfully compete with the host cell for the resources that are required for their efficient replication. They interfere with RNAPII and PABP, factors that are pivotal for the transcription and translation of polyadenylated host cell mRNAs. They steal cap-structures from cellular mRNAs in order to prime their own transcripts, thereby inducing instability of cellular mRNAs. They overwhelm the cell with the rapid transcription of their own mRNAs, which escape translational inhibition. The cumulative evidence suggests that orthobunyaviruses have adopted a gene-expression strategy that is similar to that of many positive-stranded RNA viruses, which is characterized by the exertion of a strong host cell shut-off that is circumvented by usage of an alternative translation pathway for the production of viral proteins. There is still much to uncover about how this strategy is implemented.

References

Abraham, G., and Pattnaik, A.K. (1983). Early RNA synthesis in Bunyamwera virus-infected cells. J. Gen. Virol. *64*, 1277–1290.

Acrani, G.O., Gomes, R., Proenca-Modena, J.L., da Silva, A.F., Carminati, P.O., Silva, M.L., Santos, R.I., and Arruda, E. (2010). Apoptosis induced by Oropouche virus infection in HeLa cells is dependent on virus protein expression. Virus Res. *149*, 56–63.

Afonina, E., Stauber, R., and Pavlakis, G.N. (1998). The human poly(A)-binding protein 1 shuttles between the nucleus and the cytoplasm. J. Biol. Chem. *273*, 13015–13021.

Ahn, S.H., Kim, M., and Buratowski, S. (2004). Phosphorylation of serine 2 within the RNA polymerase II C-terminal domain couples transcription and 3′ end processing. Mol. Cell *13*, 67–76.

Andrejeva, J., Childs, K.S., Young, D.F., Carlos, T.S., Stock, N., Goodbourn, S., and Randall, R.E. (2004). The V proteins of paramyxoviruses bind the IFN-inducible RNA helicase, MDA-5, and inhibit its activation of the IFN-beta promoter. Proc. Natl. Acad. Sci. U.S.A. *101*, 17264–17269.

Aquino, V.H., Moreli, M.L., and Moraes Figueiredo, L.T. (2003). Analysis of Oropouche virus L protein amino acid sequence showed the presence of an additional conserved region that could harbour an important role for the polymerase activity. Arch. Virol. *148*, 19–28.

Azevedo, R.S., Nunes, M.R., Chiang, J.O., Bensabath, G., Vasconcelos, H.B., Pinto, A.Y., Martins, L.C., Monteiro, H.A., Rodrigues, S.G., and Vasconcelos, P.F. (2007). Reemergence of Oropouche fever, northern Brazil. Emerg. Infect. Dis. *13*, 912–915.

Barr, J.N. (2007). Bunyavirus mRNA synthesis is coupled to translation to prevent premature transcription termination. RNA *13*, 731–736.

Barr, J.N., and Wertz, G.W. (2004). Bunyamwera bunyavirus RNA synthesis requires cooperation of 3′- and 5′-terminal sequences. J. Virol. *78*, 1129–1138.

Barr, J.N., Elliott, R.M., Dunn, E.F., and Wertz, G.W. (2003). Segment-specific terminal sequences of Bunyamwera bunyavirus regulate genome replication. Virology *311*, 326–338.

Barr, J.N., Rodgers, J.W., and Wertz, G.W. (2006). Identification of the Bunyamwera bunyavirus transcription termination signal. J. Gen. Virol. *87*, 189–198.

Beaty, B.J., Bishop, D.H., Gay, M., and Fuller, F. (1983). Interference between bunyaviruses in *Aedes triseriatus* mosquitoes. Virology *127*, 83–90.

Beaty, B.J., Sundin, D.R., Chandler, L.J., and Bishop, D.H. (1985). Evolution of bunyaviruses by genome reassortment in dually infected mosquitoes (*Aedes triseriatus*). Science *230*, 548–550.

Best, S.M. (2008). Viral subversion of apoptotic enzymes: escape from death row. Annu. Rev. Microbiol. *62*, 171–192.

Billecocq, A., Gauliard, N., Le May, N., Elliott, R.M., Flick, R., and Bouloy, M. (2008). RNA polymerase I-mediated expression of viral RNA for the rescue of infectious virulent and avirulent Rift Valley fever viruses. Virology *378*, 377–384.

Bishop, D.H., Calisher, C.H., Casals, J., Chumakov, M.P., Gaidamovich, S.Y., Hannoun, C., Lvov, D.K., Marshall, I.D., Oker-Blom, N., Pettersson, R.F., *et al.* (1980). *Bunyaviridae*. Intervirology *14*, 125–143.

Bishop, D.H., Gay, M.E., and Matsuoko, Y. (1983). Nonviral heterogeneous sequences are present at the 5′ ends of one species of snowshoe hare bunyavirus S complementary RNA. Nucl. Acids Res. *11*, 6409–6418.

Bishop, D.H.L. (1990). *Bunyaviridae* and their replication. I. *Bunyaviridae*. In Virology, Fields, B.N., ed. (Raven Press, New York), pp. 1155–1173.

Blakqori, G., Delhaye, S., Habjan, M., Blair, C.D., Sanchez-Vargas, I., Olson, K.E., Attarzadeh-Yazdi, G., Fragkoudis, R., Kohl, A., Kalinke, U., *et al.* (2007). La Crosse bunyavirus nonstructural protein NSs serves to suppress the type I interferon system of mammalian hosts. J. Virol. *81*, 4991–4999.

Blakqori, G., and Weber, F. (2005). Efficient cDNA-based rescue of La Crosse bunyaviruses expressing or lacking the nonstructural protein NSs. J. Virol. *79*, 10420–10428.

Blakqori, G., Kochs, G., Haller, O., and Weber, F. (2003). Functional L polymerase of La Crosse virus allows *in vivo* reconstitution of recombinant nucleocapsids. J. Gen. Virol. *84*, 1207–1214.

Blakqori, G., van Knippenberg, I., and Elliott, R.M. (2009). Bunyamwera orthobunyavirus S-segment untranslated regions mediate poly(A) tail-independent translation. J. Virol. *83*, 3637–3646.

Borucki, M.K., Chandler, L.J., Parker, B.M., Blair, C.D., and Beaty, B.J. (1999). Bunyavirus superinfection and segment reassortment in transovarially infected mosquitoes. J. Gen. Virol. *80,* 3173–3179.
Borucki, M.K., Kempf, B.J., Blitvich, B.J., Blair, C.D., and Beaty, B.J. (2002). La Crosse virus: replication in vertebrate and invertebrate hosts. Microbes Infect. *4,* 341–350.
Bouloy, M. (1991). *Bunyaviridae*: genome organization and replication strategies. Adv. Virus Res. *40,* 235–275.
Bouloy, M., and Hannoun, C. (1976). Studies on Lumbo virus replication: I. RNA-dependent RNA polymerase associated with virions. Virology *69,* 258–264.
Bouloy, M., Vialat, P., Girard, M., and Pardigon, N. (1984). A transcript from the S segment of the Germiston bunyavirus is uncapped and codes for the nucleoprotein and a nonstructural protein. J. Virol. *49,* 717–723.
Bouloy, M., Pardigon, N., Vialat, P., Gerbaud, S., and Girard, M. (1990). Characterisation of the 5′ and 3′ ends of viral messenger RNAs isolated from BHK21 cells infected with Germiston virus (Bunyavirus). Virology *75,* 50–58.
Bowen, M.D., Jackson, A.O., Bruns, T.D., Hacker, D.L., and Hardy, J.L. (1995). Determination and comparative analysis of the small RNA genomic sequences of California encephalitis, Jamestown Canyon, Jerry Slough, Melao, Keystone and Trivittatus viruses (*Bunyaviridae*, genus *Bunyavirus*, California serogroup). J. Gen. Virol. *76* 559–572.
Bowen, M.D., Trappier, S.G., Sanchez, A.J., Meyer, R.F., Goldsmith, C.S., Zaki, S.R., Dunster, L.M., Peters, C.J., Ksiazek, T.G., and Nichol, S.T. (2001). A reassortant bunyavirus isolated from acute hemorrhagic fever cases in Kenya and Somalia. Virology *291,* 185–190.
Brackney, D.E., Scott, J.C., Sagawa, F., Woodward, J.E., Miller, N.A., Schilkey, F.D., Mudge, J., Wilusz, J., Olson, K.E., Blair, C.D., *et al.* (2010). C6/36 *Aedes albopictus* cells have a dysfunctional antiviral RNA interference response. PLoS Negl. Trop. Dis. *4,* e856.
Bridgen, A., Dalrymple, D.A., Weber, F., and Elliott, R.M. (2004). Inhibition of Dugbe nairovirus replication by human MxA protein. Virus Res. *99,* 47–50.
Bridgen, A., and Elliott, R.M. (1996). Rescue of a segmented negative-strand RNA virus entirely from cloned complementary DNAs. Proc. Natl. Acad. Sci. U.S.A. *93,* 15400–15404.
Bridgen, A., Weber, F., Fazakerley, J.K., and Elliott, R.M. (2001). Bunyamwera bunyavirus nonstructural protein NSs is a nonessential gene product that contributes to viral pathogenesis. Proc. Natl. Acad. Sci. U.S.A. *98,* 664–669.
Briese, T., Rambaut, A., and Lipkin, W.I. (2004). Analysis of the medium (M) segment sequence of Guaroa virus and its comparison to other orthobunyaviruses. J. Gen. Virol. *85,* 3071–3077.
Briese, T., Bird, B., Kapoor, V., Nichol, S.T., and Lipkin, W.I. (2006). Batai and Ngari viruses: M segment reassortment and association with severe febrile disease outbreaks in East Africa. J. Virol. *80,* 5627–5630.
Brower, C.S., Sato, S., Tomomori-Sato, C., Kamura, T., Pause, A., Stearman, R., Klausner, R.D., Malik, S., Lane, W.S., Sorokina, I., *et al.* (2002). Mammalian mediator subunit mMED8 is an elongin BC-interacting protein that can assemble with Cul2 and Rbx1 to reconstitute a ubiquitin ligase. Proc. Natl. Acad. Sci. U.S.A. *99,* 10353–10358.
Buchholz, U.J., Finke, S., and Conzelmann, K.K. (1999). Generation of bovine respiratory syncytial virus (BRSV) from cDNA: BRSV NS2 is not essential for virus replication in tissue culture, and the human RSV leader region acts as a functional BRSV genome promoter. J. Virol. *73,* 251–259.
Bupp, K., Stillmock, K., and Gonzalez-Scarano, F. (1996). Analysis of the intracellular transport properties of recombinant La Crosse virus glycoproteins. Virology *220,* 485–490.
Calisher, C.H. (1994). Medically important arboviruses of the United States and Canada. Clin. Microbiol. Rev. *7,* 89–116.
Calisher, C.H. (1996). History, classification and taxonomy of viruses in the family *Bunyaviridae*. In The *Bunyaviridae*, Elliott, R.M., ed. (Plenum Press, New York), pp. 1–17.
Campbell, W.P., and Huang, C. (1999). Sequence comparisons of medium RNA segment among 15 California serogroup viruses. Virus Res. *61,* 137–144.
Casals, J. (1963). New developments in the classification of arthropod-borne animal viruses. Ann. Microbiol. *11,* 13–34.
Casals, J., and Whitman, L. (1960). A new antigenic group of arthropod-borne viruses: the Bunyamwera group. Am. J. Hyg. Trop. Med. *9,* 73–77.
Cash, P. (1982). Inhibition of La Crosse virus replication by monensin, monovalent ionophore. J. Gen. Virol. *59,* 193–196.

Chandler, L.J., Beaty, B.J., Baldridge, G.D., Bishop, D.H.L., and Hewlett, M.J. (1990). Heterologous reassortment of bunyaviruses in *Aedes triseriatus* mosquitoes and transovarial and oral transmission of newly evolved genotypes. J. Gen. Virol. *71*, 1045–1050.

Chandler, L.J., Hogge, G., Endres, M., Jacoby, D.R., Nathanson, N., and Beaty, B.J. (1991). Reassortment of La Crosse and Tahyna bunyaviruses in *Aedes triseriatus* mosquitoes. Virus Res. *20*, 181–191.

Colon-Ramos, D.A., Irusta, P.M., Gan, E.C., Olson, M.R., Song, J., Morimoto, R.I., Elliott, R.M., Lombard, M., Hollingsworth, R., Hardwick, J.M., *et al.* (2003). Inhibition of translation and induction of apoptosis by bunyaviral nonstructural proteins bearing sequence similarity to Reaper. Mol. Biol. Cell *14*, 4162–4172.

Conzelmann, K.K. (1998). Nonsegmented negative-strand RNA viruses: genetics and manipulation of viral genomes. Annu. Rev. Genet. *32*, 123–162.

Cortez, I., Aires, A., Pereira, A.M., Goldbach, R., Peters, D., and Kormelink, R. (2002). Genetic organisation of Iris yellow spot virus M RNA: indications for functional homology between the G(C) glycoproteins of tospoviruses and animal-infecting bunyaviruses. Arch. Virol. *147*, 2313–2325.

Cunningham, C., and Szilagyi, J.F. (1987). Viral RNAs synthesized in cells infected with Germiston Bunyavirus. Virology *157*, 431–439.

Delarue, M., Poch, O., Tordo, N., Moras, D., and Argos, P. (1990). An attempt to unify the structure of polymerases. Protein Eng. *3*, 461–467.

Doms, R.W., Helenius, A., and White, J. (1985). Membrane fusion activity of the influenza virus hemagglutinin. The low pH-induced conformational change. J. Biol. Chem. *260*, 2973–2981.

Drake, J.W., and Holland, J.J. (1999). Mutation rates among RNA viruses Proc. Natl. Acad. Sci. U.S.A. *96*, 13910–13913.

Dunn, E.F., Pritlove, D.C., and Elliott, R.M. (1994). The S RNA genome segments of Batai, Cache Valley, Guaroa, Kairi, Lumbo, Main Drain and Northway bunyaviruses: sequence determination and analysis. J. Gen. Virol. *75*, 597–608.

Dunn, E.F., Pritlove, D.C., Jin, H., and Elliott, R.M. (1995). Transcription of a recombinant bunyavirus RNA template by transiently expressed bunyavirus proteins. Virology *211*, 133–143.

Eifan, S.A., and Elliott, R.M. (2009). Mutational analysis of the Bunyamwera orthobunyavirus nucleocapsid protein gene. J. Virol. *83*, 11307–11317.

Elliott, R.M. (1989a). Nucleotide sequence analysis of the large (L) genomic RNA segment of Bunyamwera virus, the prototype of the family *Bunyaviridae*. Virology *173*, 426–436.

Elliott, R.M. (1989b). Nucleotide sequence analysis of the small (S) RNA segment of Bunyamwera virus, the prototype of the family *Bunyaviridae*. J. Gen. Virol. *70*, 1281–1285.

Elliott, R.M. (1990). Molecular biology of the *Bunyaviridae*. J. Gen. Virol. *71*, 501–522.

Elliott, R.M., ed. (1996). The *Bunyaviridae* (Plenum Press, New York).

Elliott, R.M. (1997). Emerging viruses: the *Bunyaviridae*. Mol. Med. *3*, 572–577.

Elliott, R.M. (2005). Negative strand RNA virus replication. In Topley and Wilson's Microbiology and Microbial Infections, Virology, Mahy, B., and ter Meulen, V., eds. (Hodder Arnold, London), pp127–134.

Elliott, R.M. (2008). Bunyaviruses: General features. In Encyclopedia of Virology, Mahy, B.W.J., and van Regenmortel, M.H., eds. (Elsevier, Oxford), pp. 390–399.

Elliott, R.M., and Wilkie, M.L. (1986). Persistent infection of *Aedes albopictus* C6/36 cells by Bunyamwera virus. Virology *150*, 21–32.

Elliott, R.M., Lees, J.F., Watret, G.E., Clark, W., and Pringle, C.R. (1984). Genetic diversity of bunyaviruses and mechanisms of genetic variation. In Mechanisms of Viral Pathogenesis: From Gene to Pathogen, Kohn, A., and Fuchs, P., eds. (Martinus Nijhoff Publishing, Boston), pp. 61–76.

Elliott, R.M., Schmaljohn, C.S., and Collett, M.S. (1991). *Bunyaviridae* genome structure and expression. Curr. Top. Microbiol. Immunol. *169*, 91–142.

Eshita, Y., Ericson, B., Romanowski, V., and Bishop, D.H. (1985). Analyses of the mRNA transcription processes of snowshoe hare bunyavirus S and M RNA species. J. Virol. *55*, 681–689.

Farrell, P.J., Balkow, K., Hunt, T., Jackson, R.J., and Trachsel, H. (1977). Phosphorylation of initiation factor elF-2 and the control of reticulocyte protein synthesis. Cell *11*, 187–200.

Fazakerley, J.K., Gonzalez-Scarano, F., Strickler, J., Dietzschold, B., Karush, F., and Nathanson, N. (1988). Organisation of the middle RNA segment of snowshoe hare bunyavirus. Virology *167*, 422–432.

Filippova, M., Johnson, M.M., Bautista, M., Filippov, V., Fodor, N., Tungteakkhun, S.S., Williams, K., and Duerksen-Hughes, P.J. (2007). The large and small isoforms of human papillomavirus type 16 E6 bind to and differentially affect procaspase 8 stability and activity. J. Virol. *81*, 4116–4129.

Flick, R., and Pettersson, R.F. (2001). Reverse genetics system for Uukuniemi virus (*Bunyaviridae*): RNA polymerase I-catalyzed expression of chimeric viral RNAs. J. Virol. *75*, 1643–1655.

Fontana, J., Lopez-Montero, N., Elliott, R.M., Fernandez, J.J., and Risco, C. (2008). The unique architecture of Bunyamwera virus factories around the Golgi complex. Cell Microbiol. *10*, 2012–2028.

Fragkoudis, R., Attarzadeh-Yazdi, G., Nash, A.A., Fazakerley, J.K., and Kohl, A. (2009). Advances in dissecting mosquito innate immune responses to arbovirus infection. J. Gen. Virol. *90*, 2061–2072.

Frese, M., Kochs, G., Feldmann, H., Hertkorn, C., and Haller, O. (1996). Inhibition of bunyaviruses, phleboviruses, and hantaviruses by human MxA protein. J. Virol. *70*, 915–923.

Fujiki, Y., Hubbard, A.L., Fowler, S., and Lazarow, P.B. (1982). Isolation of intracellular membranes by means of sodium carbonate treatment: application to endoplasmic reticulum. J. Cell Biol. *93*, 97–102.

Gao, S., von der Malsburg, A., Paeschke, S., Behlke, J., Haller, O., Kochs, G., and Daumke, O. (2010). Structural basis of oligomerization in the stalk region of dynamin-like MxA. Nature *465*, 502–506.

Garry, C.E., and Garry, R.F. (2004). Proteomics computational analyses suggest that the carboxyl terminal glycoproteins of Bunyaviruses are class II viral fusion protein (beta-penetrenes). Theor. Biol. Med. Model *1*, 10.

Gentsch, J., and Bishop, D.H. (1976). Recombination and complementation between temperature-sensitive mutants of a Bunyavirus, snowshoe hare virus. J. Virol. *20*, 351–354.

Gentsch, J.R., Rozhon, E.J., Klimas, R.A., El Said, L.H., Shope, R.E., and Bishop, D.H. (1980). Evidence from recombinant bunyavirus studies that the M RNA gene products elicit neutralizing antibodies. Virology *102*, 190–204.

Gerrard, S.R., Li, L., Barrett, A.D., and Nichol, S.T. (2004). Ngari virus is a Bunyamwera virus reassortant that can be associated with large outbreaks of hemorrhagic fever in Africa. J. Virol. *78*, 8922–8926.

Gonzales-Scarano, F., Bupp, K., and Nathanson, N. (1996). Pathogenesis of diseases caused by viruses of the *Bunyavirus* genus. In The *Bunyaviridae*, Elliott, R.M., ed. (Plenum Press, New York), pp. 227–251.

Hacker, D., Raju, R., and Kolakofsky, D. (1989). La Crosse virus nucleocapsid protein controls its own synthesis in mosquito cells by encapsidating its mRNA. J. Virol. *63*, 5166–5174.

Hacker, J.K., and Hardy, J.L. (1997). Adsorptive endocytosis of California encephalitis virus into mosquito and mammalian cells: a role for G1. Virology *235*, 40–47.

Hacker, J.K., Volkman, L.E., and Hardy, J.L. (1995). Requirement for the G1 Protein of California Encephalitis Virus in Infection *in vitro* and *in vivo*. Virology *206*, 945–953.

Haller, O., and Kochs, G. (2002). Interferon-induced Mx proteins: dynamin-like GTPases with antiviral activity. Traffic 3, 710–717.

Haller, O., Kochs, G., and Weber, F. (2006). The interferon response circuit: induction and suppression by pathogenic viruses. Virology *344*, 119–130.

Hart, T.J., Kohl, A., and Elliott, R.M. (2008). Role of the NSs protein in the zoonotic capacity of orthobunyaviruses. Zoonoses Pub. Health *56*, 285–296

Hebert, D.N., Zhang, J.-X., Chen, W., Foellmer, B., and Helenius, A. (1997). The number and location of glycans on influenza hemagglutinin determine folding and association with calnexin and calreticulin. J. Cell Biol. *139*, 613–623.

Helenius, A., and Aebi, M. (2001). Intracellular functions of N-linked glycans. Science *291*, 2364–2369.

Hentze, M.W. (1997). eIF4G: a multipurpose ribosome adapter? Science *275*, 500–501.

Hinshaw, V.S., Olsen, C.W., Dybdahl-Sissoko, N., and Evans, D. (1994). Apoptosis: a mechanism of cell killing by influenza A and B viruses. J. Virol. *68*, 3667–3673.

Huang, C., Thompson, W.H., Karabatsos, N., Grady, L., and Campbell, W.P. (1997). Evidence that fatal human infections with La Crosse virus may be associated with a narrow range of genotypes. Virus Res. *48*, 143–148.

Hutchinson, E.C., von Kirchbach, J.C., Gog, J.R., and Digard, P. (2009). Genome packaging in influenza A virus. J. Gen. Virol. *91*, 313 – 328

Igarashi, A. (1978). Isolation of a Singh's *Aedes albopictus* cell clone sensitive to Dengue and Chikungunya viruses. J. Gen. Virol. *40*, 531–544.

Ikegami, T., Won, S., Peters, C.J., and Makino, S. (2006). Rescue of infectious Rift Valley fever virus entirely from cDNA, analysis of virus lacking the NSs gene, and expression of a foreign gene. J. Virol. *80*, 2933–2940.

Iroegbu, C.U., and Pringle, C.R. (1981). Genetic interactions among viruses of the Bunyamwera complex. J. Virol. 37, 383–394.

Ishikawa, H., Ma, Z., and Barber, G.N. (2009). STING regulates intracellular DNA-mediated, type I interferon-dependent innate immunity. Nature *461*, 788–792.

Iwasaki, A., and Medzhitov, R. (2004). Toll-like receptor control of the adaptive immune responses. Nat. Immunol. *5*, 987–995.

Jacoby, D.R., Cooke, C., Prabakaran, I., Boland, J., Nathanson, N., and Gonzalez-Scarano, F. (1993). Expression of the La Crosse M segment proteins in a recombinant vaccinia expression system mediates pH-dependent cellular fusion. Virology *193*, 993 – 996.

Jin, H., and Elliott, R.M. (1991). Expression of functional Bunyamwera virus L protein by recombinant vaccinia viruses. J. Virol. *65*, 4182–4189.

Jin, H., and Elliott, R.M. (1992). Mutagenesis of the L protein encoded by Bunyamwera virus and production of monoclonal antibodies. J. Gen. Virol. *73*, 2235–2244.

Jin, H., and Elliott, R.M. (1993). Characterisation of Bunyamwera virus S RNA that is transcribed and replicated by the L protein expressed from recombinant vaccinia virus. J. Virol. *67*, 1396–1404.

Kato, H., Takeuchi, O., Sato, S., Yoneyama, M., Yamamoto, M., Matsui, K., Uematsu, S., Jung, A., Kawai, T., Ishii, K.J., *et al.* (2006). Differential roles of MDA5 and RIG-I helicases in the recognition of RNA viruses. Nature *441*, 101–105.

Kawaoka, Y., ed. (2004). Biology of negative strand RNA viruses: the power of reverse genetics. (Springer, Berlin).

Kingsford, L., Ishizawa, L.D., and Hill, D.W. (1983). Biological activities of monoclonal antibodies reactive with antigenic sites mapped on the G1 glycoprotein of La Crosse virus. Virology *129*, 443–455.

Klimas, R.A., Thompson, W.H., Calisher, C.H., Clark, G.G., Grimstad, P.R., and Bishop, D.H. (1981). Genotypic varieties of La Crosse virus isolated from different geographic regions of the continental United States and evidence for a naturally occurring intertypic recombinant La Crosse virus. Am. J. Epidem. *114*, 112–131.

Kochs, G., Janzen, C.H., Hohenberg, H., and Haller, O. (2002). Antivirally active MxA protein sequesters La Crosse virus nucleocapsid protein into perinuclear complexes. Proc. Natl. Acad. Sci. U.S.A. *99*, 3153–3158.

Kohl, A., Bridgen, A., Dunn, E., Barr, J.N., and Elliott, R.M. (2003a). Effects of a point mutation in the 3′ end of the S genome segment of naturally occurring and engineered Bunyamwera viruses. J. Gen. Virol. *84*, 789–793.

Kohl, A., Clayton, R.F., Weber, F., Bridgen, A., Randall, R.E., and Elliott, R.M. (2003b). Bunyamwera virus nonstructural protein NSs counteracts interferon regulatory factor 3-mediated induction of early cell death. J. Virol. *77*, 7999–8008.

Kohl, A., Dunn, E.F., Lowen, A.C., and Elliott, R.M. (2004a). Complementarity, sequence and structural elements within the 3′ and 5′ non-coding regions of the Bunyamwera orthobunyavirus S segment determine promoter strength. J. Gen. Virol. *85*, 3269–3278.

Kohl, A., Hart, T.J., Noonan, C., Royall, E., Roberts, L.O., and Elliott, R.M. (2004b). A Bunyamwera virus minireplicon system in mosquito cells. J. Virol. *78*, 5679–5685.

Kohl, A., Lowen, A.C., Leonard, V.H.J., and Elliott, R.M. (2006). Genetic elements regulating packaging of the Bunyamwera orthobunyavirus genome. J. Gen. Virol. *87*, 177–187.

Kolakofsky, D., Bellocq, C., and Raju, R. (1987). The translational requirement for La Crosse virus S-mRNA synthesis. Cold Spring Harb. Symp. Quant. Biol. *52*, 373–379.

Komarnitsky, P., Cho, E.J., and Buratowski, S. (2000). Different phosphorylated forms of RNA polymerase II and associated mRNA processing factors during transcription. Genes Dev. *14*, 2452–2460.

Kuismanen, E., Hedman, K., Saraste, J., and Pettersson, R.F. (1982). Uukuniemi virus maturation: accumulation of virus particles and viral antigens in the Golgi complex. Mol. Cell Biol. *2*, 1444–1458.

Kuzmine, I., Gottlieb, P.A., and Martin, C.T. (2003). Binding of the priming nucleotide in the initiation of transcription by T7 RNA polymerase. J. Biol. Chem. *278*, 2819–2823.

Lappin, D.F., Nakitare, G.W., Palfreyman, J.W., and Elliott, R.M. (1994). Localisation of Bunyamwera bunyavirus G1 glycoprotein to the Golgi requires association with G2 but not NSm. J. Gen. Virol. *75*, 3441–3451.

Lees, J.F., Pringle, C.R., and Elliott, R.M. (1986). Nucleotide sequence of the Bunyamwera virus M RNA segment: conservation of structural features in the bunyavirus glycoprotein gene product. Virology *148*, 1–14.

Leonard, V.H., Kohl, A., Osborne, J.C., McLees, A., and Elliott, R.M. (2005). Homotypic interaction of Bunyamwera virus nucleocapsid protein. J. Virol. *79*, 13166–13172.

Leonard, V.H., Kohl, A., Hart, T.J., and Elliott, R.M. (2006). Interaction of Bunyamwera orthobunyavirus NSs protein with mediator protein MED8: a mechanism for inhibiting the interferon response. J. Virol. *80*, 9667–9675.

Lim, S.I., Kweon, C.H., Yang, D.K., Tark, D.S., and Kweon, J.H. (2005). Apoptosis in Vero cells infected with Akabane, Aino and Chuzan virus. J. Vet. Sci. *6*, 251–254.

Livonesi, M.C., de Sousa, R.L., Badra, S.J., and Figueiredo, L.T. (2007). *In vitro* and *in vivo* studies of the Interferon-alpha action on distinct Orthobunyaviruses. Antiviral Res. *75*, 121–128.

Lowen, A.C., and Elliott, R.M. (2005). Mutational analyses of the nonconserved sequences in the Bunyamwera orthobunyavirus S segment untranslated regions. J. Virol. *79*, 12861–12870.

Lowen, A.C., Noonan, C., McLees, A., and Elliott, R.M. (2004). Efficient bunyavirus rescue from cloned cDNA. Virology *330*, 493–500.

Lowen, A.C., Boyd, A., Fazakerley, J.K., and Elliott, R.M. (2005). Attenuation of bunyavirus replication by rearrangement of viral coding and noncoding sequences. J. Virol. *79*, 6940–6946.

Ludwig, G.V., Christensen, B.M., Yuill, T.M., and Schultz, K.T. (1989). Enzyme processing of La Crosse virus glycoprotein G1: a bunyavirus-vector infection model. Virology *171*, 108–113.

Ludwig, G.V., Israel, B.A., Christensen, B.M., Yuill, T.M., and Schultz, K.T. (1991). Role of La Crosse virus glycoproteins in attachment of virus to host cells. Virology *181*, 564–571.

Lund, J., Sato, A., Akira, S., Medzhitov, R., and Iwasaki, A. (2003). Toll-like receptor 9-mediated recognition of Herpes simplex virus-2 by plasmacytoid dendritic cells. J. Exp. Med. *198*, 513–520.

Lund, J.M., Alexopoulou, L., Sato, A., Karow, M., Adams, N.C., Gale, N.W., Iwasaki, A., and Flavell, R.A. (2004). Recognition of single-stranded RNA viruses by Toll-like receptor 7. Proc. Natl. Acad. Sci. U.S.A. *101*, 5598–5603.

McJunkin, J.E., de los Reyes, E.C., Irazuzta, J.E., Caceres, M.J., Khan, R.R., Minnich, L.L., Fu, K.D., Lovett, G.D., Tsai, T., and Thompson, A. (2001). La Crosse encephalitis in children. N. Engl. J. Med. *344*, 801–807.

Madoff, D.H., and Lenard, J. (1982). A membrane glycoprotein that accumulates intracellularly: cellular processing of the large glycoprotein of La Crosse virus. Cell *28*, 821–829.

Milligan, J.F., Groebe, D.R., Witherell, G.W., and Uhlenbeck, O.C. (1987). Oligoribonucleotide synthesis using T7 RNA polymerase and synthetic DNA templates. Nucl. Acids Res. *15*, 8783–8798.

Miura, T.A., Carlson, J.O., Beaty, B.J., Bowen, R.A., and Olson, K.E. (2001). Expression of human MxA protein in mosquito cells interferes with La Crosse virus replication. J. Virol. *75*, 3001–3003.

Mohamed, M., McLees, A., and Elliott, R.M. (2009). Viruses in the Anopheles A, Anopheles B, and Tete serogroups in the *Orthobunyavirus* genus (family *Bunyaviridae*) do not encode an NSs protein. J. Virol. *83*, 7612–7618.

Mohl, B.-P., and Barr, J.N. (2009). Investigating the specificity and stoichiometry of RNA binding by the nucleocapsid protein of Bunyamwera virus. RNA *15*, 391–399.

Moya, A., Elena, S.F., Bracho, A., Miralles, R., and Barrio, E. (2000). The evolution of RNA viruses: A population genetics view. Proc. Natl. Acad. Sci. U.S.A. *97*, 6967–6973.

Muller, R., Poch, O., Delarue, M., Bishop, D.H., and Bouloy, M. (1994). Rift Valley fever virus L segment: correction of the sequence and possible functional role of newly identified regions conserved in RNA-dependent polymerases. J. Gen. Virol. 75, 1345–1352.

Murphy, J., and Pringle, C.R. (1987). Bunyavirus mutants: reassortment group assignment and G1 protein variants. In The Biology of Negative Strand Viruses, Mahy, B., and Kolakofsky, D., eds. (Elsevier Science Publishing, Amsterdam), pp. 357–362.

Nakitare, G.W., and Elliott, R.M. (1993). Expression of the Bunyamwera virus M genome segment and intracellular localisation of NSm. Virology *195*, 511–520.

Naslund, J., Lagerqvist, N., Habjan, M., Lundkvist, A., Evander, M., Ahlm, C., Weber, F., and Bucht, G. (2009). Vaccination with virus-like particles protects mice from lethal infection of Rift Valley fever virus. Virology *385*, 409–415.

Nelson, M.I., and Holmes, E.C. (2007). The evolution of epidemic influenza. Nat. Rev. Genet. *8*, 196–205.

Neumann, G., Watanabe, T., Ito, H., Watanabe, S., Goto, H., Gao, P., Hughes, M., Perez, D.R., Donis, R., Hoffmann, E., *et al.* (1999). Generation of influenza A viruses entirely from cloned cDNAs. Proc. Natl. Acad. Sci. U.S.A. *96*, 9345–9350.

Newton, S.E., Short, N.J., and Dalgarno, L. (1981). Bunyamwera virus replication in cultured *Aedes albopictus* (mosquito) cells: establishment of a persistent viral infection. J. Virol. *38*, 1015–1024.

Nichol, S.T. (2001). Bunyaviruses. In Fields Virology, Knipe, D.M., and Howley, P.M., ed. (Lippincott Williams and Wilkins, Philadelphia), pp. 1603–1633.

Nichol, S.T., Beaty, B.J., Elliott, R.M., Goldbach, R., Plyusnin, A., Schmaljohn, C.S., and Tesh, R.B. (2005). *Bunyaviridae*. In Virus Taxonomy, Eighth Report of the International Committee on Taxonomy of Viruses, Fauquet, C.M., Mayo, M.A., Maniloff, J., Desselberger, U., and Ball, L.A., eds. (Academic Press, Amsterdam), pp. 695–716.

Novoa, R.R., Calderita, G., Arranz, R., Fontana, J., Granzow, H., and Risco, C. (2005). Virus factories: associations of cell organelles for viral replication and morphogenesis. Biol. Cell *97*, 147–172.

Nunes, M.R., Travassos da Rosa, A.P., Weaver, S.C., Tesh, R.B., and Vasconcelos, P.F. (2005). Molecular epidemiology of group C viruses (*Bunyaviridae, Orthobunyavirus*) isolated in the Americas. J. Virol. *79*, 10561–10570.

Ogawa, Y., Kato, K., Tohya, Y., and Akashi, H. (2007a). Sequence determination and functional analysis of the Akabane virus (family *Bunyaviridae*) L RNA segment. Arch. Virol. *152*, 971–979.

Ogawa, Y., Sugiura, K., Kato, K., Tohya, Y., and Akashi, H. (2007b). Rescue of Akabane virus (family *Bunyaviridae*) entirely from cloned cDNAs by using RNA polymerase I. J. Gen. Virol. *88*, 3385–3390.

Ogg, M.M., and Patterson, J.L. (2007). RNA binding domain of Jamestown Canyon virus S segment RNAs. J. Virol. *81*, 13754–13760.

Orphanides, G., and Reinberg, D. (2002). A unified theory of gene expression. Cell *108*, 439–451.

Osborne, J.C., and Elliott, R.M. (2000). RNA binding properties of Bunyamwera virus nucleocapsid protein and selective binding to an element in the 5′ terminus of the negative-sense S segment. J. Virol. *74*, 9946–9952.

Otsuka, Y., Kedersha, N.L., and Schoenberg, D.R. (2009). Identification of a cytoplasmic complex that adds a cap onto 5′-monophosphate RNA. Mol. Cell Biol. *29*, 2155–2167.

Patterson, J.L., and Kolakofsky, D. (1984). Characterisation of La Crosse virus small genome transcripts. J. Virol. *49*, 680–685.

Patterson, J.L., Holloway, B., and Kolakofsky, D. (1984). La Crosse virions contain a primer-stimulated RNA polymerase and a methylated cap-dependent endonuclease. J. Virol. *52*, 215–222.

Pattnaik, A.K., and Abraham, G. (1983). Identification of four complementary RNA species in Akabane virus-infected cells. J. Virol. *47*, 452–462.

Pekosz, A., Phillips, J., Pleasure, D., Merry, D., and Gonzalez-Scarano, F. (1996). Induction of apoptosis by La Crosse virus infection and role of neuronal differentiation and human bcl-2 expression in its prevention. J. Virol. *70*, 5329–5335.

Pichlmair, A., Schulz, O., Tan, C.P., Rehwinkel, J., Kato, H., Takeuchi, O., Akira, S., Way, M., Schiavo, G., and Reis e Sousa, C. (2009). Activation of MDA5 requires higher-order RNA structures generated during virus infection. J. Virol. *83*, 10761–10769.

Pinheiro, F.P., Travassos da Rosa, A.P., Travassos da Rosa, J.F., Ishak, R., Freitas, R.B., Gomes, M.L., LeDuc, J.W., and Oliva, O.F. (1981). Oropouche virus. I. A review of clinical, epidemiological, and ecological findings. Am. J. Trop. Med. Hyg. *30*, 149–160.

Pitossi, F., Blank, A., Schroder, A., Schwarz, A., Hussi, P., Schwemmle, M., Pavlovic, J., and Staeheli, P. (1993). A functional GTP-binding motif is necessary for antiviral activity of Mx proteins. J. Virol. *67*, 6726–6732.

Plassmeyer, M.L., Soldan, S.S., Stachelek, K.M., MartÌn-GarcÌa, J., and Gonz·lez-Scarano, F. (2005a). California serogroup Gc (G1) glycoprotein is the principal determinant of pH-dependent cell fusion and entry. Virology *338*, 121–132.

Plassmeyer, M.L., Soldan, S.S., Stachelek, K.M., Martin-Garcia, J., and Gonzalez-Scarano, F. (2005b). California serogroup Gc (G1) glycoprotein is the principal determinant of pH-dependent cell fusion and entry. Virology *338*, 121–132.

Plassmeyer, M.L., Soldan, S.S., Stachelek, K.M., Roth, S.M., Martin-Garcia, J., and Gonzalez-Scarano, F. (2007). Mutagenesis of the La Crosse Virus glycoprotein supports a role for Gc (1066–1087) as the fusion peptide. Virology *358*, 273–282.

Poch, O., Sauvaget, I., Delarue, M., and Tordo, N. (1989). Identification of four conserved motifs among the RNA-dependent polymerase encoding elements. EMBO J. *8*, 3867–3875.

Pollitt, E., Zhao, J., Muscat, P., and Elliott, R.M. (2006). Characterization of Maguari orthobunyavirus mutants suggests the nonstructural protein NSm is not essential for growth in tissue culture. Virology *348*, 224–232.

Powers, A.M., Kamrud, K.I., Olson, K.E., Higgs, S., Carlson, J.O., and Beaty, B.J. (1996). Molecularly engineered resistance to California serogroup virus replication in mosquito cells and mosquitoes. Proc. Natl. Acad. Sci. U.S.A. *93*, 4187–4191.

Price, D.H. (2000). P-TEFb, a cyclin-dependent kinase controlling elongation by RNA polymerase II. Mol. Cell Biol. *20*, 2629–2634.

Pringle, C.R. (1996). Genetics and genome segment reassortment. In The *Bunyaviridae,* Elliott, R.M., ed. (Plenum Press, New York), pp. 189–226.

Pringle, C.R., and Iroegbu, C.U. (1982). Mutant identifying a third recombination group in a bunyavirus. J. Virol. *42*, 873–879.

Raju, R., and Kolakofsky, D. (1987). Translational requirement of La Crosse virus S-mRNA synthesis: *in vivo* studies. J. Virol. *61*, 96–103.

Randall, R.E., and Goodbourn, S. (2008). Interferons and viruses: an interplay between induction, signalling, antiviral responses and virus countermeasures. J. Gen. Virol. *89*, 1–47.

Reese, S.M., Blitvich, B.J., Blair, C.D., Geske, D., Beaty, B.J., and Black, W.C.t. (2008). Potential for La Crosse virus segment reassortment in nature. Virol. J. *5*, 164.

Reese, S.M., Beaty, M.K., Gabitzsch, E.S., Blair, C.D., and Beaty, B.J. (2009). *Aedes triseriatus* females transovarially infected with La Crosse virus mate more efficiently than uninfected mosquitoes. J. Med. Entomol. *46*, 1152–1158.

van Regenmortel, M.H.V., Fauquest, C.M., Bishop, D.H.L., Carstens, E.B., Estes, M.k., Lemon, S.M., Maniloff, J., Mayo, M.A., McGeoch, D.J., Pringle, C.R., *et al.* (2000). Virus taxonomy: classification and nomenclature of viruses. Seventh Report of the International Committee on Taxonomy of Viruses (Academic Press, San Diego).

Reguera, J., Weber, F., and Cusack, S. (2010). *Bunyaviridae* RNA polymerases (L-protein) have an N-terminal, influenza-like endonuclease domain, essential for viral cap-dependent transcription. PLoS Pathog. 6, e1001101.

Rodgers, J.W., Zhou, Q., Green, T.J., Barr, J.N., and Luo, M. (2006). Purification, crystallization and preliminary X-ray crystallographic analysis of the nucleocapsid protein of Bunyamwera virus. Acta Crystall. Sect. F *62*, 361–364.

Rossier, C., Patterson, J., and Kolakofsky, D. (1986). La Crosse virus small genome mRNA is made in the cytoplasm. J. Virol. *58*, 647–650.

Rozhon, E.J., Gensemer, P., Shope, R.E., and Bishop, D.H. (1981). Attenuation of virulence of a bunyavirus involving an L RNA defect and isolation of LAC/SSH/LAC and LAC/SSH/SSH reassortants. Virology *111*, 125–138.

Salanueva, I.J., Novoa, R.R., Cabezas, P., Lopez-Iglesias, C., Carrascosa, J.L., Elliott, R.M., and Risco, C. (2003). Polymorphism and structural maturation of Bunyamwera virus in Golgi and post-Golgi compartments. J. Virol. *77*, 1368–1381.

Santos, R.I.M., Rodrigues, A.H., Silva, M.L.c., Mortara, R.A., Rossi, M.A., Jamur, M.C., Oliver, C., and Arruda, E. (2008). Oropouche virus entry into HeLa cells involves clathrin and requires endosomal acidification. Virus Res. *138*, 139–143.

Scallan, M.F., and Elliott, R.M. (1992). Defective RNAs in mosquito cells persistently infected with Bunyamwera virus. J. Gen. Virol. *73*, 53–60.

Schmaljohn, C.S., and Hooper, J.W. (2001). *Bunyaviridae*: the viruses and their replication In Fields Virology, Knipe, D.M., and Howley, P.M., eds. (Lippincott Williams and Wilkins, Philadelphia), pp. 1581–1602.

Schmidt, A., Schwerd, T., Hamm, W., Hellmuth, J.C., Cui, S., Wenzel, M., Hoffmann, F.S., Michallet, M.C., Besch, R., Hopfner, K.P., *et al.* (2009). 5′-triphosphate RNA requires base-paired structures to activate antiviral signaling via RIG-I. Proc. Natl. Acad. Sci. U.S.A. *106*, 12067–12072.

Scott, J.C., Brackney, D.E., Campbell, C.L., Bondu-Hawkins, V., Hjelle, B., Ebel, G.D., Olson, K.E., and Blair, C.D. (2010). Comparison of dengue virus type 2-specific small RNAs from RNA interference-competent and -incompetent mosquito cells. PLoS Negl. Trop. Dis. *4*, e848.

Service, M.W., ed. (2001). The Encyclopedia of Arthropod-transmitted Infections (CABI Publishing, Wallingford).

Shi, X., and Elliott, R.M. (2004). Analysis of N-linked glycosylation of Hantaan virus glycoproteins and the role of oligosaccharide side chains on protein folding and intracellular trafficking. J. Virol. *78*, 5414–5422.

Shi, X., and Elliott, R.M. (2009). Generation and analysis of recombinant Bunyamwera orthobunyaviruses expressing V5 epitope-tagged L proteins. J. Gen. Virol. *90*, 297–306.

Shi, X., Lappin, D.F., and Elliott, R.M. (2004). Mapping the Golgi targeting and retention signal of Bunyamwera virus glycoproteins. J. Virol. *78*, 10793–10802.

Shi, X., Brauburger, K., and Elliott, R.M. (2005). Role of N-linked glycans on Bunyamwera virus glycoproteins in intracellular trafficking, protein folding, and virus infectivity. J. Virol. *79*, 13725–13734.

Shi, X., Kohl, A., Leonard, V.H.J., Li, P., McLees, A., and Elliott, R.M. (2006). Requirement of the N-terminal region of orthobunyavirus nonstructural protein NSm for virus assembly and morphogenesis. J. Virol. *80*, 8089–8099.

Shi, X., Kohl, A., Li, P., and Elliott, R.M. (2007). Role of the cytoplasmic tail domains of Bunyamwera orthobunyavirus glycoproteins Gn and Gc in virus assembly and morphogenesis. J. Virol. *81*, 10151–10160.

Shi, X., Goli, J., Clark, G., Brauburger, K., and Elliott, R.M. (2009). Functional analysis of the Bunyamwera orthobunyavirus Gc glycoprotein. J. Gen. Virol. *90*, 2483–2492.

Shi, X., van Mierlo, j., French, A., and Elliott, R.M. (2010). Visualizing the replication cycle of Bunyamwera orthobunyavirus expressing fluorescent protein-tagged Gc glycoprotein. J. Virol. *84*, 8460–8469.

Smithburn, K.C., Haddow, A.J., and Mahaffy, A.F. (1946). A neurotropic virus isolated from Aedes mosquitoes caught in the Semliki forest. Am. J. Tropl. Med. *26*, 189–208.

Soldan, S.S., Plassmeyer, M.L., Matukonis, M.K., and Gonzalez-Scarano, F. (2005). La Crosse virus nonstructural protein NSs counteracts the effects of short interfering RNA. J. Virol. 79, 234–244.

Soldan, S.S., Hollidge, B.S., Wagner, V., Weber, F., and Gonz·lez-Scarano, F. (2010). La Crosse virus (LACV) Gc fusion peptide mutants have impaired growth and fusion phenotypes, but remain neurotoxic. Virology *404, 139–147.*

Southern, J.A., Young, D.F., Heaney, F., Baumgartner, W.K., and Randall, R.E. (1991). Identification of an epitope on the P and V proteins of simian virus 5 that distinguishes between two isolates with different biological characteristics. J. Gen. Virol. *72*, 1551–1557.

Streitenfeld, H., Boyd, A., Fazakerley, J.K., Bridgen, A., Elliott, R.M., and Weber, F. (2003). Activation of PKR by Bunyamwera virus is independent of the viral interferon antagonist NSs. J. Virol. *77*, 5507–5511.

Sundin, D.R., Beaty, B.J., Nathanson, N., and Gonzalez-Scarano, F. (1987). A G1 glycoprotein epitope of La Crosse virus: a determinant of infection of *Aedes triseriatus*. Science *235*, 591–593.

Tabeta, K., Georgel, P., Janssen, E., Du, X., Hoebe, K., Crozat, K., Mudd, S., Shamel, L., Sovath, S., Goode, J., *et al.* (2004). Toll-like receptors 9 and 3 as essential components of innate immune defense against mouse cytomegalovirus infection. Proc. Natl. Acad. Sci. U.S.A. *101*, 3516–3521.

Takeuchi, O., and Akira, S. (2009). Innate immunity to virus infection. Immunol. Rev. *227*, 75–86.

Thomas, D., Blakqori, G., Wagner, V., Banholzer, M., Kessler, N., Elliott, R.M., Haller, O., and Weber, F. (2004). Inhibition of RNA polymerase II phosphorylation by a viral interferon antagonist. J. Biol. Chem. *279*, 31471–31477.

Urquidi, V., and Bishop, D.H. (1992). Non-random reassortment between the tripartite RNA genomes of La Crosse and snowshoe hare viruses. J. Gen. Virol. *73*, 2255–2265.

Utz, J.T., Apperson, C.S., MacCormack, J.N., Salyers, M., Dietz, E.J., and McPherson, J.T. (2003). Economic and social impacts of La Crosse encephalitis in western North Carolina. Am. J. Trop. Med. Hyg. *69*, 509–518.

Verbruggen, P., Ruf, M., Blakqori, G., Oeverby, A.K., Heidemann, M., Eick, D., and Weber, F. (2011). Interferon antagonist NSs of La Crosse virus triggers a DNA-damage response-like degradation of transcribing RNA polymerase II. J. Biol. Chem. *286*, 3681–3692.

Vezza, A.C., Repik, P.M., Cash, P., and Bishop, D.H.L. (1979). *In vivo* transcription and protein synthesis capabilities of bunyaviruses: wild-type snowshoe hare virus and its temperature-sensitive group I, group II, and group I/II mutants. J. Virol. *31*, 426–436.

Vialat, P., and Bouloy, M. (1992). Germiston virus transcriptase requires active 40S ribosomal subunits and utilizes capped cellular RNAs. J. Virol. *66*, 685–693.

Walter, C.T., Costa Bento, D.F., Guerrero Alonso, A., and Barr, J.N. (2011). Amino acid changes within the Bunyamwera virus nucleocapsid protein differentially affect mRNA transcription and RNA replication activity of assembled ribonucleoprotein templates. J. Gen. Virol. *92*, 80–84.

Watts, S., Garcia-Maruniak, A., and Maruniak, J. (2009). Tensaw virus genome sequence and its relation to other *Bunyaviridae*. Virus Genes *39*, 309–318.

Weber, F., Dunn, E.F., Bridgen, A., and Elliott, R.M. (2001). The Bunyamwera virus nonstructural protein NSs inhibits viral RNA synthesis in a minireplicon system. Virology *281*, 67–74.

Weber, F., Bridgen, A., Fazakerley, J.K., Streitenfeld, H., Randall, R.E., and Elliott, R.M. (2002). Bunyamwera bunyavirus nonstructural protein NSs counteracts the induction of alpha/beta interferon. J. Virol. *76*, 7949–7955.

Widlak, P., and Garrard, W.T. (2005). Discovery, regulation, and action of the major apoptotic nucleases DFF40/CAD and endonuclease G. J. Cell Biochem. *94*, 1078–1087.

Yanase, T., Kato, T., Yamakawa, M., Takayoshi, K., Nakamura, K., Kokuba, T., and Tsuda, T. (2006). Genetic characterization of Batai virus indicates a genomic reassortment between orthobunyaviruses in nature. Arch. Virol. *151*, 2253–2260.

Yoneyama, M., Kikuchi, M., Natsukawa, T., Shinobu, N., Imaizumi, T., Miyagishi, M., Taira, K., Akira, S., and Fujita, T. (2004). The RNA helicase RIG-I has an essential function in double-stranded RNA-induced innate antiviral responses. Nat. Immunol. *5*, 730–737.

Zeidner, N.S., Higgs, S., Happ, C.M., Beaty, B.J., and Miller, B.R. (1999). Mosquito feeding modulates Th1 and Th2 cytokines in flavivirus susceptible mice: an effect mimicked by injection of sialokinins, but not demonstrated in flavivirus resistant mice. Parasite Immunol. *21*, 35–44.

Zeller, H., and Bouloy, M. (2000). Infections by viruses of the families *Bunyaviridae* and Filoviridae. Rev. Sci. Tech. *19*, 79–91.

Molecular Biology of Hantavirus Infection

2

Christina F. Spiropoulou

Abstract

Hantaviruses are mainly rodent-borne human pathogens with a worldwide distribution and mortality rates among infected individuals of up to 40%. Members of the *Hantavirus* genus can be broadly divided into the Old World hantaviruses, which cause haemorrhagic fever with renal syndrome (HFRS), and the New World hantaviruses, which cause hantavirus pulmonary syndrome (HPS). The virus life cycle with few exceptions is very similar to that of other members of the *Bunyaviridae* family. What is really unique to hantaviruses is the unusual mechanisms by which they cause severe disease in humans and establish persistence in their rodent reservoir hosts. In this chapter, features of the hantavirus life cycle and the virus host cell interaction are highlighted in an attempt to understand what makes hantaviruses among the most important human pathogens of the family *Bunyaviridae*.

Introduction

Hantaviruses are mainly rodent-borne human pathogens although an increasing number of shrew-associated hantaviruses have also been discovered. Rodent-borne members of the *Hantavirus* genus are broadly split into the Old World hantaviruses, which cause haemorrhagic fever with renal syndrome (HFRS), and the more recently discovered New World hantaviruses, which cause hantavirus pulmonary syndrome (HPS). The prototype strain for the Old World hantaviruses is Hantaan virus (HTNV). Sin Nombre virus (SNV) and Andes virus (ANDV) are the major representatives of the New World hantaviruses in North and South America, respectively, and the most common aetiological agents of HPS. Case fatalities of approximately 1–15% and 40% are associated with pathogenic Old World and New World hantavirus infections, respectively. Hantaviruses are enveloped viruses with a tripartite negative-sense RNA genome. The viral genome encodes the nucleocapsid N protein, the glycoprotein precursor GPC (which is cotranslationally cleaved into the two viral glycoproteins, Gn and Gc), and the RNA polymerase L protein.

The focus of this chapter is the cellular biology of hantaviruses and their interaction with their host. The chapter is divided into sections describing different aspects of these features.

Virus particle morphology

Hantavirus virions possess an outer lipid envelope in which are embedded the two surface glycoproteins Gn and Gc, appearing as fuzzy surface projections, approximately 7 nm in length (Fig. 2.1). Inside the virion there are three genomic ribonucleoprotein complexes arranged as circular coils. Viewed by electron microscopy (EM) negative staining, hantavirus virions appear to have a grid like pattern on the surface (Martin *et al.*, 1985). Based on this grid-like appearance and recent extensive biochemical studies, a model has been proposed in which the glycoprotein complex in the virions is thought to consist of Gn tetramers interconnected with Gc dimers (Hepojoki *et al.*, 2010a). In general, the virus particles adopt a spherical shape, although with quite broad variation in the size ranging from 70 to 210 nm (Goldsmith *et al.*, 1995; Tao *et al.*, 1987). The pleiomorphic size of the virion particles is thought to be the result of some virion particles possessing additional copies of genomic segments (Rodriguez *et al.*, 1998). EM studies have also revealed elongated virus-like particles narrower than mature virus particles (approx. 50 nm) budding into the cisternae of the Golgi apparatus in cells infected with New and Old World hantaviruses (Goldsmith *et al.*, 1995). Unusually elongated tubular projections have been also observed at the infected VeroE6 cell surface, most prominently in the New World hantaviruses (Goldsmith *et al.*, 1995). Occasionally these projections are also associated with virus particles. Structurally, they composed of two concentric rings measuring up to 550 nm long but only 28 nm wide, and they can be stained for the virus glycoproteins but not the nucleocapsid. Their function, if any, is not yet known.

Virus genome structure

Hantaviruses possess three genome segments termed small (S), medium (M), and large (L) (Schmaljohn and Dalrymple, 1983). Hantavirus S segment varies in size from 1600 to 2004 nt and encodes the N protein (Schmaljohn *et al.*, 1986b; Spiropoulou *et al.*, 1994). The length variation observed reflects mainly differences in the 3′ (relative to the antigenomic RNA) non-translated region which contains a series of imprecise repeated sequences. These

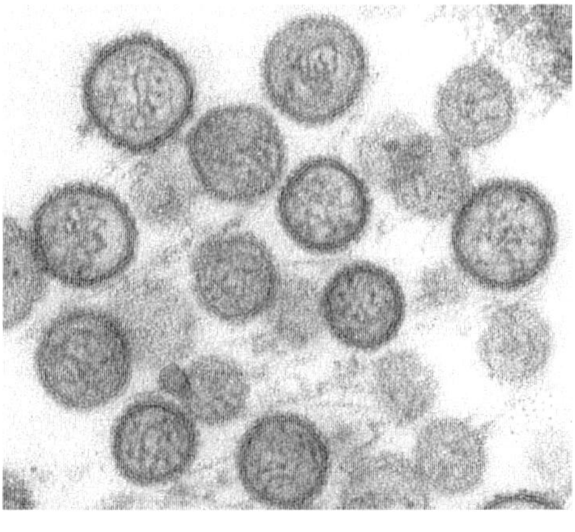

Figure 2.1 Electron micrograph thin section image of Sin Nombre virus particles released from Vero-E6 cells. The internal filamentous or 'coiled bead' structures are likely viral ribonucleocaspid cores (courtesy of C.S. Goldsmith and L.H. Elliott).

repeats may be due to virus RNA polymerase slippage or stuttering (Spiropoulou *et al.*, 1994). An overlapping second open reading frame (ORF) encoding a NSs protein may exist within the N coding region of some hantaviruses (Spiropoulou *et al.*, 1994; Plyusnin, 2002). The hantavirus M segment is ~3600 nt in length and encodes in a single ORF the glycoprotein precursor for two glycoproteins, Gn and Gc (Schmaljohn *et al.*, 1986a, 1987; Kamrud and Schmaljohn, 1994). The hantavirus M segment coding strategy is simpler than that seen in members of some other genera, in which an additional non-structural protein, NSm, is also encoded (Nakitare and Elliott, 1993). The hantavirus L segment is approximately 6500 nt in length and encodes the virus RNA polymerase, L protein (Schmaljohn, 1990). The functional genome is composed of three circularized RNA segments encapsidated by the N protein, forming the ribonucleocapsid cores (RNP).

Virus attachment and entry

Hantaviruses have been shown to utilize β3 and β1 integrins for their cell attachment (Gavrilovskaya *et al.*, 1998, 1999; ; Hall *et al.*, 2008; Hall *et al.*, 2009; Larson *et al.*, 2005). In particular, β3 integrin has been implicated as a receptor of the pathogenic hantaviruses and β1 as receptor of the non-pathogenic hantaviruses. It is unclear if integrins are also used for hantavirus cell entry. It would be expected that the use of integrins for virus entry would result in integrin-dependent cell-signalling activation; however, this has not been observed during hantavirus infection (Stewart and Nemerow, 2007). Since β3 integrins are present on platelets and endothelial cells and they can regulate vascular permeability, platelet activation and adhesion, the use of these receptors by hantaviruses has been suggested to have implications in pathogenesis (Mackow and Gavrilovskaya, 2001). However, cell tropism and pathogenicity could not be explained by the use of different integrins as both pathogenic and non-pathogenic hantaviruses readily infect the macrovascular endothelium.

Additional cellular receptors such us decay-accelerating factor (DAF/CD55) and gC1qR/p32 (the receptor of complement C1q) also have been reported to be used as receptors for HTNV (Choi *et al.*, 2008; Krautkramer and Zeier, 2008). Virus attachment to cells is mediated by the virus surface glycoproteins Gn and Gc. Virion uptake by the infected cells is followed by low pH-dependent fusion of the virus and endosomal membranes and release of the viral RNP cores into the cytoplasm. The low pH fusion between the virus and endosomal cellular membranes is mediated by the hantavirus glycoproteins, in particular Gc (Ogino *et al.*, 2004; Tischler *et al.*, 2005).

Virus transcription and replication

After the release of the virus RNP cores into the cytoplasm, each negative-sense RNA segment is transcribed by the few virion-associated L polymerase molecules, and functional virus mRNAs are produced. A prime-and-realign model has been proposed for the initiation of hantavirus RNA transcription (Fig. 2.2) (Garcin *et al.*, 1995). According to this model a short capped cellular oligonucleotide with a terminal guanine residue snatched by the viral polymerase would align to the third position of the viral RNA template in order to initiate transcription. During the initial process of elongation, the viral RNA will realign by slipping backward in such way that the 5′ terminal G residue will be exposed as an extension. In this manner, the 5′ ends of all three mRNAs (N, GPC and L) are capped and contain

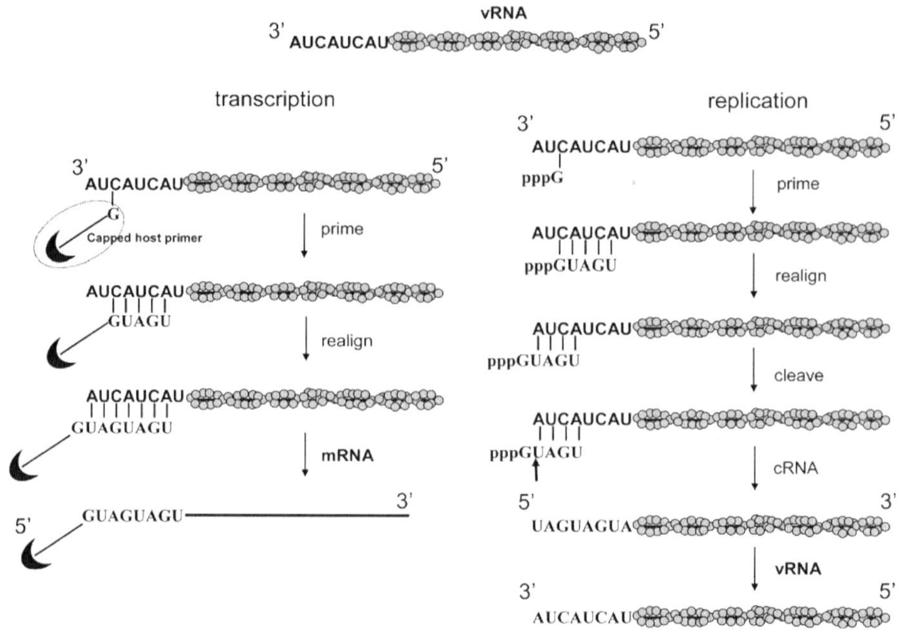

Figure 2.2 A schematic representation of the proposed model of hantavirus transcription replication. A prime and realign model has been proposed for the initiation of hantavirus RNA transcription. In this model a capped host oligonucleotide will prime with a G residue at third position of the vRNA. After a short extension, the nascent RNA realigns by slipping back two nucleotides. The resulting mRNA is capped with 10–16 nucleotides derived from the host mRNAs, and it is shorter than the full-length vRNA segment at the 3′ end. In hantavirus replication, full-length copies of the antigenomic RNA templates are synthesized based on the realign model. Encapsidated antigenomic positive sense RNAs subsequently serve as template for the synthesis of the negative sense viral RNA genome.

a non-templated G. Each mature mRNA is shorter than the corresponding full-length genomic segment and (with the possible exception the GPC mRNA) is not polyadenylated. The precise transcription termination signal for each RNA segment is unknown. A GC-rich motif has been suggested as transcription termination signal for the S segment (Hutchinson *et al.*, 1996).

The relative amounts of the mRNAs synthesized are inversely proportional to their length. Thus, N mRNA is the most abundant and the L mRNA the least (Hutchinson *et al.*, 1996). A simple explanation for this may be that the transcription elongation process is discontinuous, with the polymerase pausing and possibly prematurely terminating at specific sites along the genome. The longer the gene, the higher the possibility for the presence of pausing sites. The three mRNAs are transcribed in P bodies (Mir *et al.*, 2008). The GPC mRNA is translated by membrane-bound ribosomes whereas the S and L mRNAs are translated by free ribosomes. The ratio of the individual virus proteins synthesized reflects the relative abundance of the virus mRNAs, with the N the most abundant protein.

The exact mechanism by which the virus switches from the transcription to the replication mode remains unclear. The main switch is thought to involve increasing concentrations of N protein leading to more efficient RNA encapsidation. The signals for encapsidation have been mapped at the 3′ and 5′ regions of the each genomic RNA segment (Severson *et al.*, 2001; Mir and Panganiban, 2006). The termini of each viral genome RNA form a panhandle-like structure. Each panhandle is at least 23 nucleotides long and is bound with high affinity by the trimeric N (Mir and Panganiban, 2004, 2005). This interaction is required for encapsidation of the viral RNAs, and allows the polymerase to read through the transcription termination signals and synthesize the full-length copies of the antigenomic RNA templates. Encapsidated antigenomic RNAs subsequently serve as template for the synthesis of the negative sense viral RNA genome.

Translation of the viral proteins

The nucleocapsid N protein

In hantaviruses, N mRNA is the first viral RNA to accumulate during infection, with detectable amounts appearing as early as 2 h post infection (Hutchinson *et al.*, 1996; Kariwa *et al.*, 2003). Accordingly, the N protein (~50 kD in size) also appears shortly after, as has been shown in immunofluorescence and immunoprecipitation studies, and is the most dominant viral protein synthesized early in infections. In EM studies using immunogold labelling techniques, the majority of N protein is seen as granular and filamentous formations which are not membrane-bound (Goldsmith *et al.*, 1995). These formations, called inclusion bodies, are very characteristic of hantavirus infections. A portion of hantaviral N protein is membrane-bound and co-localizes with cis-medial and trans-Golgi membranes, and recycling endosomes (Ravkov and Compans, 2001; Ramanathan *et al.*, 2007; Rowe *et al.*, 2008). The main role of the N protein is to protect the virus genomic RNA from nuclease degradation, and to form the viral RNPs. Thus N encapsidates the viral negative sense genomic RNA as well as the positive sense antigenomic RNA, but not viral mRNA. As mentioned above each RNA segment terminal panhandle is necessary for efficient and specific binding of the N protein (Mir and Panganiban, 2004, 2005, 2006; Mir *et al.*, 2006). Many studies have shown that the N protein involved in the viral RNA encapsidation is in a trimeric form. Trimerization involves the interaction of a coiled-coil structure mainly present in the first 75 N-terminal amino acids of N (Alfadhli *et al.*, 2001; Kaukinen *et al.*, 2004; Mir and Panganiban, 2004; Alminaite *et al.*, 2006; Wang *et al.*, 2008). This is a region which is highly conserved among all hantaviruses (Alminaite *et al.*, 2006). More recently N has also been shown to play an active role in virus transcription, by binding, protecting and sequestering cellular 5′ mRNA caps which are stored in the P bodies (Mir and Panganiban, 2008). Not only does N enrich the accumulation of 5′ caps in the P bodies in order to enhance transcription, but it is also capable of replacing the whole eIF4F cap-binding complex to initiate translation (Mir and Panganiban, 2008). Finally, the N protein has also been shown to play and active role in virus assembly.

The virus Gn/Gc glycoproteins

The Gn and Gc proteins are formed by cotranslational cleavage of the GPC by cellular signal peptidases in the lumen of the ER. The cleavage site for all hantaviruses is located immediately downstream of the conserve amino acid motif WAASA (Lober *et al.*, 2001). Both Gn and Gc glycoproteins are type I integral transmembrane proteins with the C-terminal hydrophobic anchor domains; their N-termini facing towards the ER lumen and their C-termini into the cytoplasm. Gn possesses an unusually long cytoplasmic tail (Gn-CT) approximately 150 aa in length, whereas Gc has a short but heavily charged cytoplasmic tail ~approximately 10 aa in length. Conserved signals such as those for nitrosylation and possible redox regulation are present in the hantavirus Gn-CT. Most importantly, conserved YRTL motif is present within the Gn-CT of all known hantavirus species; this is predicted to be a glycoprotein trafficking signal and a late domain motif (Spiropoulou, 2001). The cytoplasmic tail of Gc possesses the ER-retention signal KKXX that is functional when Gc is expressed alone, but is masked during Gc interaction with Gn. This masking allows the Gn/Gc heterodimer to exit the ER and enter the Golgi complex (Pensiero and Hay, 1992).

Unlike many other negative strand RNA viruses, the hantaviruses lack a matrix (M) protein. It has been speculated that perhaps the unusually long Gn-CT seen in hantaviruses may substitute for the M protein function. Both Gn and Gc glycoproteins when coexpressed in transfected or infected cells, have the capacity to cause fusion under low pH conditions (Ogino *et al.*, 2004), and Gc is predicted to direct this fusion activity (Tischler *et al.*, 2005). The Gn and Gc proteins are both modified by N-linked glycosylation. There are four potential N-glycosylation signals with consensus sequences N-X-S/T, conserved among all hantaviruses. There are three sites located in Gn (e.g. at aa positions 142, 358 and 410 in SNV) and one in Gc (position 939) (Spiropoulou *et al.*, 1994). Based on point mutation experiments, all four sites are used for glycosylation (Shi and Elliott, 2004). Gn and Gc acquire their sugars cotranslationally in the ER. It seems that each N-linked glycosylation site plays a role in the correct folding and trafficking of the glycoproteins, with the first being the most crucial (Shi and Elliott, 2004). An additional feature also conserved in all hantaviruses is the high number of cysteine residues present in both glycoproteins (57 in total).

Initially both glycoproteins are associated with ER chaperones for correct folding and their heterodimerization. Pulse-chase analysis has shown that Gc folds more rapidly than Gn, and that heterodimerization occurs over 10–60 mins (Shi and Elliott, 2004). Endo-H treatment, which cleaves high mannose N-linked oligosaccharides, is frequently used to monitor whether glycoproteins are transported out of the ER and reach the medial Golgi complex. However, it appears that the oligosaccharides linked to the hantavirus glycoproteins remain in the high mannose form, despite their apparent transport through the Golgi complex to the cell surface (Antic *et al.*, 1992; Spiropoulou *et al.*, 2003; Ogino *et al.*, 2004; Shi and Elliott, 2004).

As already mentioned, one of the hallmarks of the members of the family *Bunyaviridae* is the unusual property of maturing in the Golgi apparatus (Pettersson RF, 1996). The maturation of bunyaviruses in Golgi is thought to be due to the retention and accumulation of the two viral glycoproteins Gn and Gc in the Golgi complex (Matsuoka *et al.*, 1988; Pettersson RF, 1996). In general, it is thought that the glycoprotein closest to the N-terminus of the GPC precursor (i.e. Gn) possesses the signal for Golgi localization (Pettersson, 1996). It is difficult to predict the Golgi retention signal based on observations from other members of the family *Bunyaviridae*, as differences can be seen even between viruses within the same genus. For instance, within the genus *Phlebovirus*, the Golgi retention signal of Punta Toro virus is mapped to the Gn transmembrane domain and the first 10 aa residues of the cytoplasmic tail, whereas the Uukuniemi virus signal maps only to the cytoplasmic tail (Matsuoka *et al.*, 1994; Andersson *et al.*, 1997). For hantaviruses, the precise amino acid sequence required for Golgi retention remains unclear. Based on extensive deletion mutant analysis it has been suggested that the Golgi retention signal may be conformational and dependent on Gn/Gc heterodimerization (Shi and Elliott, 2002). This would be consistent with the fact that Gn and Gc co-expression is required for their transport and retention to Golgi (Deyde *et al.*, 2005; Shi and Elliott, 2002). The Gc protein expressed alone is retained in the ER, whereas Gn expressed alone is located in aggresomes, where excess misfolded proteins accumulate (Shi and Elliott, 2002; Spiropoulou *et al.*, 2003).

It is clear for both Old and New World hantaviruses that, when viral Gn and Gc proteins are expressed together, they accumulate in the Golgi and at later time points are readily detectable on the cell surface (Spiropoulou *et al.*, 2003; Ogino *et al.*, 2004). Gn and Gc proteins of Andes virus (ANDV) have been also seen in recycling endosomes, where they partially co-localize with ANDV N protein (Rowe *et al.*, 2008). Experimental data indicate that the Gn protein, when expressed from GPC, can follow at least two distinct pathways: one leading to the formation of a non-covalent Gn/Gc complex and its Golgi localization, and the other leading to the delivery of Gn glycoprotein to the lysosomes (Spiropoulou *et al.*, 2003). It appears plausible that the first is the productive pathway which leads to virion formation, whereas the second is a non-productive pathway leading to degradation of the virus glycoproteins. Similarly, Gn-CT has also been shown to be ubiquitinated and proteosomally degraded (Geimonen *et al.*, 2003a; Wang *et al.*, 2009). These may represent mechanisms by which hantaviruses down-regulate glycoprotein levels and virus maturation.

The L protein

The virus L protein is approximately 250 kDa in size. L functions as an RNA transcriptase, replicase and endonuclease. These functions of L are fundamental for virus mRNA transcription, viral replication and the acquisition of 5′end capped primers from host mRNAs. Hantavirus L polymerase possesses the five conserved amino acid motifs seen in polymerases of other RNA viruses that fold into a palm-like structure (Kukkonen *et al.*, 2005). The significance of these motifs to the L polymerase functions is not known. Compared with N protein and the Gn and Gc glycoproteins, very few studies have been done to examine L polymerase function. The main obstacles to L analysis have been its low abundance in virus infected cells (there are ~2000-times more molecules of N relative to L), and the lack of L-specific antibodies. Recently, specific antibodies raised against Tula virus (TULV) L protein have allowed basic studies on the detection of L and demonstration of its membrane association in TULV-infected cells (Kukkonen *et al.*, 2005). However, the precise location of the L protein in hantavirus-infected cells is still unclear. Studies with expressed L protein fused with green fluorescent protein (GFP), have shown that L exhibits a punctuated perinuclear staining pattern and partially co-localizes with expressed N protein consistent with the interaction of the two proteins (Kukkonen *et al.*, 2004; Zhang *et al.*, 2008).

Hantavirus assembly and release

The last critical steps in the hantavirus life cycle are virus assembly and release. For hantavirus assembly, two viral components are required, the RNP core (containing the vRNA, the N protein and the L polymerase) and the acquired host cell membrane containing the embedded viral Gn and Gc glycoproteins. Hantavirus virion assembly requires the transport of the viral components to the assembly site and interaction between the viral components. It is generally believed that the viral glycoproteins determine the site of virus assembly and budding. This is based on the observation that viral glycoproteins can accumulate at the site of virus budding even when expressed in the absence of RNPs. As already mentioned, the general feature of Golgi maturation of bunyaviruses is thought to be due to the retention and accumulation of their Gn and Gc glycoproteins in the Golgi complex (Matsuoka *et al.*, 1988; Pettersson RF, 1996). For a long time, the dogma was that, similar to members of other Bunyaviridae genera, the hantaviruses also mature only in the Golgi. With the discovery of the New World hantaviruses, came the observation that they could also mature at the cell surface. EM studies showed that with New World hantaviruses most of the virus particles were present extracellularly, with only rare virus-like particles being observed inside the cell (Goldsmith *et al.*, 1995; Ravkov *et al.*, 1997). More recent studies have suggested an even broader spectrum of virus maturation sites that can include also the ER-Golgi compartment in addition and the plasma membranes (Xu *et al.*, 2007).

As stated above, many other negative strand RNA viruses have an matrix (M) protein, and expression of this protein alone will result in maturation of virus-like particles. With the hantaviruses, expression of N or Gn/Gc alone is insufficient for virus maturation. Only when both the N and Gn/Gc were co-expressed in mammalian cells were they able to form virus like particles (VLPs) (Betenbaugh *et al.*, 1995). The long cytoplasmic tail of Gn is thought to substitute for the lack of an M protein, and mediate the interaction of the Gn/Gc multimers with N for virus assembly to occur. A predicted late domain amino acid motif, YXXL, known to interact with cellular factors to facilitate virus budding, is present in the Gn-CT. For other members within the family *Bunyaviridae* it has been shown that the Gn tail is required for packaging of the viral genome or for formation of VLPs, indicating that the Gn tail plays a critical role in the virus assembly (Overby *et al.*, 2007a,b; Shi *et al.*, 2007). Most recently, a direct interaction between the hantaviral Gn-CT and the N protein has been documented as well (Hepojoki *et al.*, 2010b, Wang *et al.*, 2010).The N protein has been implicated to have an additional role in hantavirus assembly, in that the protein has been shown to be associated with actin filaments. Treatment of Black Creek Canal virus (BCCV) infected cells with cytochalasin D, an actin depolymerizing agent, resulted in reduction of virus yields, indicating that intact actin microfilaments are important in hantavirus morphogenesis (Ravkov *et al.*, 1998). However, different results have been obtained in ANDV infected cells, where the intracellular distribution of the N protein is dependent on microtubules and not on microfilaments (Ramanathan *et al.*, 2007). Little is known about other host cellular factors involved in hantavirus assembly and release. The N protein has been shown to interact with SUMO-1 and Ubc9 cellular proteins (Kaukinen *et al.*, 2003; Lee *et al.*, 2003). Mutations of the N protein motifs which mediate the interaction with these host proteins abolished the Golgi localization of N, suggesting that these host proteins play a role in the transport of N to a possible assembly site.

Virus release from polarized cells may differ among different hantaviruses. BCCV, a New World pathogenic hantavirus, has been shown to be released at the apical membrane during infection in polarized epithelial cells (Ravkov *et al.*, 1997). Considering that hantavirus infection of humans is believed to be through aerosol transmission, apical release from cells could facilitate cell-to-cell spread of the virus in the respiratory tract. However, it would not explain how the virus crosses the epithelium to infect the macrovascular endothelium. In contrast, both apical and basolateral release of virus has been demonstrated in ANDV infection of primary hamster epithelium. This bidirectional mode of release would allow the virus to get access from the epithelium to the respiratory endothelium (Rowe and Pekosz, 2006).

In vitro studies of host cell responses to hantaviral infection

Cytopathic effects – apoptosis

In general, hantaviruses do not induce a cytopathic effect in infected cells. Studies on possible induction of apoptosis by hantaviruses have yielded contradictory results. One study suggested bystander apoptosis occurred in HEK-293 cells infected with hantaviruses, but the mechanism of induction was not determined (Markotic *et al.*, 2003). Further reports have shown that hantaviruses may induce a low level apoptosis in dividing cells but not in confluent cells (Hardestam *et al.*, 2005). Also, TULV, which is not pathogenic in humans, is reported to induce significant levels of apoptosis in infected cells. In these studies, the induction of apoptosis has been attributed to ER stress and involved caspase 8 activation (Li *et al.*, 2005). The significance of these findings is unclear given that little evidence of infected cell apoptosis is seen in HPS or HFRS cases (see later section).

Induction of host cell innate immune responses

In general, type I interferons (IFNs) are considered the first line of host defence to RNA virus infections. The host innate immune system has evolved to recognize pathogen-associated molecular patterns leading to activation of innate immune responses such as the type I IFNs and pro-inflammatory cytokines responsible for impairment of viral replication, and induction of adaptive immune responses that contribute to pathogen elimination (Garcia-Sastre and Biron, 2006; Takeuchi and Akira, 2009). Recognition of viral RNA by pathogen sensors such as Retinoic Acid Inducible Gene-I (RIG-I) like receptors (RLRs) or Toll-like receptors (TLRs) is critical in the activation of the type I IFN-dependent antiviral innate immune response. 5′PPP-ssRNA or short dsRNA, as well as poly uridine motifs with 5′PPP-termini in some RNA viruses, are known to serve as ligands that activate RIG-I. Activation of RIG-I initiates a signalling cascade leading to phosphorylation of pre-existing transcription factors, such as interferon regulatory factor 3 (IRF-3). After phosphorylation, IRF-3 dimerizes and translocates from the cytoplasm to the nucleus. In the nucleus, IRF-3 recruits the transcriptional co-activator CREB-binding protein and initiates IFN-β mRNA synthesis. In addition, other transcription factors such as NF-κB and AP-1 can help to enhance IFN-β synthesis. Secreted IFN-β binds to its receptor and initiates the activation of the IFN signalling pathway and an antiviral state within infected and neighbouring cells. Through the JAK-STAT pathway, phosphorylated Stat-1 and Stat-2 recruit IRF-9 to form the IFN-stimulated gene factor 3, which translocates to the nucleus to activate the expression of IFN-stimulated genes. Induction of IFN results in the activation of many genes with antiviral activities, including MxA and PKR. Induced MxA protein has been shown to inhibit hantavirus replication (Frese *et al.*, 1996; Kanerva *et al.*, 1996; Nam *et al.*, 2003; Khaiboullina *et al.*, 2005; Oelschlegel *et al.*, 2007). TLR3 and TLR4 have been implicated in the IFN induction by hantavirus infection (Jiang *et al.*, 2008; Handke *et al.*, 2009).

Hantaviruses, like other RNA viruses, have evolved several mechanisms to evade the host IFN response. These include both ways to avoid or block IFN induction and inhibition of the IFN signalling pathway. Hantaviruses avoid activation the IFN pathway by modification of their RNA 5′ ends from a 5′PPP to 5′monophosphate, using the prime-and-realigned mechanism for the virus RNA transcription (Garcin *et al.*, 1995; Habjan *et al.*, 2008). This is consistent with the fact that RIG-I is not activated at early times post infection (Handke *et al.*, 2009). The Gn and Gc glycoproteins have also been found to down-regulate the IFN response by directly inhibiting IRF-3 phosphorylation, dimerization and nuclear translocation, and impairing the function of IKKe and TBK-1 kinases (Alff *et al.*, 2006, 2008; Spiropoulou *et al.*, 2007).

Often it is the virus non-structural proteins that are involved in down-regulation of the host innate immune response. However, hantaviruses are known to encode only the four structural proteins, N, Gn, Gc and L. Some studies have predicted the presence in some hantaviruses of a second potential open reading frame within the N ORF capable of encoding an NSs-like protein (Spiropoulou *et al.*, 1994; Plyusnin, 2002). Although there is no link between the presence of NSs and hantavirus pathogenicity, overexpression of the putative NSs protein of Tula and Puumala viruses resulted in a modest suppression of IFN induction (Jaaskelainen *et al.*, 2007). Additional studies are needed to more precisely define the role of NSs, if any, in hantavirus infection.

In general, there is a clear difference between pathogenic and non-pathogenic hantaviruses in their ability to induce early interferon responses. The non-pathogenic hantaviruses, such as Prospect Hill virus (PHV), induce a robust IFN-β response early after infection, whereas the pathogenic hantaviruses do not (Geimonen *et al.*, 2002; Kim *et al.*, 2007; Prescott *et al.*, 2007; Spiropoulou *et al.*, 2007; Jiang *et al.*, 2008; Handke *et al.*, 2009). However, despite these differences, both pathogenic and non-pathogenic hantaviruses were found to inhibit the interferon signalling pathway by down-regulating phosphorylation and nuclear translocation of STAT 1 and 2. Surprisingly, it was the virus glycoproteins that were found to be involved in this IFN-signalling inhibition (Spiropoulou *et al.*, 2007). This rather unusual property of hantavirus glycoproteins may be related to the unique long Gn-CT and in particular in the presence of a possible ITAM (with the consensus sequence YxxL/Ix 6–8YxxL/I), which is conserved among the pathogenic hantaviruses (Geimonen *et al.*, 2003b). The precise hantavirus glycoprotein regions that are required for the downregulation of the interferon signalling are not known so far.

In addition to the viral glycoproteins, the N protein has also been implicated in the inhibition of inflammatory responses by hantaviruses. It has been reported that in HTNV virus-infected cells activated with TNF-alpha, the N was able to block the import of NF-κB to the nucleus by binding to importin-alpha (Taylor *et al.*, 2009a). This characteristic is not seen with all pathogenic hantaviruses and it seems to mostly be a feature of the Old but not the New World pathogenic hantaviruses (Taylor *et al.*, 2009b).

Hantavirus induced immunopathology in humans

The major hantavirus target is the microvascular endothelium, and severe human disease has been attributed to microvascular leakage. Hantaviruses are transmitted to humans by inhalation of infectious aerosol from rodent excreta, and virus entry involves the use of the β3 integrin receptor (Larson *et al.*, 2005; Medina *et al.*, 2007). In general, the host cell responds to viral infection with cellular and humoral immune responses. Initially, the innate immune system recognizes the virus and as a result pro-inflammatory cytokines and chemokines are induced. This is resulting in responses of the adaptive immune system, which consists of cytotoxic T cells and of B cells that produce virus specific antibodies. Production of additional cytokines and chemokines could attract pro-inflammatory cells such as monocytes-macrophages to the sites of infection. Although these responses are crucial for virus clearing, if deregulated they can also cause damage to the host. It is believed that the induction of an uncontrolled immune response to the hantavirus infection, rather than the viral infection per se, is the cause of the microvascular leakage and ultimately HPS and HFRS (Zaki *et al.*, 1995; Terajima *et al.*, 2007). It is widely accepted today, that hantavirus-associated diseases in humans are immune-modulated. The evidence for this is outlined below.

- Presence of large immunoblasts in the circulation and in necropsy tissues at the onset of the disease – large immunoblasts are present in the circulation at the onset of HPS (Koster *et al.*, 2001). In addition, immunoblasts are primarily concentrated in the lungs of fatal HPS cases (Zaki *et al.*, 1995). Large mononuclear cells with the appearance of immunoblasts have also been seen within the red pulp of the spleen and other lymphoid organs in both HPS and HFRS cases (Nolte *et al.*, 1995; Zaki *et al.*, 1995; Temonen *et al.*, 1996).
- At the time of acute disease presentation, all HPS and HFRS patients have mounted an active immune response (IgM detectable in all cases and IgG in most) – hantaviruses induce a long lasting humoral immune response. Antibodies to the hantaviral N protein, are present in the HPS patient serum soon after the onset of disease, quickly followed by neutralizing antibodies directed against Gn and Gc (Jenison *et al.*, 1994; Valdivieso *et al.*, 2006). However, antibodies do not appear to contribute to the immunopathology seen and, to the contrary, the presence of high titre neutralizing antibodies in the patient serum correlates with the milder disease outcome (Bharadwaj *et al.*, 2000).
- Despite prominent accumulation of viral antigen in the infected vascular endothelium, no evidence of cell destruction has been observed – capillary endothelial cells are the major target cells in both HPS and HFRS. A uniform infection of the lung macrovascular endothelium and high viral titres, are seen in lung material from HPS cases (Zaki *et al.*, 1995). In addition to the lung endothelium, endothelial cells in various tissues throughout the body are also infected. In kidney biopsy samples from HFRS patients, some cell destruction has been observed in tubular epithelial cells, but not any gross architectural changes in the kidney. In addition to endothelial cells, hantaviruses can also infect other human cell types including monocytes/macrophages (Temonen *et al.*, 1993; Markotic *et al.*, 2007) and dendritic cells (Raftery *et al.*, 2002), without apparent cytopathic effect. Evidence of apoptosis is not always lacking. For instance, elevated levels of granzyme B and apparent epithelial apoptosis have been observed in Puumala virus-infected patients (Klingström *et al.*, 2006).
- High levels of cytokine/chemokine-producing cells are present in target tissues – high levels of cytokines have been detected in both HPS and HFRS patients. In lung tissues from HPS cases, several cytokines, including those potentially produced by T cells, such as TNF-α, IL-α, IL-6 and IFN-γ are detected (Mori *et al.*, 1999). Moreover, immunohistological analyses of kidney biopsies from patients during acute PUUV-associated HFRS have shown infiltrating lymphocytes consisting primarily of CD8+ T cells (Temonen *et al.*, 1996). The high levels of activated T lymphocytes could produce high amounts of TNF-α, contributing to microvascular leakage. In addition, *in vitro*, human primary microvascular endothelial cells infected with hantaviruses have shown to secrete high levels of chemokines including IP-10 and RANTES, known to attract T cells (Sundstrom *et al.*, 2001).

- Specific HLA alleles have been correlated with either mild or severe disease – genetic factors associated with the clinical outcome in hantavirus infection have been also investigated in a few studies. Specific HLA alleles are linked with either mild or severe disease. In HPS patients, the HLA-B*3501 allele has been shown to be associated with severe disease (Kilpatrick *et al.*, 2004). The severity of PUUV induced HFRS (often referred to as nephropathia epidemica), has been also linked to specific HLA alleles. HLA haplotypes B8 and DR3, have been associated with severe disease, whereas the HLA-B27 allele is associated with mild disease (Mustonen *et al.*, 1998; Makela *et al.*, 2002). It is possible that this could reflect a linkage between specific HLA haplotypes and polymorphisms in the promoter region of the TNF-α gene.
- Correlation between the frequency of virus-specific $CD8^+$ T cells and disease severity – a robust T-cell response is generated in humans during hantavirus infection and this is followed by a long-lived memory T-cell response (Van Epps *et al.*, 2002). Several $CD4^+$ and $CD8^+$ T-cell clones obtained from the blood of acute and convalescent patients recognize epitopes identified in the N, Gn and Gc viral proteins (Ennis *et al.*, 1997; Van Epps *et al.*, 1999; Terajima *et al.*, 2002; Kilpatrick *et al.*, 2004). Higher frequencies of SNV-specific $CD8^+$ T cells were detected in patients with severe HPS than in patients with less severe symptoms (Kilpatrick *et al.*, 2004).

Model of hantavirus pathogenesis

Understanding of the sequence of events leading to hantavirus-induced disease is still limited. However, current data suggest a cascade of events. First, hantavirus infects either the respiratory epithelium and/or the dendritic cells (DCs) residing in the respiratory epithelium (Raftery *et al.*, 2002). After that the virus can get access to its major target, the respiratory endothelium, either by its basolateral release from the infected epithelium or by transport with the infected DCs to the lymph nodes, where infection of monocytes and macrophages could occur (Raftery *et al.*, 2002; Rowe and Pekosz, 2006). Early post infection the virus inhibits induction of type 1 IFN in infected cells which allows for its efficient replication and spread. Hantavirus-infected endothelial cells produce pro-inflammatory cytokines and chemokines and upregulate adhesion molecules on their cell surface, attracting monocytes/macrophages and T cells (Sundström *et al.*, 2001). In addition, hantavirus-infected monocytes, macrophages attracted to the infected endothelium will produce additional chemokines/cytokines and other pro-inflammatory molecules (Temonen *et al.*, 1995, 1996; Markotic *et al.*, 2003, 2007; Cebalo and Markotic, 2007). This cytokine storm will result in the migration of monocytes and T cells into the macrovascular endothelium and through factors like TNF-α, reactive oxygen/nitrogen species (RONS) and possibly VEGF, increase vascular permeability (Fig. 2.3) (Krakauer *et al.*, 1995; Linderholm *et al.*, 1996a,b; Davis *et al.*, 2002; Gavrilovskaya *et al.*, 2008). This abnormally strong induced immune response could be the underlying cause of the microvascular endothelium immunopathology observed.

Hantavirus infection in rodent reservoir

Hantaviruses persistently infect their rodent reservoirs causing an asymptomatic lifelong infection. During the viral infection in rodents there is an initial short 2–3 week viraemic

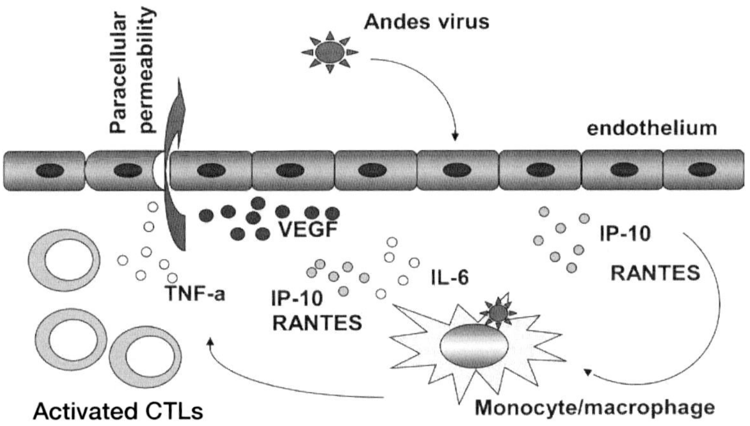

Figure 2.3 Proposed model for hantavirus-induced immunopathology. Hantavirus infection of the microvascular endothelium resorts in the induction of pro-inflammatory chemokines such us IP-10 and RANTES. These chemokines will attract infected monocytes, macrophages and activated T cells to the vicinity of the infected endothelium. A cytokine storm will be created with the release of increased amounts of proinflammatory cytokines–chemokines and permeability factors such as TNF-α and VEGF, which results in increased vascular permeability.

phase where high viral titres and viral RNA load could be detected. During this phase viral antigens can be detected in the blood and different organs, mainly lungs. Viral shedding in urine, faeces and saliva is prominent in the acute phase of infection, and persists for the lifetime of the infected rodent (Kariwa *et al.*, 1998). This acute phase is followed by viral persistence in which the virus is maintained at low cycling levels despite the presence of high levels of neutralizing antibodies (Hutchinson *et al.*, 1998, 2000; Botten *et al.*, 2000, 2003).

Hantavirus infection of the rodent hosts is in striking contrast to that seen in humans. Since the disease in humans has been attributed to the a dysregulated overactivation of the immune response, particularly the T-cell response, virus persistence in rodents may result from a much less robust immune response. Two studies have implicated regulatory T cells (Treg) in the establishment of hantavirus persistence in rodents. In the acute phase of infection, $CD4^+$ T cells from SNV-infected deer mice were found to produce a broad spectrum of cytokines, whereas only TGF-β1 was found to be expressed by $CD4^+$ T cells obtained during the persistent infection phase. Increased expression of TGF-β1 and the transcription factor forkhead box protein 3 (FoxP3), is indicative of a regulatory T-cell phenotype (Schountz *et al.*, 2007). In similar experiments, an increased percentage of $CD4^+CD25^+FoxP3^+$ Treg cells were also detected in rats persistently infected with SEOV. Functional regulatory T cells suppress the expression of pro-inflammatory cytokines such as TNF-α and IL-6 (Easterbrook *et al.*, 2007). The results from both studies suggest that the activation of Treg cells may be a common mechanism mediating persistence of hantaviruses in natural rodent reservoirs (Easterbrook *et al.*, 2007). On the question on whether or not the induction of pro-inflammatory responses are suppressed during hantavirus infection of its rodent reservoir host, a recent study found no generalized immunosuppression occurred, but only a selective reduction of pro-inflammatory responses in the lung (Easterbrook and Klein, 2008). Clearly there is a subtle balance between sufficient immune response to limit virus replication, and a massive dysregulated response which results in disease.

Animal model of hantavirus disease

The development of animal models of hantavirus is a key to the study of pathogenesis and to evaluate the possible effectiveness of antiviral drugs, immunotherapy, and vaccines. Hantaviruses infection is lethal in newborn suckling mice, but without any evidence of HFRS or HPS disease symptoms (Kim and McKee, 1985). Non-human primates can also be infected with hantaviruses, but also without human disease-like symptoms (Groen *et al.*, 1995; McElroy *et al.*, 2002; Hooper *et al.*, 2006). There is, however, one study, where HFRS-like illness was observed in cynomolgus macaques infected with a PUUV strain which was not tissue culture-adapted but propagated in colonized host rodents (Klingström *et al.*, 2002; Sironen *et al.*, 2008). Currently, the only model for HPS disease is infection of Syrian golden hamsters with either of two South America hantaviruses, ANDV and Maporal virus (Hooper *et al.*, 2001; Milazzo *et al.*, 2002). Infection of hamsters with these viruses results in the development of the disease with the characteristics similar to those of HPS, including interstitial pneumonitis, microvascular leakage and high mortality rates. This represented a major breakthrough for studies of hantavirus pathogenesis. Unfortunately, the Syrian hamster model is applicable only to HPS and not to HFRS causing hantaviruses. We are still lacking a small animal model of HFRS.

Concluding remarks

An increasing number of rodent and shrew hantaviruses continue to be discovered. Significant advances have been made in recent years regarding understanding of the molecular and cellular mechanisms of virus entry, replication, maturation and release. For example, we now know that hantaviruses utilize integrins as cellular receptors, and that virus maturation can occur both intracellularly and/or at the cell surface. The structure and function of the viral structural proteins and the basic mechanisms on virus transcription and replication are also largely discerned. We are beginning to gain insight into the basis of induction of immunopathology in humans versus persistence in rodents. However, there is still much work to be done. Core questions such as the basis of the switch from virus RNA transcription to replication, or precise trigger initiating development of HPS or HFRS remain to be answered. Development of a hantavirus reverse genetic system similar to those developed for several other negative-stranded RNA viruses would greatly facilitate virus transcription and replication studies. Unfortunately, this goal has been elusive despite multiple concerted efforts in several different laboratories. On the pathogenesis front, although the hamster HPS model was a big step forward, it has a number of limitations, including the low availability of the specific reagents to allow characterization of the hamster response to infection. Further exploitation of this model for HPS, and the development of a small animal model for HFRS in addition to the existing macaque model, would greatly improve our understanding of these serious public health problems.

References

Alfadhli, A., Love, Z., Arvidson, B., Seeds, J., Willey, J., and Barklis, E. (2001). Hantavirus nucleocapsid protein oligomerization. J. Virol. *75*, 2019–2023.

Alff, P.J., Gavrilovskaya, I.N., Gorbunova, E., Endriss, K., Chong, Y., Geimonen, E., Sen, N., Reich, N.C., and Mackow, E.R. (2006). The pathogenic NY-1 hantavirus G1 cytoplasmic tail inhibits RIG-I- and TBK-1-directed interferon responses. J. Virol. *80*, 9676–9686.

Alff, P.J., Sen, N., Gorbunova, E., Gavrilovskaya, I.N., and Mackow, E.R. (2008). The NY-1 hantavirus Gn cytoplasmic tail coprecipitates TRAF3 and inhibits cellular interferon responses by disrupting TBK1–TRAF3 complex formation. J. Virol. *82*, 9115–9122.

Alminaite, A., Halttunen, V., Kumar, V., Vaheri, A., Holm, L., and Plyusnin, A. (2006). Oligomerization of hantavirus nucleocapsid protein: analysis of the N-terminal coiled-coil domain. J. Virol. *80*, 9073–9081.

Andersson, A.M., Melin, L., Bean, A., and Pettersson, R.F. (1997). A retention signal necessary and sufficient for Golgi localization maps to the cytoplasmic tail of a *Bunyaviridae* (Uukuniemi virus) membrane glycoprotein. J. Virol. *71*, 4717–4727.

Antic, D., Wright, K.E., and Kang, C.Y. (1992). Maturation of Hantaan virus glycoproteins G1 and G2. Virology *189*, 324–328.

Betenbaugh, M., Yu, M., Kuehl, K., White, J., Pennock, D., Spik, K., and Schmaljohn, C. (1995). Nucleocapsid- and virus-like particles assemble in cells infected with recombinant baculoviruses or vaccinia viruses expressing the M and the S segments of Hantaan virus. Virus Res. *38*, 111–124.

Bharadwaj, M., Nofchissey, R., Goade, D., Koster, F., and Hjelle, B. (2000). Humoral immune responses in the hantavirus cardiopulmonary syndrome. J. Infect. Dis. *182*, 43–48.

Botten, J., Mirowsky, K., Kusewitt, D., Bharadwaj, M., Yee, J., Ricci, R., Feddersen, R.M., and Hjelle, B. (2000). Experimental infection model for Sin Nombre hantavirus in the deer mouse (Peromyscus maniculatus). Proc. Natl. Acad. Sci. U.S.A. *97*, 10578–10583.

Botten, J., Mirowsky, K., Kusewitt, D., Ye, C., Gottlieb, K., Prescott, J., and Hjelle, B. (2003). Persistent Sin Nombre virus infection in the deer mouse (Peromyscus maniculatus) model: sites of replication and strand-specific expression. J. Virol. *77*, 1540–1550.

Cebalo, L., and Markotic, A. (2007). Chemokine production predominates in human monocytes infected with Tula virus. Viral Immunol. *20*, 206–213.

Choi, Y., Kwon, Y.C., Kim, S.I., Park, J.M., Lee, K.H., and Ahn, B.Y. (2008). A hantavirus causing hemorrhagic fever with renal syndrome requires gC1qR/p32 for efficient cell binding and infection. Virology *381*, 178–183.

Davis, I.C., Zajac, A.J., Nolte, K.B., Botten, J., Hjelle, B., and Matalon, S. (2002). Elevated generation of reactive oxygen/nitrogen species in hantavirus cardiopulmonary syndrome. J. Virol. *76*, 8347–8359.

Deyde, V.M., Rizvanov, A.A., Chase, J., Otteson, E.W., and St Jeor, S.C. (2005). Interactions and trafficking of Andes and Sin Nombre Hantavirus glycoproteins G1 and G2. Virology *331*, 307–315.

Easterbrook, J.D., and Klein, S.L. (2008). Seoul virus enhances regulatory and reduces pro-inflammatory responses in male Norway rats. J. Med. Virol. *80*, 1308–1318.

Easterbrook, J.D., Zink, M.C., and Klein, S.L. (2007). Regulatory T cells enhance persistence of the zoonotic pathogen Seoul virus in its reservoir host. Proc. Natl. Acad. Sci. U.S.A. *104*, 15502–15507.

Ennis, F.A., Cruz, J., Spiropoulou, C.F., Waite, D., Peters, C.J., Nichol, S.T., Kariwa, H., and Koster, F.T. (1997). Hantavirus pulmonary syndrome: CD8+ and CD4+ cytotoxic T lymphocytes to epitopes on Sin Nombre virus nucleocapsid protein isolated during acute illness. Virology *238*, 380–390.

Frese, M., Kochs, G., Feldmann, H., Hertkorn, C., and Haller, O. (1996). Inhibition of bunyaviruses, phleboviruses, and hantaviruses by human MxA protein. J. Virol. *70*, 915–923.

Garcia-Sastre, A., and Biron, C.A. (2006). Type 1 interferons and the virus–host relationship: a lesson in detente. Science *312*, 879–882.

Garcin, D., Lezzi, M., Dobbs, M., Elliott, R.M., Schmaljohn, C., Kang, C.Y., and Kolakofsky, D. (1995). The 5′ ends of Hantaan virus (*Bunyaviridae*) RNAs suggest a prime-and-realign mechanism for the initiation of RNA synthesis. J. Virol. *69*, 5754–5762.

Gavrilovskaya, I.N., Shepley, M., Shaw, R., Ginsberg, M.H., and Mackow, E.R. (1998). beta3 Integrins mediate the cellular entry of hantaviruses that cause respiratory failure. Proc. Natl. Acad. Sci. U.S.A. *95*, 7074–7079.

Gavrilovskaya, I.N., Brown, E.J., Ginsberg, M.H., and Mackow, E.R. (1999). Cellular entry of hantaviruses which cause hemorrhagic fever with renal syndrome is mediated by beta3 integrins. J. Virol. 73, 3951–3959.

Gavrilovskaya, I.N., Gorbunova, E.E., Mackow, N.A., and Mackow, E.R. (2008). Hantaviruses direct endothelial cell permeability by sensitizing cells to the vascular permeability factor VEGF, while angiopoietin 1 and sphingosine 1-phosphate inhibit hantavirus-directed permeability. J. Virol. *82*, 5797–5806.

Geimonen, E., Neff, S., Raymond, T., Kocer, S.S., Gavrilovskaya, I.N., and Mackow, E.R. (2002). Pathogenic and nonpathogenic hantaviruses differentially regulate endothelial cell responses. Proc. Natl. Acad. Sci. U.S.A. *99*, 13837–13842.

Geimonen, E., Fernandez, I., Gavrilovskaya, I.N., and Mackow, E.R. (2003a). Tyrosine residues direct the ubiquitination and degradation of the NY-1 hantavirus G1 cytoplasmic tail. J. Virol. *77*, 10760–10868.

Geimonen, E., LaMonica, R., Springer, K., Farooqui, Y., Gavrilovskaya, I.N., and Mackow, E.R. (2003b). Hantavirus pulmonary syndrome-associated hantaviruses contain conserved and functional ITAM signaling elements. J. Virol. *77*, 1638–1643.

Goldsmith, C.S., Elliott, L.H., Peters, C.J., and Zaki, S.R. (1995). Ultrastructural characteristics of Sin Nombre virus, causative agent of hantavirus pulmonary syndrome. Arch. Virol. *140*, 2107–2122.

Groen, J., Gerding, M., Koeman, J.P., Roholl, P.J., van Amerongen, G., Jordans, H.G., Niesters, H.G., and Osterhaus, A.D. (1995). A macaque model for hantavirus infection. J. Infect. Dis. *172*, 38–44.

Habjan, M., Andersson, I., Klingstrom, J., Schumann, M., Martin, A., Zimmermann, P., Wagner, V., Pichlmair, A., Schneider, U., Muhlberger, E., *et al.* (2008). Processing of genome 5′ termini as a strategy of negative-strand RNA viruses to avoid RIG-I-dependent interferon induction. PloS One *3*, e2032.

Hall, P.R., Hjelle, B., Brown, D.C., Ye, C., Bondu-Hawkins, V., Kilpatrick, K.A., and Larson, R.S. (2008). Multivalent presentation of antihantavirus peptides on nanoparticles enhances infection blockade. Antimicrob. Agents Chemother. *52*, 2079–2088.

Hall, P.R., Hjelle, B., Njus, H., Ye, C., Bondu-Hawkins, V., Brown, D.C., Kilpatrick, K.A., and Larson, R.S. (2009). Phage display selection of cyclic peptides that inhibit Andes virus infection. J. Virol. *83*, 8965–8969.

Handke, W., Oelschlegel, R., Franke, R., Kruger, D.H., and Rang, A. (2009). Hantaan virus triggers TLR3-dependent innate immune responses. J. Immunol. *182*, 2849–2858.

Hardestam, J., Klingstrom, J., Mattsson, K., and Lundkvist, A. (2005). HFRS causing hantaviruses do not induce apoptosis in confluent Vero E6 and A-549 cells. J. Med. Virol. *76*, 234–240.

Hepojoki, J., Strandin, T., Vaheri, A., and Lankinen, H. (2010a). Interactions and oligomerization of hantavirus glycoproteins. J. Virol. *84*, 227–242.

Hepojoki, J., Strandin, T., Wang, H., Vaheri, A., and Lankinen, H. (2010b). Cytoplasmic tails of hantavirus glycoprotein interact with the nucleocapsid. J. Gen. Virol. *91*, 4889–4897.

Hooper, J.W., Larsen, T., Custer, D.M., and Schmaljohn, C.S. (2001). A lethal disease model for hantavirus pulmonary syndrome. Virology *289*, 6–14.

Hooper, J.W., Custer, D.M., Smith, J., and Wahl-Jensen, V. (2006). Hantaan/Andes virus DNA vaccine elicits a broadly cross-reactive neutralizing antibody response in nonhuman primates. Virology *347*, 208–216.

Hutchinson, K.L., Peters, C.J., and Nichol, S.T. (1996). Sin Nombre virus mRNA synthesis. Virology *224*, 139–149.

Hutchinson, K.L., Rollin, P.E., and Peters, C.J. (1998). Pathogenesis of a North American hantavirus, Black Creek Canal virus, in experimentally infected Sigmodon hispidus. Am. J. Trop. Med. Hyg. *59*, 58–65.

Hutchinson, K.L., Rollin, P.E., Shieh, W.J., Zaki, S., Greer, P.W., and Peters, C.J. (2000). Transmission of Black Creek Canal virus between cotton rats. J. Med. Virol. *60*, 70–76.

Jaaskelainen, K.M., Kaukinen, P., Minskaya, E.S., Plyusnina, A., Vapalahti, O., Elliott, R.M., Weber, F., Vaheri, A., and Plyusnin, A. (2007). Tula and Puumala hantavirus NSs ORFs are functional and the products inhibit activation of the interferon-beta promoter. J. Med. Virol. *79*, 1527–1536.

Jenison, S., Yamada, T., Morris, C., Anderson, B., Torrez-Martinez, N., Keller, N., and Hjelle, B. (1994). Characterization of human antibody responses to four corners hantavirus infections among patients with hantavirus pulmonary syndrome. J. Virol. *68*, 3000–3006.

Jiang, H., Wang, P.Z., Zhang, Y., Xu, Z., Sun, L., Wang, L.M., Huang, C.X., Lian, J.Q., Jia, Z.S., Li, Z.D., *et al.* (2008). Hantaan virus induces toll-like receptor 4 expression, leading to enhanced production of beta interferon, interleukin-6 and tumor necrosis factor-alpha. Virology *380*, 52–59.

Kamrud, K.I., and Schmaljohn, C.S. (1994). Expression strategy of the M genome segment of Hantaan virus. Virus Res. *31*, 109–121.

Kanerva, M., Melen, K., Vaheri, A., and Julkunen, I. (1996). Inhibition of puumala and tula hantaviruses in Vero cells by MxA protein. Virology *224*, 55–62.

Kariwa, H., Fujiki, M., Yoshimatsu, K., Arikawa, J., Takashima, I., and Hashimoto, N. (1998). Urine-associated horizontal transmission of Seoul virus among rats. Arch. Virol. *143*, 15–24.

Kariwa, H., Tanabe, H., Mizutani, T., Kon, Y., Lokugamage, K., Lokugamage, N., Iwasa, M.A., Hagiya, T., Araki, K., Yoshimatsu, K., *et al.* (2003). Synthesis of Seoul virus RNA and structural proteins in cultured cells. Arch. Virol. *148*, 1671–1685.

Kaukinen, P., Vaheri, A., and Plyusnin, A. (2003). Mapping of the regions involved in homotypic interactions of Tula hantavirus N protein. J. Virol. *77*, 10910–10916.

Kaukinen, P., Kumar, V., Tulimaki, K., Engelhardt, P., Vaheri, A., and Plyusnin, A. (2004). Oligomerization of Hantavirus N protein: C-terminal alpha-helices interact to form a shared hydrophobic space. J. Virol. *78*, 13669–13677.

Khaiboullina, S.F., Rizvanov, A.A., Deyde, V.M., and St Jeor, S.C. (2005). Andes virus stimulates interferon-inducible MxA protein expression in endothelial cells. J. Med. Virol. *75*, 267–275.

Kilpatrick, E.D., Terajima, M., Koster, F.T., Catalina, M.D., Cruz, J., and Ennis, F.A. (2004). Role of specific CD8+ T cells in the severity of a fulminant zoonotic viral hemorrhagic fever, hantavirus pulmonary syndrome. J. Immunol. *172*, 3297–3304.

Kim, G.R., and McKee, K.T., Jr. (1985). Pathogenesis of Hantaan virus infection in suckling mice: clinical, virologic, and serologic observations. Am. J. Trop. Med. Hyg. *34*, 388–395.

Kim, I.W., Hwang, J.Y., Kim, S.K., Kim, J.K., and Park, H.S. (2007). Interferon-stimulated genes response in endothelial cells following Hantaan virus infection. J. Korean Med. Sci. *22*, 987–992.

Klingström, J., Plyusnin, A., Vaheri, A., and Lundkvist, A. (2002). Wild-type Puumala hantavirus infection induces cytokines, C-reactive protein, creatinine, and nitric oxide in cynomolgus macaques. J. Virol. *76*, 444–449.

Klingström, J., Hardestam, J., Stoltz, M., Zuber, B., Lundkvist, A., Linder, S., and Ahlm, C. (2006). Loss of cell membrane integrity in puumala hantavirus-infected patients correlates with levels of epithelial cell apoptosis and perforin. J. Virol. *80*, 8279–8282.

Koster, F., Foucar, K., Hjelle, B., Scott, A., Chong, Y.Y., Larson, R., and McCabe, M. (2001). Rapid presumptive diagnosis of hantavirus cardiopulmonary syndrome by peripheral blood smear review. Am. J. Clin. Pathol. *116*, 665–672.

Krakauer, T., Leduc, J.W., and Krakauer, H. (1995). Serum levels of tumor necrosis factor-alpha, interleukin-1, and interleukin-6 in hemorrhagic fever with renal syndrome. Viral Immunol. *8*, 75–79.

Krautkramer, E., and Zeier, M. (2008). Hantavirus causing hemorrhagic fever with renal syndrome enters from the apical surface and requires decay-accelerating factor (DAF/CD55). J. Virol. *82*, 4257–4264.

Kukkonen, S.K., Vaheri, A., and Plyusnin, A. (2004). Tula hantavirus L protein is a 250 kDa perinuclear membrane-associated protein. J. Gen. Virol. *85*, 1181–1189.

Kukkonen, S.K., Vaheri, A., and Plyusnin, A. (2005). L protein, the RNA-dependent RNA polymerase of hantaviruses. Arch. Virol. *150*, 533–556.

Larson, R.S., Brown, D.C., Ye, C., and Hjelle, B. (2005). Peptide antagonists that inhibit Sin Nombre virus and hantaan virus entry through the beta3-integrin receptor. J. Virol. *79*, 7319–7326.

Lee, B.H., Yoshimatsu, K., Maeda, A., Ochiai, K., Morimatsu, M., Araki, K., Ogino, M., Morikawa, S., and Arikawa, J. (2003). Association of the nucleocapsid protein of the Seoul and Hantaan hantaviruses with small ubiquitin-like modifier-1-related molecules. Virus Res. *98*, 83–91.

Li, X.D., Lankinen, H., Putkuri, N., Vapalahti, O., and Vaheri, A. (2005). Tula hantavirus triggers pro-apoptotic signals of ER stress in Vero E6 cells. Virology 333, 180–189.

Linderholm, M., Ahlm, C., Settergren, B., Waage, A., and Tarnvik, A. (1996a). Elevated plasma levels of tumor necrosis factor (TNF)-alpha, soluble TNF receptors, interleukin (IL)–6, and IL-10 in patients with hemorrhagic fever with renal syndrome. J. Infect. Dis. *173*, 38–43.

Linderholm, M., Groeneveld, P.H., and Tarnvik, A. (1996b). Increased production of nitric oxide in patients with hemorrhagic fever with renal syndrome – relation to arterial hypotension and tumor necrosis factor. Infection *24*, 337–340.

Lober, C., Anheier, B., Lindow, S., Klenk, H.D., and Feldmann, H. (2001). The Hantaan virus glycoprotein precursor is cleaved at the conserved pentapeptide WAASA. Virology *289*, 224–229.

McElroy, A.K., Bray, M., Reed, D.S., and Schmaljohn, C.S. (2002). Andes virus infection of cynomolgus macaques. J. Infect. Dis. *186*, 1706–1712.

Mackow, E.R., and Gavrilovskaya, I.N. (2001). Cellular receptors and hantavirus pathogenesis. Curr. Top. Microbiol. Immunol. *256*, 91–115.

Makela, S., Mustonen, J., Ala-Houhala, I., Hurme, M., Partanen, J., Vapalahti, O., Vaheri, A., and Pasternack, A. (2002). Human leukocyte antigen-B8-DR3 is a more important risk factor for severe Puumala hantavirus infection than the tumor necrosis factor-alpha(–308) G/A polymorphism. J. Infect. Dis. *186*, 843–846.

Markotic, A., Hensley, L., Geisbert, T., Spik, K., and Schmaljohn, C. (2003). Hantaviruses induce cytopathic effects and apoptosis in continuous human embryonic kidney cells. J. Gen. Virol. *84*, 2197–2202.

Markotic, A., Hensley, L., Daddario, K., Spik, K., Anderson, K., and Schmaljohn, C. (2007). Pathogenic hantaviruses elicit different immunoreactions in THP-1 cells and primary monocytes and induce differentiation of human monocytes to dendritic-like cells. Coll. Antropol. *31*, 1159–1167.

Martin, M.L., Lindsey-Regnery, H., Sasso, D.R., McCormick, J.B., and Palmer, E. (1985). Distinction between *Bunyaviridae* genera by surface structure and comparison with Hantaan virus using negative stain electron microscopy. Arch. Virol. *86*, 17–28.

Matsuoka, Y., Ihara, T., Bishop, D.H., and Compans, R.W. (1988). Intracellular accumulation of Punta Toro virus glycoproteins expressed from cloned cDNA. Virology *167*, 251–260.

Matsuoka, Y., Chen, S.Y., and Compans, R.W. (1994). A signal for Golgi retention in the bunyavirus G1 glycoprotein. J. Biol. Chem. *269*, 22565–22573.

Medina, R.A., Mirowsky-Garcia, K., Hutt, J., and Hjelle, B. (2007). Ribavirin, human convalescent plasma and anti-beta3 integrin antibody inhibit infection by Sin Nombre virus in the deer mouse model. J. Gen. Virol. *88*, 493–505.

Milazzo, M.L., Eyzaguirre, E.J., Molina, C.P., and Fulhorst, C.F. (2002). Maporal viral infection in the Syrian golden hamster: a model of hantavirus pulmonary syndrome. J. Infect. Dis. *186*, 1390–1395.

Mir, M.A., and Panganiban, A.T. (2004). Trimeric hantavirus nucleocapsid protein binds specifically to the viral RNA panhandle. J. Virol. *78*, 8281–8288.

Mir, M.A., and Panganiban, A.T. (2005). The hantavirus nucleocapsid protein recognizes specific features of the viral RNA panhandle and is altered in conformation upon RNA binding. J. Virol. *79*, 1824–1835.

Mir, M.A., and Panganiban, A.T. (2006). Characterization of the RNA chaperone activity of hantavirus nucleocapsid protein. J. Virol. *80*, 6276–6285.

Mir, M.A., and Panganiban, A.T. (2008). A protein that replaces the entire cellular eIF4F complex. EMBO J. *27*, 3129–3139.

Mir, M.A., Brown, B., Hjelle, B., Duran, W.A., and Panganiban, A.T. (2006). Hantavirus N protein exhibits genus-specific recognition of the viral RNA panhandle. J. Virol. *80*, 11283–11292.

Mir, M.A., Duran, W.A., Hjelle, B.L., Ye, C., and Panganiban, A.T. (2008). Storage of cellular 5′ mRNA caps in P bodies for viral cap-snatching. Proc. Natl. Acad. Sci. U.S.A. *105*, 19294–19299.

Mori, M., Rothman, A.L., Kurane, I., Montoya, J.M., Nolte, K.B., Norman, J.E., Waite, D.C., Koster, F.T., and Ennis, F.A. (1999). High levels of cytokine-producing cells in the lung tissues of patients with fatal hantavirus pulmonary syndrome. J. Infect. Dis. *179*, 295–302.

Mustonen, J., Partanen, J., Kanerva, M., Pietila, K., Vapalahti, O., Pasternack, A., and Vaheri, A. (1998). Association of HLA B27 with benign clinical course of nephropathia epidemica caused by Puumala hantavirus. Scand. J. Immunol. *47*, 277–279.

Nakitare, G.W., and Elliott, R.M. (1993). Expression of the Bunyamwera virus M genome segment and intracellular localization of NSm. Virology *195*, 511–520.

Nam, J.H., Hwang, K.A., Yu, C.H., Kang, T.H., Shin, J.Y., Choi, W.Y., Kim, I.B., Joo, Y.R., Cho, H.W., and Park, K.Y. (2003). Expression of interferon inducible genes following Hantaan virus infection as a mechanism of resistance in A549 cells. Virus Genes *26*, 31–38.

Nolte, K.B., Feddersen, R.M., Foucar, K., Zaki, S.R., Koster, F.T., Madar, D., Merlin, T.L., McFeeley, P.J., Umland, E.T., and Zumwalt, R.E. (1995). Hantavirus pulmonary syndrome in the United States: a pathological description of a disease caused by a new agent. Hum. Pathol. *26*, 110–120.

Oelschlegel, R., Kruger, D.H., and Rang, A. (2007). MxA-independent inhibition of Hantaan virus replication induced by type I and type II interferon *in vitro*. Virus Res. *127*, 100–105.

Ogino, M., Yoshimatsu, K., Ebihara, H., Araki, K., Lee, B.H., Okumura, M., and Arikawa, J. (2004). Cell fusion activities of Hantaan virus envelope glycoproteins. J. Virol. *78*, 10776–10782.

Overby, A.K., Pettersson, R.F., and Neve, E.P. (2007a). The glycoprotein cytoplasmic tail of Uukuniemi virus (*Bunyaviridae*) interacts with ribonucleoproteins and is critical for genome packaging. J. Virol. *81*, 3198–3205.

Overby, A.K., Popov, V.L., Pettersson, R.F., and Neve, E.P. (2007b). The cytoplasmic tails of Uukuniemi Virus (*Bunyaviridae*) G(N) and G(C) glycoproteins are important for intracellular targeting and the budding of virus-like particles. J. Virol. *81*, 11381–11391.

Pensiero, M.N., and Hay, J. (1992). The Hantaan virus M-segment glycoproteins G1 and G2 can be expressed independently. J. Virol. *66*, 1907–1914.

Pettersson RF, M.L. (1996). Synthesis, assembly and intracellular transport of *Bunyaviridae* membrane proteins. In The *Bunyaviridae*, Elliott, R.M., ed. (Plenum Press, New York), pp. 159–188.

Plyusnin, A. (2002). Genetics of hantaviruses: implications to taxonomy. Arch. Virol. *147*, 665–682.

Prescott, J.B., Hall, P.R., Bondu-Hawkins, V.S., Ye, C., and Hjelle, B. (2007). Early innate immune responses to Sin Nombre hantavirus occur independently of IFN regulatory factor 3, characterized pattern recognition receptors, and viral entry. J. Immunol. *179*, 1796–1802.

Raftery, M.J., Kraus, A.A., Ulrich, R., Kruger, D.H., and Schonrich, G. (2002). Hantavirus infection of dendritic cells. J. Virol. *76*, 10724–10733.

Ramanathan, H.N., Chung, D.H., Plane, S.J., Sztul, E., Chu, Y.K., Guttieri, M.C., McDowell, M., Ali, G., and Jonsson, C.B. (2007). Dynein-dependent transport of the hantaan virus nucleocapsid protein to the endoplasmic reticulum-Golgi intermediate compartment. J. Virol. *81*, 8634–8647.
Ravkov, E.V., and Compans, R.W. (2001). Hantavirus nucleocapsid protein is expressed as a membrane-associated protein in the perinuclear region. J. Virol. *75*, 1808–1815.
Ravkov, E.V., Nichol, S.T., and Compans, R.W. (1997). Polarized entry and release in epithelial cells of Black Creek Canal virus, a New World hantavirus. J. Virol. *71*, 1147–1154.
Ravkov, E.V., Nichol, S.T., Peters, C.J., and Compans, R.W. (1998). Role of actin microfilaments in Black Creek Canal virus morphogenesis. J. Virol. *72*, 2865–2870.
Rodriguez, L.L., Owens, J.H., Peters, C.J., and Nichol, S.T. (1998). Genetic reassortment among viruses causing hantavirus pulmonary syndrome. Virology *242*, 99–106.
Rowe, R.K., and Pekosz, A. (2006). Bidirectional virus secretion and nonciliated cell tropism following Andes virus infection of primary airway epithelial cell cultures. J. Virol. *80*, 1087–1097.
Rowe, R.K., Suszko, J.W., and Pekosz, A. (2008). Roles for the recycling endosome, Rab8, and Rab11 in hantavirus release from epithelial cells. Virology *382*, 239–249.
Schmaljohn, C.S. (1990). Nucleotide sequence of the L genome segment of Hantaan virus. Nucleic Acids Res. *18*, 6728.
Schmaljohn, C.S., and Dalrymple, J.M. (1983). Analysis of Hantaan virus RNA: evidence for a new genus of bunyaviridae. Virology *131*, 482–491.
Schmaljohn, C.S., Hasty, S.E., Rasmussen, L., and Dalrymple, J.M. (1986a). Hantaan virus replication: effects of monensin, tunicamycin and endoglycosidases on the structural glycoproteins. J. Gen. Virol. *67 (Pt 4)*, 707–717.
Schmaljohn, C.S., Jennings, G.B., Hay, J., and Dalrymple, J.M. (1986b). Coding strategy of the S genome segment of Hantaan virus. Virology *155*, 633–643.
Schmaljohn, C.S., Schmaljohn, A.L., and Dalrymple, J.M. (1987). Hantaan virus M RNA: coding strategy, nucleotide sequence, and gene order. Virology *157*, 31–39.
Schountz, T., Prescott, J., Cogswell, A.C., Oko, L., Mirowsky-Garcia, K., Galvez, A.P., and Hjelle, B. (2007). Regulatory T cell-like responses in deer mice persistently infected with Sin Nombre virus. Proc. Natl. Acad. Sci. U.S.A. *104*, 15496–15501.
Severson, W.E., Xu, X., and Jonsson, C.B. (2001). cis-Acting signals in encapsidation of Hantaan virus S-segment viral genomic RNA by its N protein. J. Virol. *75*, 2646–2652.
Shi, X., and Elliott, R.M. (2002). Golgi localization of Hantaan virus glycoproteins requires co-expression of G1 and G2. Virology *300*, 31–38.
Shi, X., and Elliott, R.M. (2004). Analysis of N-linked glycosylation of hantaan virus glycoproteins and the role of oligosaccharide side chains in protein folding and intracellular trafficking. J. Virol. *78*, 5414–5422.
Shi, X., Kohl, A., Li, P., and Elliott, R.M. (2007). Role of the cytoplasmic tail domains of Bunyamwera orthobunyavirus glycoproteins Gn and Gc in virus assembly and morphogenesis. J. Virol. *81*, 10151–10160.
Sironen, T., Klingstrom, J., Vaheri, A., Andersson, L.C., Lundkvist, A., and Plyusnin, A. (2008). Pathology of Puumala hantavirus infection in macaques. PloS One 3, e3035.
Spiropoulou, C.F. (2001). Hantavirus maturation. Curr. Top. Microbiol. Immunol. *256*, 33–46.
Spiropoulou, C.F., Morzunov, S., Feldmann, H., Sanchez, A., Peters, C.J., and Nichol, S.T. (1994). Genome structure and variability of a virus causing hantavirus pulmonary syndrome. Virology *200*, 715–723.
Spiropoulou, C.F., Goldsmith, C.S., Shoemaker, T.R., Peters, C.J., and Compans, R.W. (2003). Sin Nombre virus glycoprotein trafficking. Virology *308*, 48–63.
Spiropoulou, C.F., Albarino, C.G., Ksiazek, T.G., and Rollin, P.E. (2007). Andes and Prospect Hill hantaviruses differ in early induction of interferon although both can downregulate interferon signaling. J. Virol. *81*, 2769–2776.
Stewart, P.L., and Nemerow, G.R. (2007). Cell integrins: commonly used receptors for diverse viral pathogens. Trends Microbiol. *15*, 500–507.
Sundstrom, J.B., McMullan, L.K., Spiropoulou, C.F., Hooper, W.C., Ansari, A.A., Peters, C.J., and Rollin, P.E. (2001). Hantavirus infection induces the expression of RANTES and IP-10 without causing increased permeability in human lung microvascular endothelial cells. J. Virol. *75*, 6070–6085.
Takeuchi, O., and Akira, S. (2009). Innate immunity to virus infection. Immunol. Rev. *227*, 75–86.
Tao, H., Xia, S.M., Chan, Z.Y., Song, G., and Yanagihara, R. (1987). Morphology and morphogenesis of viruses of hemorrhagic fever with renal syndrome. II. Inclusion bodies--ultrastructural markers of hantavirus-infected cells. Intervirology *27*, 45–52.

Taylor, S.L., Frias-Staheli, N., Garcia-Sastre, A., and Schmaljohn, C.S. (2009a). Hantaan virus nucleocapsid protein binds to importin alpha proteins and inhibits tumor necrosis factor alpha-induced activation of nuclear factor kappa B. J. Virol. *83*, 1271–1279.

Taylor, S.L., Krempel, R.L., and Schmaljohn, C.S. (2009b). Inhibition of TNF-alpha-induced activation of NF-kappaB by hantavirus nucleocapsid proteins. Ann. N. Y. Acad. Sci. *1171 (Suppl 1)*, E86–93.

Temonen, M., Vapalahti, O., Holthofer, H., Brummer-Korvenkontio, M., Vaheri, A., and Lankinen, H. (1993). Susceptibility of human cells to Puumala virus infection. J. Gen. Virol. *74 (Pt 3)*, 515–518.

Temonen, M., Lankinen, H., Vapalahti, O., Ronni, T., Julkunen, I., and Vaheri, A. (1995). Effect of interferon-alpha and cell differentiation on Puumala virus infection in human monocyte/macrophages. Virology *206*, 8–15.

Temonen, M., Mustonen, J., Helin, H., Pasternack, A., Vaheri, A., and Holthofer, H. (1996). Cytokines, adhesion molecules, and cellular infiltration in nephropathia epidemica kidneys: an immunohistochemical study. Clin. Immunol. Immunopathol. *78*, 47–55.

Terajima, M., Van Epps, H.L., Li, D., Leporati, A.M., Juhlin, S.E., Mustonen, J., Vaheri, A., and Ennis, F.A. (2002). Generation of recombinant vaccinia viruses expressing Puumala virus proteins and use in isolating cytotoxic T cells specific for Puumala virus. Virus Res. *84*, 67–77.

Terajima, M., Hayasaka, D., Maeda, K., and Ennis, F.A. (2007). Immunopathogenesis of hantavirus pulmonary syndrome and hemorrhagic fever with renal syndrome: Do CD8+ T cells trigger capillary leakage in viral hemorrhagic fevers? Immunol. Lett. *113*, 117–120.

Tischler, ND, Gonzalez, A., Perez-Acle, T., Rosemblatt, M., and Valenzuela, P.D. (2005). Hantavirus Gc glycoprotein: evidence for a class II fusion protein. J. Gen. Virol. *86*, 2937–2947.

Valdivieso, F., Vial, P., Ferres, M., Ye, C., Goade, D., Cuiza, A., and Hjelle, B. (2006). Neutralizing antibodies in survivors of Sin Nombre and Andes hantavirus infection. Emerg. Infect. Dis. *12*, 166–168.

Van Epps, H.L., Schmaljohn, C.S., and Ennis, F.A. (1999). Human memory cytotoxic T-lymphocyte (CTL) responses to Hantaan virus infection: identification of virus-specific and cross-reactive CD8(+) CTL epitopes on nucleocapsid protein. J. Virol. 73, 5301–5308.

Van Epps, H.L., Terajima, M., Mustonen, J., Arstila, T.P., Corey, E.A., Vaheri, A., and Ennis, F.A. (2002). Long-lived memory T lymphocyte responses after hantavirus infection. J. Exp. Med. *196*, 579–588.

Wang, H., Strandin, T., Hepojoki, J., Lankinen, H., and Vaheri, A. (2009). Degradation and Aggresome Formation of the Gn Tail of the Apathogenic Tula Hantavirus. J. Gen. Virol. *90*, 2995–3001.

Wang, H., Alminaite, A., Vaheri, A., and Plyusnin, A. (2010). Interaction between hantaviral nucleocapsid protein and the cytoplasmic tail of surface glycoprotein Gn. Virus Res. *151*, 205–212.

Wang, Y., Boudreaux, D.M., Estrada, D.F., Egan, C.W., St Jeor, S.C., and De Guzman, R.N. (2008). NMR structure of the N-terminal coiled coil domain of the Andes hantavirus nucleocapsid protein. J. Biol. Chem. *283*, 28297–28304.

Xu, F., Yang, Z., Wang, L., Lee, Y.L., Yang, C.C., Xiao, S.Y., Xiao, H., and Wen, L. (2007). Morphological characterization of hantavirus HV114 by electron microscopy. Intervirology *50*, 166–172.

Zaki, S.R., Greer, P.W., Coffield, L.M., Goldsmith, C.S., Nolte, K.B., Foucar, K., Feddersen, R.M., Zumwalt, R.E., Miller, G.L., Khan, A.S., *et al.* (1995). Hantavirus pulmonary syndrome. Pathogenesis of an emerging infectious disease. Am. J. Pathol. *146*, 552–579.

Zhang, Y., Li, X.H., Jiang, H., Huang, C.X., Wang, P.Z., Mou, D.L., Sun, L., Xu, Z., Wei, X., and Bai, X.F. (2008). Expression of L protein of Hantaan virus 84FLi strain and its application for recovery of minigenomes. Apmis *116*, 1089–1096.

Genetics and Evolution of Hantaviruses

3

Tarja Sironen and Alexander Plyusnin

Abstract

Hantaviruses are globally important human pathogens. They are emerging viruses, and both the number of different hantavirus species discovered and the amount of human infections are increasing. The hantaviral sequence data set is rapidly growing, and several fully sequenced genomes have been reported. Until recently, rodents have been regarded as the main reservoir and evolutionary scene of hantaviruses. These viruses have indeed proven to be an excellent model to study virus evolution and co-evolution of a virus with its host. During the last couple of years, however, majority of novel hantaviruses have been discovered from insectivore hosts instead of rodents, thus challenging some of the earlier views on hantavirus evolution. In the following paragraphs we summarize the current knowledge on the genome organization, genetic diversity, and evolution of hantaviruses from the major groups associated with Cricetidae and Muridae rodents and Soricomorph insectivores.

Introduction

Since some hantaviruses are important, even if not major, human pathogens, they have been studied quite intensively in recent years and there was a steady growth of the hantaviral sequence database that by June 2011 contained over 3800 entries. During the 1980s and 1990s, most of the currently recognized 23 hantavirus species have been described, and for five of them, Hantaan, Seoul, Puumala, Sin Nombre, and Tula viruses (HTNV, SEOV, PUUV, SNV, and TULV), complete genome sequences have been determined. Basic mechanisms and driving forces of hantavirus evolution have been established. Practically all studies were performed on rodent-borne (ROBO) hantaviruses. It was found that the genetic diversity of hantaviruses is generated primarily by genetic drift, that is accumulation of point mutations and insertions/deletions of one or several nucleotides. Evidence for genetic shift in hantaviruses via reassortment of genome RNA segments or intra-segment recombination has been collected as well. The leading role of stabilizing (or negative) selection in hantavirus evolution became apparent and it seems that the evolution keeps to a neutral (or quasi-neutral) mode thus the ever changing quasispecies swarms of viral genomes in individual hosts move around selected adaptive peaks. It is a widely shared opinion that the evolution of hantaviruses follows the evolution of their natural hosts. Consequently, specific features of distinct hantavirus species, or groups of species, are thought to be determined primarily by the nature of their hosts. Occasional host-switching events present unique opportunity to study processes of hantavirus–host co-speciation and co-evolution in greater detail.

During the last ten years, there has been a rapid increase of the hantaviral sequence data. Complete genomes have been recovered from Dobrava, Saaremaa, Andes, Prospect Hill, Rio Mamore, Maporal, Amur/Soochong, Da Bie Shan, Thottopalayam, Imjin, and Asama viruses (DOBV, SAAV, ANDV, PHV, RIOMV, MAPV, ASV, DBSV, TPMV, MJNV, ASAV). Large number of partial sequences became available as well. Most recently, the input from insectivore-borne (INBO) hantaviruses has been increasing especially fast. In general, the newly available information fitted well into the previously established views on hantavirus genetics and evolution. However, some challenges were noticed, too. In the following paragraphs we summarize the current knowledge on the genome structure, genetic diversity, and evolution of hantaviruses from the major groups associated with Cricetidae and Muridae rodents and Soricomorph insectivores.

Genome organization and evolution of hantaviruses

Hantavirus genome

Hantaviruses have three molecules of negative ssRNA as their genome (Nichol *et al.*, 2005). These segments are designated L (large), M (medium) and S (small), and their total length is 11–13 kb. The L, M and S segments encode, respectively, the viral RNA-dependent RNA polymerase (L protein), two envelope glycoproteins (Gn and Gc) and the nucleocapsid (N) and non-structural (NSs) proteins. Each RNA segment is non-covalently circular and the terminal nucleotides are complementary and form a base-paired panhandle-like structure.

The L segment of different hantaviruses is 6474–6581 nucleotides (nt) long, and the reverse complement of the vRNA contains a single open reading frame (ORF) encoding for the L protein of 2149–2156 amino acid (aa) residues. The non-coding regions (NCRs) at the segment termini are shorter than for the two other segments. The L segment is on average the most conserved part of the genome sequence for hantaviruses, the diversity on the protein level is up to 40% for the more distantly related ones. There are five conserved regions detected in all RNA-dependent RNA-polymerases, and these are present in the hantavirus L protein (Kukkonen *et al.*, 2005). Owing to the conserved sequence regions, the L segment has turned out to be an excellent choice for recovering sequences of novel hantaviruses.

The M segment is 3613–3801 nt long, and there is a single ORF encoding for a glycoprotein precursor that is cotranslationally cleaved to yield the two mature proteins Gn and Gc (Löber *et al.*, 2001). The 5′NCR sequence is of similar length to those found on L and S segments, while the 3′NCR sequence is approximately 200–250 nt long. Potential glycosylation sites are found in the Gn and Gc sequences, as well as a highly conserved WAASA motif at the cleavage site. Some bunyaviruses have a second ORF in the M segments encoding for a non-structural protein NSm, but no such ORF is found on hantaviruses. The glycoprotein sequences are highly variable: the diversity is up to 60%.

The S segment sequence is 1530–2078 nt long; the shortest is found from hantaviruses carried by insectivores and Muridae rodents, while Cricetidae-borne hantaviruses have clearly longer S segment. All hantavirus S segment sequences have a short 5′NCR

(~40–70 nt), and a 3′NCR of highly variable length (~200–800 nt). In contrast to L and M segments, the S segment 3′NCR sequences cannot be properly aligned, except for the last ~100 nt. The S segment encodes for the N protein of 428–434 aa. The diversity of the N protein sequences is up to 57% on the aa level. A highly variable region around nt 700–900 is found in the S segment with more conserved regions before and after this site. The Cricetidae-borne hantaviruses contain also a second ORF encoding for the non-structural protein NSs. The NSs protein sequences are 88–95 aa long for the Arvicolinae-borne hantaviruses, while the Sigmodontinae- and Neotominae-borne viruses have a slightly shorter NSs protein of approximately 60 aa residues.

Mechanisms of hantavirus evolution

The persistently infected rodents are the evolutionary scene for hantaviruses (Plyusnin and Morzunov, 2001; Nemirov *et al.*, 2004). Genetic studies have shown that the main mechanism leading to hantavirus diversification is genetic drift, i.e. the gradual accumulation of genomic changes such as nucleotide substitutions, deletions and insertions (Plyusnin *et al.*, 1995; Rowe *et al.*, 1995; Lundkvist *et al.*, 1998). Similarly to many other RNA viruses, hantaviruses appear as populations of closely related genetic variants, i.e. quasispecies (Plyusnin *et al.*, 1995, Plyusnin *et al.*, 1996b; Lundkvist *et al.*, 1997; Feuer *et al.*, 1999; Chung *et al.*, 2007; Sironen *et al.*, 2008). Likewise, the error rate per replication cycle is as high for hantaviruses as it is for RNA viruses in general (Severson *et al.*, 2003; Sironen *et al.*, 2008). In contrast, the evolutionary rate of hantaviruses seems to be slow (Hughes and Friedemann, 2000; Sironen *et al.*, 2001), and these viruses have adapted to their rodent hosts a long time ago, and thus little selection pressure for change remains in these animals (Lundkvist *et al.*, 1997). Strong purifying selection is apparent as high level of strain variation at the nt level is translated only to a low variation at the aa level (Hjelle *et al.*, 1995b). Hantavirus evolution may also be driven by positive selection in situations requiring fast adaptation. This has been shown in the adaptation of PUUV to cell culture, which was accomplished through an expanded quasispecies diversity and fixation of nt substitutions in the non-coding regions of the S segment (Lundkvist *et al.*, 1997).

The life-long hantavirus infection of the carrier rodents enables co-infections and thus segment reassortment or intra-segment recombination between different viral strains. These mechanisms are called genetic shift. In nature, inter-species hantavirus reassortment was first demonstrated for SNV (Li *et al.*, 1995), and recently also for PUUV (Razzauti *et al.*, 2008; 2009). The situation with intra-species reassortment appears a bit more complicated. *In vitro* generated reassortants between SNV and BCCV were infrequent and, importantly, unstable and deleterious (Rodriquez *et al.*, 1998). On the other hand, natural (therefore supposedly stable) reassortants between HTNV and SEOV have been observed in wild rats (*Rattus norvegicus*) in China (Zou *et al.*, 2008b). In addition, conflicting phylogenies for the S and M segment suggest that CADV is a reassortant as well (see below).

Genetic shift in hantaviruses may also happen through recombination. This was initially shown for TULV (Sibold *et al.*, 1999, Plyusnin *et al.*, 2002b), and later also for PUUV (Sironen *et al.*, 2001), HNTV (Chare *et al.*, 2003) and ANDV (Medina *et al.*, 2009). The recombination rate of hantaviruses appears to be low (Chare *et al.*, 2003), in a similar fashion to other negative-stranded RNA viruses.

Taxonomy and phylogenetic relationships

The International Committee on Taxonomy of Viruses has formulated four criteria to define hantavirus species (Nichol *et al.*, 2005; Plyusnin *et al.*, 2011). First, the different species are usually found in a different primary reservoir, i.e. they are in a unique ecological niche. Second, the aa difference of the complete glycoprotein precursor protein and the nucleocapsid protein sequences exceeds 7%. Third, a 4-fold difference in two-way cross-neutralization test is observed, and fourth, different species do not reassort naturally. It is important to follow the ICTV policy on defining virus species as polythetic entities, and this is especially true for hantaviruses. Many examples exist in which one or the other rule is not fulfilled, and yet the sequences discovered probably represent truly unique hantavirus species. With the current speed of sequence recovery, it should be kept in mind that at least two genes should be characterized for species definition, preferably the complete genome. Nearly full-length sequences have been obtained even without virus isolation, and complete S segment sequences are recovered routinely (Sironen *et al.*, 2001). Although there are examples of successful recovery of complete M or even complete L segments without isolation (e.g. ELMCV, TULV, ASAV, OXBV), these will not become routine procedure for quite some time. Thus isolation is required in order to fully characterize a new hantavirus species, especially when the sequence diversity is less than 7%, but the viruses are biologically clearly distinct and represent unique species (Plyusnin *et al.*, 2002a).

In general, the evolution of hantaviruses follows that of their hosts, and phylogenetic reconciliation analysis has provided evidence for virus–host co-divergence (Jackson *et al.*, 2004). Some recent data have challenged this view (Ramsden *et al.*, 2009); yet even that study found evidence for clade-defining amino acid replacements, compatible with host-specific adaptation. This co-evolutionary adaptation is reflected on phylogenetic trees: hantaviruses form groups according to their reservoir hosts. The rodent-borne hantaviruses are divided into three groups (Fig. 3.1). First group is associated with family Muridae rodents of the subfamily Murinae, which are mice and rats of the Old World. The two other groups are associated with family Cricetidae rodents, one with subfamily Arvicolinae (voles and lemmings of Eurasia and North America), and the other with subfamilies Neotominae and Sigmodontinae (mice and rats of the New World). The virus–host association is usually stable, yet sometimes hantaviruses have apparently spread and colonized new hosts. Such host-switching events are seen in phylogenetic trees, when the topology of the host tree and the viral tree do not match. For rodent-borne hantaviruses, host-switching has been suggested for e.g. MGLV/NYV (Morzunov *et al.*, 1998), TOPV/KHAV (Vapalahti *et al.*, 1999), and DOBV/SAAV (Nemirov *et al.*, 2002). For insectivore-borne viruses, host-switching events may have taken place even between more distantly related lineages. The insectivore-borne hantaviruses are divided into two groups, the first one being ancestral to all other hantaviruses, and the other one sharing common ancestry with Murinae-borne viruses. Insectivores of the subfamily Crocidurinae (family Soricidae) carry viruses of the first group, while subfamily Soricinae insectivores carry viruses that belong to the second group. Talpidae-borne hantaviruses are found in both of these groups, and even in a third place sharing ancestry with Sigmodontinae-borne hantaviruses, thus suggesting deep host-switching events for e.g. NVAV/OXBV/ASAV (Arai *et al.*, 2008b; Kang *et al.*, 2009b,c). It may even be speculated that the whole second group consisting mainly of Soricinae-borne viruses, is a result of host-switch early in the evolutionary history of hantaviruses.

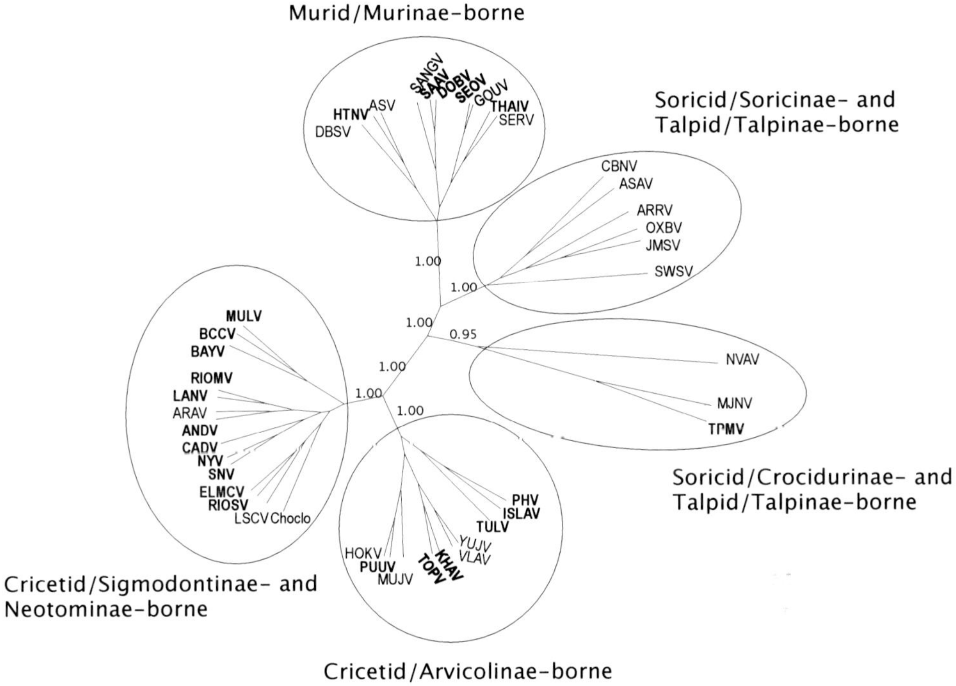

Figure 3.1 Phylogenetic tree of S segment coding sequences of representatives of different hantavirus species (in bold) as well as tentative species. The tree was reconstructed using the Bayesian Markov Chain Monte Carlo method in BEAST (http://beast.bio.ed.ac.uk/). A maximum clade credibility tree with an arbitrary root is shown with mean branch lengths (substitutions per site), and Bayesian posterior probabilities given at the nodes.

Main hantavirus lineages and genetic description of different hantavirus species

Cricetidae/Arvicolinae-borne hantaviruses

The Arvicolinae-borne hantaviruses (family Cricetidae) have been found in the rodent genera *Myodes, Microtus* and *Lemmus,* from all over the world (Table 3.1). These viruses represent interesting twists in their evolutionary history not following the strict host-virus co-evolution. Phylogenetically, these viruses are separated into three different clusters. First cluster is formed by *Myodes*-carried viruses: PUUV and two tentative species MUJV and HOKV. Only PUUV is firmly associated with disease, nephropathia epidemica that is a mild form of haemorrhagic fever with renal syndrome. The *Microtus*-carried hantaviruses are separated into two clusters, the first of these includes KHAV, two tentative species VLAV and YUJV, and TOPV carried by lemmings. This cluster is phylogenetically closer to the *Myodes*-carried hantaviruses than to the other *Microtus*-borne viruses that form the third cluster: TULV, PHV and ISLAV. This cluster of *Microtus*-borne hantaviruses includes the important non-pathogenic model virus TULV from Europe, another model virus PHV from USA, and the poorly characterized ISLAV from USA. All these *Microtus*- and *Lemmus*-carried viruses are apparently non-pathogenic to humans; the basis for this is yet unknown.

Table 3.1 List of hantaviruses carried by rodents from the family Cricetidae, subfamily Arvicolinae. In ***bold italics*** are the hantavirus species currently recognized by the International Committee on Taxonomy of Viruses; others are tentative species

Virus	Abbreviation	Host	Reference	Full-length sequences		
				S	M	L
Puumala	PUUV	*Myodes glareolus*, bank vole	Brummer-Korvenkontio *et al.* (1980)	NC_005224	NC_005223	NC_005225
Tula	TULV	*Microtus arvalis*, European common vole	Plyusnin *et al.* (1994)	NC_005227	NC_005228	NC_005226
Topografov	TOPV	*Lemmus sibiricus*, lemming	Plyusnin *et al.* (1996)	AJ011646	AJ011647	
Khabarovsk	KHAV	*Microtus maximowiczii*, Maximowicz vole	Hörling *et al.* (1996)	U35255	AJ011648	
		Microtus fortis, reed vole				
Prospect Hill	PHV	*Microtus pennsylvanicus*, meadow vole	Lee *et al.* (1982, 1985)	Z49098	X55129	EF646763
Isla Vista	ISLAV	*Microtus californicus*, Californian vole	Song *et al.* (1995)	U19303		
Vladivostok	VLAV	*Microtus fortis*, reed vole	Kariwa *et al.* (1999)	FJ170796		
Muju	MUJV	*Myodes regulus*, royal vole	Song *et al.* (2007)	DQ138142	EF198413	
Hokkaido	HOKV	*Myodes rufocanus*, red bank vole	Kariwa *et al.* (1995)	AB010730		
Yuanjiang	YUJV	*Microtus fortis*, reed vole	Zou *et al.* (2008)	FJ170795		

Puumala virus (PUUV) was discovered in Finland in 1980 (Brummer-Korvenkontio *et al.*, 1980). PUUV is the champion when it comes to the number of sequences available with over 600 deposits in the GenBank. These sequences show high diversity that even exceeds the limits for hantavirus species definition. Five strains have been fully sequenced: prototype strain Sotkamo from Finland, three Russian strains (Kazan, DTK/Ufa-97 and CG1820), and one from Sweden (Umeå/hu). PUUV is found in most of Europe and in European parts of Russia, and phylogenetically it forms geographical lineages. The overall phylogeny is star-like, suggesting early split of lineages and independent evolution of the lineages thereafter (Sironen *et al.*, 2001).

The full-length L segment sequences are 6550 nt long with diversity of 13–19%, except for the two closely related Russian strains DTK/Ufa-97 and CG1820 that differ only by 0.3%. The ORF encodes the RNA-dependent RNA polymerase that is 2156 aa long, with the exception that the human isolate Umeå/hu has a deletion of 8 aa in the C-terminal part of the protein. The full-length aa sequences display diversity of 0–7%. The currently available full-length sequences are all from a geographically narrow area, and the level of diversity could be much higher if sequences from other PUUV lineages were known.

Altogether, eight full-length M segment sequences are available with eight additional full-length coding sequences. The M segments sequences are 3682 nt long, with a 5′NCR region of 30 nt and 3′NCR of 195 nt. The 3′NCR is longer in the M segment than in the L, but it can still be easily aligned. The full-length sequences differ by 10–20% and the three closely related Russian strains by 0.3%. The 16 full-length coding sequences are all 3444 nt long, differing from each by 7–20%, while the very closely related Russian sequences differ by only 1%. The ORF encodes for a precursor for the two glycoproteins, and the diversity of the 1148 aa long sequences is 0–11%. Some of the PUUV strains are thus more divergent from each other than the limit set for taxonomically different hantavirus species that is more than a 7% difference in the aa sequence of the glycoproteins.

Most of the PUUV sequences available are from the S segment with 47 full-length sequences (1784–1882 nt) and altogether 87 complete and unique coding sequences. In addition, there are close to 300 partial S segment sequences available. The NCRs are 42 nt long at the 5′ and 440–538 nt long at the 3′-end. These long 3′-NCRs are highly divergent and cannot be reliably aligned except for the last ~100 nucleotides. The diversity of the full-length sequences is up to 26%, and for the complete coding sequences it is up to 19%. The aa sequences of the N protein (433 aa) are 92–100% identical, and thus these too exceed the species definition limit of 7%.

PUUV strains form seven geographical lineages, and between these lineages the nt diversity is 11–19%. The intra-lineage diversity is 0–9% for those lineages that originate from a narrow geographical area. When the area widens, the nt diversity between strains of a particular lineage increases to 15% as seen for the Russian and Southern Scandinavian lineages. Within a local rodent population the sequence diversity is 0–5% (Plyusnin *et al.*, 1995; Razzauti *et al.*, 2008, 2009). Interestingly, in Konnevesi, Finland, the L segment diversity was up to 9.7% within the local rodent population (Razzauti *et al.*, 2008). Within an individual rodent the quasispecies diversity of PUUV is 0–1% (Plyusnin *et al.*, 1995; Sironen *et al.*, 2008). PUUV NSs sequences differ by 0–15% on the nt level, and thus they are slightly more conserved than the full-length S segment sequences. The translated NSs protein sequences differ by 0–36%.

Muju virus (MUJV) is a tentative new hantavirus species carried by *Myodes regulus,* the royal vole. MUJV was discovered from vole tissue samples collected from various regions in Korea between 1996 and 2007 (Song *et al.,* 2007a). Earlier it has been noted that more than 5% of the HFRS cases in Korea exhibit higher antibody titre to PUUV than to HTNV, suggesting that a PUUV-like hantavirus could be the causative agent of these cases. Whether MUJV is pathogenic to humans may only be confirmed after successful isolation of this virus in cell culture.

Full-length S and M segment sequences are available with additional partial sequences, and these show geographical clustering. The M segment sequence (1142 nt) has NCR's of similar length to PUUV and the sequence diversity is 13–15% on the nt level and on the aa level 10–12%. In the N-terminal part of the glycoprotein Gn, MUJV has a deletion of 6 aa. The analysis of the S segment sequences (1857 nt) shows a slightly lower level of diversity from PUUV with 4–8% on the aa level. Even though this difference is less than the general limit for species definition, MUJV is most probably a distinct species, since it has a distinct host, it is phylogenetically clearly different from PUUV and the glycoprotein difference is high enough.

Hokkaido virus (HOKV) is another tentative hantavirus species carried by *Myodes*-rodents. The first sequences were discovered from *Myodes rufocanus* in Japan (Kariwa *et al.,* 1995). Later HOKV was also found in China (Zhang *et al.,* 2007) and in Buryatia, Russia (Plyusnina *et al.,* 2008). Serological data suggest that HOKV might be pathogenic to humans (Kariwa *et al.,* 2000), but virus isolation and neutralization studies are needed to confirm this.

The phylogenetic position of HOKV is somewhat uncertain, and either there is no monophyly or the support for monophyly is low. HOKV strains from China seem to be highly divergent from those found in Japan and Russia. It might very well be that the Fusong-strains from China belong to yet another hantavirus species distinct from HOKV. MUJV is ancestral to PUUV and HOKV that share a more recent common ancestor. The HOKV N protein sequences (433 aa) are 5–7% different from MUJV and 4–9% different from PUUV sequences. One full-length M segment sequence (3639 nt) is available from HOKV in China, and similarly to MUJV, there is a deletion in the N-terminal part of the Gn protein.

Topografov virus (TOPV) is carried by *Lemmus sibiricus* (Plyusnin *et al.,* 1996a; Vapalahti *et al.,* 1999) lemmings. These are arvicoline rodents that are phylogenetically ancestral to both *Myodes*- and *Microtus*-rodents. TOPV is thought to be apathogenic to humans, although serologically it is not possible to distinguish antibody reactivity between PUUV/TOPV/KHAV antigens. Full-length M segment sequence is 3728 nt long, encoding a glycoprotein precursor of 1142 aa, and the full S sequence is 1951 nt long, encoding nucleocapsid protein of 433 aa. TOPV is closely related to the microtine-carried KHAV, and these two differ at the amino acid level by only 4% for the S segment, while the M segment sequences differ by 12% at the amino acid level. Host-switching scenarios have been proposed to explain the phylogenetic position of TOPV (Vapalahti *et al.,* 1999) that later seems to have jumped from lemmings to microtine rodents to yield KHAV. Long deletions in the NCRs of KHAV S and M segment sequences suggest that TOPV is the more ancestral one of this pair.

Khabarovsk virus (KHAV) was discovered initially from the reed vole, *Microtus fortis,* in the Far East Russia (Hörling *et al.,* 1996). Later, KHAV strains (named Yakeshi-strains) were found in China from *Microtus maximowiczii* (Zou *et al.,* 2008c), and this rodent could in fact be the true host for KHAV. The different strains from Russia and China are very similar, they differ in the N protein amino acid sequences by only 1.5%. Full M and S segment sequences are available: M is 3695 nt long (1144 aa), and S 1843 nt (433 aa).

Vladivostok virus (VLAV) is a tentative hantavirus species that was originally discovered in Vladivostok, Far East Russia, in the same host as the first KHAV sequences, *M. fortis* (Kariwa *et al.*, 1999). Later, more sequences were recovered in China (Fusong-strains) from the same host, and these were highly similar to the original VLAV strain (Zou *et al.*, 2008c), thus strengthening the view that VLAV is a distinct hantavirus species. VLAV strains have been identified also in the Shenyang city in China (Zou *et al.*, 2008d) and in Buryatia, Russia (Plyusnina *et al.*, 2008). Analysis of the S segment amino acid sequences (433 aa) reveals that the strains differ by less than 5%. In Buryatia, one VLAV strain was recovered from *Microtus oeconomus*, and it appeared highly identical to other VLAV sequences. This could represent a case of spill-over from the host to another sympatric rodent species, which is a phenomenon known to take place for hantaviruses. Alternatively, VLAV could be carried by two different hosts. Yuanjiang strains (YUJV) from China present one more twist to the story of *Microtus fortis*-carried hantaviruses (Zou *et al.*, 2008d). These appear to be highly divergent from VLAV strains, especially in the M segment amino acid sequences that differ by more than 8%. The S segment sequences are slightly closer with a 5% divergence. The analysis of hosts showed that there are two distinct lineages of *M. fortis*, one of these carries VLAV strains and the other one the Yuanjiang strains. YUJV might be a distinct novel hantavirus species or alternatively it is a lineage of VLAV. In any case, hantaviruses associated with *M. fortis* demonstrate great complexity in China. This region is endemic for HFRS, and it will be important to study the pathogenicity of these different hantaviruses to humans.

Tula virus (TULV) is a widely used non-pathogenic model hantavirus that was originally discovered in two sibling species of the European common vole, *Microtus arvalis* and *Microtus levis/rossiaemeridionalis* (Plyusnin *et al.*, 1994). TULV has also been demonstrated in other species such as *M. agrestis* (Scharninghausen *et al.*, 2002) and *M. subterraneus* (Song *et al.*, 2002), but the main reservoir is thought to be *M. arvalis*. No clinical disease is associated with TULV, although some evidence of human infection have been provided (Vapalahti *et al.*, 1996; Klempa *et al.*, 2003).

TULV was first described from voles in central Russia, and later it has been found in several countries in Central and Eastern Europe. Phylogenetically, TULV strains are monophyletic with distinct, geographical lineages. TULV strains originating from Slovakia were the first recombinant hantavirus sequences to be discovered (Sibold *et al.*, 1999), and this mechanism was later confirmed *in vitro* (Plyusnin *et al.*, 2002b, 2005). Quasispecies nature of hantavirus sequences was also first described with TULV sequences (Plyusnin *et al.*, 1996b).

The number of TULV sequences in the GenBank is over 150, including one fully sequenced strain Tula/Moravia/5302v/95. In addition to the prototype strain, full-length M segment sequence is available for one TULV strain (Tula/Moravia/5286Ma/94), and full-length S segment sequence for 21 strains. The TULV L segment sequence is 6541 nt long, and it contains one ORF that encodes a 2153 aa long RNA-dependent RNA-polymerase with typical arrangement of conserved motifs. The two full M segment sequences (3694 nt) differ by 2% (nt), and they both contain an ORF encoding a 1141 aa glycoprotein precursor protein.

The S segment sequences are 1824–1852 nt long, and in contrast to PUUV, also the NCRs can be reliably aligned. The sequence diversity is up to 16%. The S segment contains two overlapping ORFs encoding a nucleocapsid protein and a non-structural protein NSs. The N protein is 429 or 430 aa long, and the diversity is up to 6%. The NSs sequences display up to 14% diversity on the nt and up to 37% on the aa level.

Prospect Hill virus (PHV) is another widely used non-pathogenic model virus. It was the first hantavirus discovered in the American continent (Lee *et al.*, 1985), and it was found from meadow voles (*Microtus pennsylvanicus*). Related strains named Bloodland Lake virus (BLLV) have been discovered from *M. ochrogaster*, the prairie vole, and these are regarded as a subtype of PHV.

The fully sequenced strain PH-1 has an L segment that is 6559 nt long, including an ORF for a 2155-aa long RNA-dependent RNA polymerase. The NCRs of PHV L segment are highly similar to those in other Arvicolinae-borne hantaviruses, with the exception of a 12-nt insertion in the 3′NCR. The 3707-nt long M segment includes one ORF that encodes for a 1142 aa long glycoprotein precursor. The S segment sequences are 1675 nt long, and that of BLLV is 1722 nt with insertions in the 3′NCR. The diversity is up to 8% for the PHV strains, and BLLV differs from these by 21% on the nt level. The S segment encodes for a 433 aa nucleocapsid protein with 0–2% diversity for the PHV strains, and 7% difference from these to BLLV. The S segment encodes also for another ORF for putative NSs proteins. In this region, the sequence diversity is up to 15% on the nt and up to 34% on the aa level.

Isla vista virus (ISLAV) is another North American *Microtus*-borne hantavirus. It was described initially from California voles (*Microtus californicus*) (Song *et al.*, 1995). Spill-over of ISLAV has been detected in California mouse (*Peromyscus californicus*), in which also SNV has been reported. No L segment sequences are available and only very short partial M segment sequences. Three full-length S segment sequences are available, these are 1720 nt long, and their diversity is only 1–2%. The N protein sequences are 433 aa long, and they differ only by one or two amino acids. An ORF for putative NSs protein is present, and the sequences are all identical.

Cricetidae/Sigmodontinae-borne hantaviruses

The Sigmodontinae-borne hantaviruses originate mainly from the South America with few North American ones (Table 3.2). The relationships of these viruses are complicated, and many discrepancies between S and M segment phylogenies exist suggesting multiple reassortment events as well as host-switching events. Many of the Sigmodontinae-borne hantaviruses are not isolated nor fully characterized with only few sequences available, and thus final species specification still has to wait for further analysis. In fact, there is only one full L segment sequence, that of ANDV. Furthermore, the diversity of the reservoir rodents in South America is an impressive example of broad adaptive radiation, and thus hantaviruses carried by these rodents also show high diversity as they follow the on-going speciation events of their hosts.

The Sigmodontinae-borne hantaviruses can be roughly split in three different clusters. One of these contains ANDV and numerous ANDV-like strains, as well as the tentative species ARAV. The ANDV-cluster shares common ancestry with the second cluster that includes RIOMV, LANV and Choclo as well as some other tentative species. The third cluster is formed by BAYV, BCCV and MULV, and these are HCPS-causing viruses from

Table 3.2 List of hantaviruses carried by rodents from the family Cricetidae, subfamily Sigmodontinae. In ***bold italics*** are the hantavirus species currently recognized by the International Committee on Taxonomy of Viruses; others are tentative species

				Full-length sequences		
Virus	Abbreviation	Host	Reference	S	M	L
Andes	ANDV	*Oligoryzomys longicaudatus,* long-tailed pygmy rice rat	Lopez *et al*. (1996); Meissner *et al*. (2002)	NC_003466	NC_003467	NC_003468
Bayou	BAYV	*Oryzomys palustris,* marsh rice rat	Morzunov *et al*. (1995)	L36929	L36930	
Black Creek Canal	BCCV	*Sigmodon hispidus,* hispid cotton rat	Rollin *et al*. (1995); Ravkov *et al*. (1995)	L39949	L39950	
Cano Delgadito	CADV	*Sigmodon alstoni,* Alston's cotton rat	Fulhorst *et al*. (1997)	DQ285566	DQ284451	
Laguna Negra	LANV	*Calomys laucha,* vesper mouse	Johnson *et al*. (1997)	AF005727	AF005728	
Muleshoe	MULV	*Sigmodon hispidus,* hispid cotton rat	Rawlings *et al*. (1996)	U54575		
Rio Mamore	RIOMV	*Oligoryzomys microtis,* small-eared pygmy rice rat	Bharadwaj *et al*. (1997)	U52136		
Araraquara	ARAV	*Bolomys lasiurus,* hairy-tailed bolo mouse	Johnson *et al*. (1999); Suzuki *et al*. (2004)			
Calabazo		*Zygodontomys brevicauda,* short-tailed cane mouse	Vincent *et al*. (2000)			
Castelo dos Sonhos		Not known	Johnson *et al*. (1999)			
Choclo		*Oligoryzomys fulvescens,* fulvous pygmy rice rat	Vincent *et al*. (2000)	DQ285046	DQ285047	

North America. CADV is a special case: its S segment may have originated from Neotominae-associated hantavirus lineage while the M segment stands alone in phylogeny.

Andes virus (ANDV) is the most important aetiological agent of HCPS in southern South America. It was the first hantavirus reported from South America (Lopez *et al.*, 1996), and the prototype strain from Chile was described later in detail (Meissner *et al.*, 2002). Ever since new sequences have been reported with high speed, and currently there are almost 400 ANDV sequences deposited in the GenBank. Two strains are fully sequenced (Chile-9717869 and Chile-7913), and in addition to these, eight full S and four full M segment sequences are known.

The predominant reservoir of ANDV is the long-tailed pygmy rice rat, *Oligoryzomys longicaudatus* (Levis *et al.*, 1998). Spillover into other species seems to happen relatively frequently, and strains of ANDV have indeed been described from other rodent species as well. Whether these hantavirus strains represent novel species or variants of ANDV, is not yet determined.

One monophyletic subcluster of ANDV-like strains have been described from the black-footed pygmy rice rat, *Oligoryzomys nigripes*, from different countries in South America. These include, for example, Itapua37 and Itapua38 from Paraguay (Chu *et al.*, 2006), and Juquitiba strains from Uruguay, Paraguay, and Brazil (Suzuki *et al.*, 2004). These sequences are closely related to Araucaria sequences from HCPS-patients in Brazil (Raboni *et al.*, 2005).

Another subcluster is composed of hantavirus strains hosted by akodontine rodents: Pergamino carried by *Akodon azarae* (Levis *et al.*, 1998, Bohlman *et al.*, 2002) in Argentina, and Maciel carried by *Necromys benefactus* (formerly *Bolomys obscurus*). This subcluster is sometimes (especially M-tree) combined with Araraquara-like strains that also have akodontine-rodents as hosts.

A third cluster is composed of *Oligoryzomys*-carried strains. These include the HCPS-associated Oran, carried by *Oligoryzomys longicaudatus* (Levis *et al.*, 1997) and found in Argentina; similarly HCPS-associated Lechiguanas carried by *Oligoryzomys flavescens*, from Argentina and Uruguay; as well as Bermejo, carried by *Oligoryzomys chacoensis* in Argentina, and Bermejo-Neembucu from Paraguay and Argentina (Chu *et al.*, 2006). A fourth cluster is composed of the prototype ANDV with sequences derived from HCPS-patients.

The two ANDV L segment sequences are 6562 nt long, and they are highly identical differing only by 1% on the aa level. The encoded RNA-dependent RNA-polymerase is 2153 aa long with all typical motifs present. The M segment sequences are 3634–3671 nt long encoding a 1138 aa glycoprotein precursor. The diversity of the M segment coding regions is up to 21% on the nt level, and up to 8% on the aa level. The full-length S segment sequences are 1869–1938 nt long, and the diversity of the coding regions is up to 19% on the nt level. The diversity of the encoded 428 aa N proteins is up to 6%. Castelo dos Sonhos virus is a tentative hantavirus species that was recovered from a patient in Brazil with partial sequences only (Johnson *et al.*, 1999). The rodent host has not been described, and based on the aa similarities, this sequence could represent a variant of ANDV.

Araraquara virus (ARAV) is a tentative hantavirus species first described from a fatal HCPS case in Brazil (Johnson *et al.*, 1999). The host was identified later, and it is the hairy-tailed bolo mouse (*Bolomys lasiurus*) (Suzuki *et al.*, 2004). One full-length S segment sequence is available (strain P5/Cajuru), and partial sequences are available from patients (e.g. strain Paranoa) (Melo-Silva *et al.*, 2009). Based on the aa similarities to ANDV, ARAV could be defined as a lineage of ANDV. On the other hand, all sequences from *B. lasiurus* clearly cluster together to form their own lineage separate from ANDV. They do share a common ancestor with ANDV strains derived from akodontine rodents, and neutralization assays are needed in order to establish ARAV as a hantavirus species.

Rio Mamore virus (RIOMV) was first described from the small-eared pygmy rice rat, *Oligoryzomys microtis*, captured in Bolivia (Bharadwaj *et al.*, 1997). Later, RIOMV was isolated in cell culture from rodents collected in Peru (Powers *et al.*, 1999). No human cases of RIOMV have yet been described. One full-length S segment sequence of 1975 nt is available (strain OM-556), and it encodes for a 428-aa N protein and a putative NSs protein. Very recently, a full sequence of RIOMV was reported from strain HTN-007, isolated from a small-eared pygmy rice rat captured in Peru (Richter *et al.*, 2010). RIOMV shares common ancestry with LANV, from which it differs by 7% on the aa level (N protein). A closely related strain, Alto Paraguay, has been described from *Holochilus chacarius* with N protein aa diversity of only 4%. Similarly, closely related sequences with 4–6% aa diversity have been found in Brazil: Anajatuba from *Oligoryzomys fornesi* and Rio Mearim from *Holochilus sciureus* (Rosa *et al.*, 2005).

Laguna Negra virus (LANV) was discovered as the causative agent of a large HCPS-outbreak in Paraguay (Johnson *et al.*, 1997). The virus was found both in patients and in tissue samples of the small vesper mouse, *Calomys laucha*. Full-length S (1904 nt) and M (3698 nt) segment sequences are available, and these encode a 428 aa N protein and a 1138 aa glycoprotein precursor. LANV has also been found in *C. callosus*, the large vesper mouse, captured in Bolivia (Carroll *et al.*, 2005).

Choclo virus (CHOV) is a tentative hantavirus species known to cause HCPS in Panama, Central America. It was discovered from HCPS-patients and the reservoir was identified to be the fulvous pygmy rice rat, *Oligoryzomys fulvescens* (Vincent *et al.*, 2000). CHOV has been isolated, and its infectivity for Syrian hamsters has been tested, but unlike ANDV, CHOV causes a non-pathogenic infection in this hamster-model (Eyzaguirre *et al.*, 2008). Full-length S and M segment sequences are available, the S segment is 1972 nt long, and it encodes for a 428 aa N protein, as well as a NSs protein. The M segment is 3685 nt long encoding for a 1138 aa glycoprotein precursor. CHOV sequences are clearly distinct, the closest hantaviruses being ANDV and ARAV from which CHOV differs by with 9–11% for the N protein, and for the glycoprotein precursor by 17%.

Maporal virus has been identified and isolated from the same host as CHOV, the fulvous pygmy rice rat, but Maporal is found from Venezuela (Fulhorst *et al*, 2004). The full-length S segment (strain HV-97021050) is 1859 nt encoding for a 428 aa N protein, and a putative NSs protein. The full-length M segment is 3684 nt long encoding for a1139 aa glyprotein precursor. The full-length L segment sequence has been reported very recently (Richter *et al.*, 2010). Maporal appears to be distinct from all other hantaviruses, including CHOV, with at least 8% aa diversity for the N protein, and 15% diversity for the glycoprotein precursor. There is also another report of Maporal virus isolation from an arboreal rice rat *(Oecomys bicolor)* captured in central Venezuela, and it has been used to infect Syrian hamsters (Milazzo *et al.*, 2002). Unlike CHOV, the infection was mimicking HCPS in a similar fashion to ANDV. Sequences from this report are not available in the GenBank, and thus it is not clear whether Maporal from the two different hosts, is the same virus or not. The cluster RIOMV/LANV/CHOV also contains a hantavirus strain Ape Aime-Itapúa (AAI) from *Akodon montensis* (Chu *et al.*, 2006). Based on the level of sequence diversity of the S segments, AAI could represent a novel hantavirus species. AAI might also be a reassortant virus, since the S segment is related to RIOMV and LANV, while M segment clusters with other Akodon-carried ANDV strains Pergamino and Maciel.

Bayou virus (BAYV) is, after SNV, the second most important causative agent of HCPS in North America. It was originally discovered from a HCPS-patient in Louisiana (Morzunov *et al.*, 1995). The reservoir host was later identified to be *Oryzomys palustris*, the marsh rice rat (Torrez-Martinez *et al.*, 1998). Full-length S (1958 nt) and M (3677 nt) segment sequences are available, and they encode a 428 aa N protein and a 1141 aa glycoprotein precursor.

A closely related strain, Catacamas, was detected in *Oryzomys couesi* in Honduras (Milazzo *et al.*, 2006). It differs from BAYV by 5% for the N protein sequences and by 6% for the glycoprotein, and thus it could be a variant of BAYV. In another subspecies of *O. couesi*, a variant named Playa de Oro was described from Mexico (Chu *et al.*, 2008). The N protein sequence of this strain differs from BAYV by 7%, and from Catacamas by 8%.

Black Creek Canal virus (BCCV) has been isolated in cell culture, and it has been used as a model hantavirus both in cell culture and in an experimental animal model. It was discovered from cotton rats (*Sigmodon hispidus*) in Florida (Ravkov *et al.*, 1995), and it has been associated with HCPS (Khan *et al.*, 1996). The only sequences available are those from the original isolate: full-length S (1989 nt) and M (3668 nt) segment sequences encode for a 428 aa N protein and a 1141 aa glycoprotein precursor. BCCV shares common ancestry with BAYV, MULV, and BAYV-like strains, and it differs from them by 7–10% for the N protein, and by 11% for the glycoprotein sequences.

Muleshoe virus (MULV) was discovered in Texas from the same host as BCCV, the cotton rat (Rawlings *et al.*, 1996). One full-length S segment sequence (1989 nt) is available, and the N protein sequence differs from BAYV and BCCV by 7–10%.

Cano Delgadito virus (CADV) was isolated from *Sigmodon alstoni*, Alston's cotton rat, in Venezuela (Fulhorst *et al.*, 1997), and it is the first hantavirus discovered from rodents indigenous to South America. The position of CADV in phylogenetic trees is uncertain suggestive of reassortment: the S segment shares ancestry with Neotominae-borne hantaviruses while the M segment is placed separately from any major lineages. The M segment sequence is 3675 nt long, and it encodes for a 1138 aa long glycoprotein precursor. The S segment is 1986 nt long encoding for a 428 aa N protein and putative 63 aa NSs protein. Both S and M segment sequences are clearly distinct from all other Sigmodontinae-borne hantaviruses.

Calabazo virus is a tentative hantavirus species discovered from *Zygodontomys brevicauda*, the short-tailed cane mouse, in Panama (Vincent *et al.*, 2000). Only short partial sequences are available, but phylogenetically it holds an interesting position between Sigmodontinae- and Neotominae-borne hantaviruses.

Cricetidae/Neotominae-borne hantaviruses

Neotominae-borne hantaviruses come in two groups, the HPS-causing ones carried by Peromyscus-mice (SNV and NYV), and the apparently non-pathogenic ones carried by Reithrodontomys rodents (RIOSV and ELMCV) (Table 3.3). In addition, there is one in between: although LSCV is carried by *Peromyscus leucopus*, it is more closely related to the *Reithrodontomys*-carried hantaviruses. These viruses are all found only in North America.

Table 3.3 List of hantaviruses carried by rodents from the familiy Cricetidae, subfamily Neotominae. In ***bold italics*** are the hantavirus species currently recognized by the International Committee on Taxonomy of Viruses; others are tentative species

				Full-length sequences		
Virus	Abbreviation	Host	Reference	S	M	L
Sin Nombre	SNV	*Peromyscus maniculatus,* deer mouse	Nichol *et al*. (1993)	NC_005216	NC_005215	NC_005217
New York	NYV	*Peromyscus leucopus,* white-footed mouse	Hjelle *et al*. (1995b)	U09488	U36801	
El Moro Canyon	ELMCV	*Reithrodontomys megalotis,* western harvest mouse	Hjelle *et al*. (1994)	U11427	U26828	
Rio Segundo	RIOSV	*Reithrodontomys mexicanus,* Mexican harvest mouse	Hjelle *et al*. (1995)	U18100		
Limestone canyon	LSCV	*Peromyscus boylii,* brush mouse	Sanchez *et al*. (2001)	AF307322	AF307323	

The genetic relationships of hantaviruses carried by *Peromyscus*-rodents are complex, most likely as a result of the rapid radiation and speciation of New World Neotominae rodents and occasional host-switching events. Furthermore, phylogenetically Neotominae- and Sigmodontinae-borne cannot be separated, and it looks like reassortment could also happen between these lineages of hantaviruses.

Sin Nombre virus (SNV) is the prototype North American hantavirus that was discovered as the causative agent of an HCPS outbreak in 1993 (Nichol *et al.*, 1993). 146 sequences are currently deposited in the GenBank, with three strains fully sequenced. These include the prototype strain NM H10 from an HPS-patient, the same strain after adaptation to cell culture (NM R11), and the strain Convict Creek 107 isolated in cell culture from deer mice captured in California. The reservoir host of SNV is the deer mouse (*Peromyscus maniculatus*). Closely related sequences have been found in *P. leucopus*, the host of NYV, and these strains are named Blue River virus. SNV-like sequences named Monongahela virus have been discovered from the forest form of deer mice, *P. maniculatus nubiterrae* (Song *et al.*, 1996).

SNV has been widely used both in cell culture models, but also as a model to study hantavirus properties in the host animal, in this case the deer mouse. It has been shown that *P. maniculatus* forms several divergent phylogeographical lineages, and similarly, different variants of SNV are detected in the different host clades: e.g. SNV in the western USA and Monongahela in the eastern USA and Canada (Monroe *et al.*, 1999; Dragoo *et al.*, 2006). Whether these different Peromyscine-carried viruses (NYV, Monongahela and Blue River) are distinct hantavirus species or not, can only be determined after neutralization studies. Based on the nt and aa diversities, these strains are below the species demarcation cut-off line, and similarly is NYV.

The full-length L segment sequences are 6562 nt long encoding a 2153 aa long RNA-dependent RNA-polymerase. The NM H10 strain shows only minimal genetic changes after cell-culture adaptation, and the difference to Convict Creek 107 is 12% on the nt and 1% on the aa level. The M segment sequences are 3696 nt encoding a glycoprotein precursor of 1140 aa. The diversity is 11–20% for the nt coding sequences and 1–7% for the aa sequences. The full-length S segment sequences are 2047–2083 nt long, encoding for a 428 aa N protein, and a putative NSs protein of 63 aa. The diversity of the S segment coding sequences is 3–18% on the nt and 0–7% on the aa level.

New York virus (NYV) is another HCPS-associated hantavirus from the United States, it is carried by the white-footed mouse, *Peromyscus leucopus* (Hjelle *et al.*, 1995b; 1995c). NYV was first found in patients, and soon after highly identical sequences were determined from the white-footed mice. NYV has been isolated and thereafter used as a model virus in cell culture studies. Only a few sequences are available in the GenBank including one full-length S segment sequence of 2078 nt encoding a 428-aa long N protein, and with an unusually long 3′NCR. The three full-length M segment sequences of 3668 nt encode a 1140-aa long glycoprotein precursor. The amino acid identity to SNV is 7–8% for the N protein, and 4–7% for the glycoprotein precursor, and these values are below the 7% cut-off line. These two viruses are nevertheless antigenically distinct, and thus they are regarded as different virus species.

El Moro Canyon virus (ELMCV) was discovered during hantavirus surveillance in the United States, from the most common *Reithrodontomys*-species rodent, the western harvest mouse (*R. megalotis*) (Hjelle *et al.*, 1994). Full-length S and M segment sequences are known from the strain RM-97. The M segment sequence is 3801 nt long, and the 3′NCR is longer than on the average for hantavirus. The M segment encodes for a glycoprotein precursor of 1139 aa. The diversity from other Neotominae-borne hantaviruses is 29% on the nt and 21% on the aa level. ELMCV S segment sequence is 1896 nt long, and it encodes for a 428 aa N protein. The ORF for a putative NSs protein is present. The S segment sequence differs from the other Neotominae-borne hantaviruses by 24% on the nt and 16% on the aa level.

Rio Segundo virus (RIOSV) was discovered while screening *Reithrodontomys*-rodents from the USA and Mexico for hantaviruses after the discovery of ELMCV. RIOSV was recovered from the Mexican harvest mouse (*Reithrodontomys mexicanus*) in Costa Rica (Hjelle *et al.*, 1995a). Only one sequence is available (strain HMV-2), and that is a full-length S segment sequence of 1749 nucleotides. It is most closely related to ELMCV with a 21% diversity on the nt and 9% diversity on the aa level. The ORF for putative NSs protein can be found in the sequence, but it contains an earlier stop codon, and the protein would be 11 aa shorter.

Limestone Canyon virus (LSCV) is a tentative hantavirus species carried by the brush mouse, *Peromyscus boylii* (Sanchez *et al.*, 2001). Seropositive mice were found during mark-recapture trapping studies in the USA. Nearly full-length M segment and partial S segment sequences are available, and the S sequences are most closely related to *Reithrodontomys*-carried ELMCV and RIOSV, while the M sequences are equally distant from all other Neotominae-borne viruses.

Muridae/Murinae-borne hantaviruses

The Murinae-borne hantaviruses (Table 3.4) are the causative agents of HFRS, and they are responsible for most of the human hantavirus infections. Muridae rodents are a large group of animals, and since hantaviruses have been recovered only from one subfamily (Murinae), it may be speculated that many hantaviruses are yet to be discovered. In fact, there are already patient-derived sequences in the GenBank holding such positions in phylogeny that these most probably represent novel hantaviruses. Some of these misidentified strains have already been further characterized, including the tentative species DBSV.

The Murinae-borne hantaviruses have interesting host–virus relationships including multiple host-switching events, and reassortment possibly even between different hantavirus species. Three major lineages can be found, first one of these is composed of the prototype hantavirus, HTNV, and two tentative species DBSV and ASV. These all have closely related sequences, but they nevertheless display unique neutralization patterns, and they also have unique reservoir hosts. The second lineage includes viruses carried by rats: the cosmopolitan SEOV, tentative GOUV, as well as THAIV and Serang virus. The third lineage includes European hantaviruses carried by *Apodemus* mice: DOBV and SAAV. The first African hantavirus, Sangassou, seems to be somewhat related to the DOBV and SAAV.

The Murinae-borne hantaviruses are clearly distinct from the Cricetidae-borne ones, but interestingly some of the insectivore-borne hantaviruses seem to share the same ancestry. A clear difference is found in the S segment: Cricetidae-borne hantaviruses have an additional ORF encoding for an NSs protein while neither Murinae- nor insectivore-borne hantaviruses have this ORF.

Hantaan virus (HTNV) was discovered in the 1970s: first sera from Korean haemorrhagic fever (KHF) patients were found to react with tissue sections of striped field mice (Lee and Lee 1976), next HTNV was isolated in laboratory-colonized field mice (Lee and

Table 3.4 List of hantaviruses carried by rodents from the family Muridae, subfamily Murinae. In ***bold italics*** are the hantavirus species currently recognized by the International Committee on Taxonomy of Viruses; others are tentative species

				Full-length sequences		
Virus	Abbreviation	Host	Reference	S	M	L
Hantaan	HTNV	*Apodemus agrarius coreae,* dark-striped field mouse	Lee *et al*. (1978)	NC_005218	NC_005219	NC_005222
Dobrava	DOBV	*Apodemus flavicollis,* yellow-necked mouse	Avsic-Zupanc *et al*. (1992, 1995)	NC_005233	NC_005234	NC_005235
Saaremaa	SAAV	*Apodemus agrarius agrarius,* striped field mouse	Plyusnin *et al.* (1997); Nemirov *et al*. (1999)	AJ009773	AJ009774	AJ410618
Thailand	THAIV	*Bandicota indica,* great bandicoot rat	Hörling *et al.* (1996)	AB186420	L08756	
Seoul	SEOV	*Rattus norvegicus*, *Rattus rattus,* Norway rat, black rat	Elwell *et al*. (1985)	NC_005236	NC_005237	NC_005238
Amur/ Soochong	ASV	*Apodemus peninsulae,* Korean field mouse	Liang *et al*. (1994), Yashina *et al*. (2001), Lokugamage *et al*. (2004), Baek *et al*. (2006)	AY675349	AY675353	DQ056292
Da Bie Shan	DBSV	*Niviventer confucianus,* Chinese white-bellied rat	Wang *et al*. (2000)	AB027523	AB027115	DQ989237
Gou	GOUV	*Rattus rattus,* black rat[a]	Wang *et al*. (2000)	AF184988	AB027521	
Sangassou	SANGV	*Hylomyscus simus,* African wood mouse	Klempa *et al*. (2006)			
Serang	SERV	*Rattus tanezumi,* Tanezumi rat/Asian house rat	Plyusnina *et al*. (2009)			

[a]In the original publication (most likely *R. tanezumi*).

Lee 1978), and finally in cell culture (French *et al.*, 1981). HTNV is the most important cause of HFRS. The virus is endemic in China, and it is found also in other countries in eastern Asia (reviewed by Kariwa *et al.*, 2007).

Molecular biology of HTNV has been studied widely in cell culture models, and many epidemiological surveys have also been reported. Thus, it is not unexpected that HTNV is well represented in the GenBank with 316 deposited sequences. Seven strains have been fully described: the protype strain 76–118, two cell culture-adapted variants derived from the protype (cl-1 and cl-2), LR1 and Z10 used for vaccine production and two slightly more divergent strains 84FLi and Q32. There are altogether 53 full-length S segment sequences, of which 47 coding sequences are unique, with diversity up to 19% on the nt level. The encoded N protein sequences are 429 aa long with diversity of less than 5%. There is no ORF for an NSs that is found in the Cricetidae-borne hantaviruses. 50 full-length M segment sequences are available (3615–3618 nt long), of which 46 have unique coding regions with diversity up to 21%. The encoded glycoprotein precursor sequences are 1135 aa long with diversity up to 10%. Six full-length L segment sequences (6533 nt) are available, and the nt diversity of them is up to 16%, and the encoded RNA-dependent RNA-polymerase sequences (2151 aa) present diversity of 0–3%.

The main host of HTNV is the striped field mouse, *Apodemus agrarius*, and HTNV sequences are mainly host dependent, and to some extent show geographical clustering (Zou *et al.*, 2008a). Multiple phylogroups of HTNV can circulate simultaneously in one location, as was shown in Guizhou, China. Spillover of HTNV from *A. Agrarius* to *Rattus* rats has been detected, resulting potentially into reassortment between SEOV and HTNV (Zou *et al.*, 2008b). HTNV has also been detected in China in *Apodemus peninsulae*, the host of the tentative hantavirus species ASV (Zhang *et al.* 2007). These *Apodemus*-species are closely related and they often inhabit the same forest, and thus both might be able to act as a HTNV reservoir. One HTNV strain (A16) in the GenBank seems to represent reassortment between HTNV and ASV: the S segment clusters clearly with HTNV while the M segment is placed within the ASV cluster. Furthermore, one HTNV-designated strain Maaji-1 derived from a HFRS patient displays such diversity from HTNV that it most probably represents a novel hantavirus species.

Amur/Soochong virus (ASV) is an HFRS-causing hantavirus present at least in Russia, China and Korea. The first ASV sequences were isolated from HFRS patients in China, but these were designated as HTNV strains H5 and B78 (Liang *et al.*, 1994). The name Amur was proposed when distinct genotypes were described from HFRS patients in Russia (Yashina *et al.*, 2000). The reservoir host was identified, when ASV sequences were recovered from the Korean field mouse, *Apodemus peninsulae*, captured near Amur river in the far east of Russia (Yashina *et al.*, 2001). Further evidence was provided by another survey in Russia, in which ASV was found in *A. peninsulae* (Lokugamage *et al.*, 2002). ASV isolates from patients display distinct neutralizing pattern in comparison to HTNV, and thus ASV has been tentatively named as a novel hantavirus species (Lokugamage *et al.*, 2004). The first isolates from *A. peninsulae* were derived from mice captured in Korea and these viral strains were initially designated as Soochong virus (Baek *et al.*, 2006). Later, it was suggested that these viral strains should be called collectively ASV (Jiang *et al.*, 2007).

Two cell culture-isolated ASV strains have been fully sequenced: Soochong-1 and -2. These have L segment sequences of 6533 nt that differ from each other by 1.4% on the nt level; the encoded L protein sequences of 2151 aa differ by 0.6%. The diversity from HTNV is 19% on the nt and 4–5% on the aa level. Nine full M segment sequences are available with the length of 3615 nt, encoding a glycoprotein precursor of 1135 aa. The diversity within ASV lineage is 2–14% (nt) or 1–4% (aa), and diversity from HTNV is 19–21% (nt) or 7–11% (aa). Altogether, 11 full S segment sequences are available with lengths of approximately 1700 nt encoding a 429 aa nucleocapsid protein. These sequences differ from each other by 1–14% (nt) and 0–3% (aa), and from HTNV by 14–19% (nt) and 2–5% (aa). Thus, the sequence diversity levels alone are not enough to recognize ASV as a novel hantavirus species. Further support by a unique neutralizing pattern, and a unique reservoir host nevertheless, confirms the status.

Da Bie Shan virus (DBSV) is a tentative hantavirus species carried by *Niviventer confucianus*, the Chinese white-bellied rat (Wang *et al.*, 2000). DBSV was discovered in China and it was first classified as HTNV. The virus was isolated, and it was noted that the virus isolate (strain NC167) displayed a unique neutralizing pattern in antigenic characterization when compared with HTNV, supporting the definition of DBSV as novel hantavirus species. It has been speculated that DBSV could represent a host-switch event from *A. agrarius* to *N. confucianus*.

The strain NC167 has been fully sequenced. The L segment sequence is 6533 nt long, encoding a 2151-aa RNA-dependent RNA-polymerase. Sequence diversity to HTNV is 22% on the nt and 8% on the aa level. Two M segment sequences are available, these are 3625 nt long, encoding for a glycoprotein precursor of 1134 aa. These two sequences are highly identical with 3% aa difference between them, and the diversity from HTNV is 24% (nt) or 14–17% (aa). Three full-length S segment sequences are available (NC167, AH09, AH211), encoding for a 429-aa N protein. These sequences are highly identical, and they diverge from HTNV by 19–23% on the nt and by 7–9% on the aa level.

Dobrava virus (DOBV) is the causative agent of severe HFRS in Europe. DOBV sequences were first described from the yellow-necked mouse, *Apodemus flavicollis*, captured in Slovenia (Avsic-Zupanc *et al.*, 1992), and the virus isolate was characterized later (Avisc-Zupanc *et al.*, 1995). Earlier studies had already described HFRS-cases from Yugoslavia (Gligic *et al.*, 1989a; 1989b; 1992), and the viral isolates from these patients were designated as Belgrade virus. These sequences were reported later (Taller *et al.*, 1993), when DOBV had already been described. The name Dobrava-Belgrade virus has been used since, but the current taxonomy designates this virus as DOBV.

DOBV is well represented in the GenBank with 198 sequences. Yet only one strain is fully sequenced, DOBV/Ano-Poroia/Afl9/1999, from Greece. The L segment is 6532 nt long, encoding a 2151-aa RNA-dependent RNA-polymerase. Three full M segment sequences are available, and they are 3635–3648 nt long, encoding a glycoprotein precursor of 1133 aa. Sequence diversity is 6–7% on the nt level, and 0–1% on the aa level. DOBV M coding sequences differ from the closely related SAAV by 14–18% on the nt, and by 6% on the aa level. Four full S sequences are available, these are up to 1700 nt long, encoding a 429 aa nucleocapsid protein. Sequence diversity of the coding region is low: 2–5% on the nt level and 0–1% on the aa level.

DOBV-like sequences have also been reported from *Apodemus ponticus*, captured in Russia (Klempa *et al.*, 2008). Similar sequence from an HFRS-patient is also deposited in the GenBank. Both full M and S sequences are available, and these differ on the aa level from DOBV by 6–7% and 2–3%. These strains may represent a lineage of DOBV or even a novel hantavirus species.

Saaremaa virus (SAAV) is the most recently established hantavirus species. It causes moderate HFRS, and it has been described from many European countries and Russia. SAAV was first misidentified as a variant of DOBV carried by the western form of *Apodemus agrarius* (Plyusnin *et al.*, 1997). When SAAV was isolated and characterized, the neutralizing pattern was noticed to be different from the prototype DOBV (Nemirov *et al.*, 1999). Furthermore, both SAAV and DOBV can co-circulate simultaneously at the same region, and yet remain separate, i.e. these two viruses are reproductively isolated (Avsic-Zupanc *et al.*, 2000). Later it was confirmed that SAAV and DOBV are indeed different hantavirus serotypes (Sjölander *et al.*, 2002), and they seem to display different pathogenicity to humans (Golovljova *et al.*, 2007).

SAAV not only shares interesting ancestry with DOBV, but also HTNV that is carried by the eastern subspecies of the same host, *A. agrarius*. A host-switch of the common ancestor of DOBV and SAAV from *A. flavicollis* to the western form of *A. agrarius* has been suggested to explain discrepancy between virus and host phylogenies (Nemirov *et al.*, 2002).

One strain of SAAV is fully sequenced: Saaremaa-160V. The full L segment is 6532 nt long, encoding a 2151 aa RNA-dependent RNA-polymerase that differs from DOBV by 2% (aa). Four full M segment sequences of 3643–3645 nt encode a 1133 aa glycoprotein precursor; sequence diversity is 2–13% on the nt level, and 0–4% on the aa level. The diversity from DOBV is 6% (aa) and from DOBV in *A. ponticus* 9–10%. Fifteen full S sequences are approximately 1700 nt long, encoding a 429 aa nucleocapsid protein. Sequence diversity of the coding region is 0–14% on the nt level and 0–4% on the aa level; diversity from DOBV is 2–4% on the aa level.

Sangassou virus (SANGV) was the first African hantavirus ever discovered. It was found in an African wood mouse, *Hylomyscus simus*, captured in Guinea (Klempa *et al.*, 2006). Earlier serological surveys had reported possible human HTNV-like infections, but since *Apodemus agrarius*, the host of HTNV is not present in Africa, other rodent species were studied for the presence of hantaviruses. Partial S, M, and L segment sequences were recovered, and found to share common ancestry with other Murinae-borne hantaviruses.

Seoul virus (SEOV) is the true cosmopolitan among the hantaviruses. SEOV strains have spread around the world, sailing with its host, the brown rat (*Rattus norvegicus*). SEOV was first isolated from urban rats in Korea (Lee *et al.*, 1982), and human cases of SEOV-caused HFRS are mainly found in Asia. The prototype strain 80–39 has been fully sequenced, and so are also four other Chinese isolates, ZT10, ZT71, Z37, and L99 (Shi *et al.*, 2003). The two last ones are vaccine strains, and L99 was isolated from *Rattus losea*. Altogether 280 SEOV sequences are deposited in the GenBank, covering the worldwide distribution with sequences from, for example, Japan, Korea, Cambodia, Singapore, Indonesia, USA, Brazil, France and Belgium. The sequence diversity of SEOV-strains is low, and in striking contrast to many other hantavirus species. Furthermore, no geographical clustering of lineages can be observed.

The L segment sequences are 6530 nt long, encoding a 2151-aa RNA-dependent RNA-polymerase. The sequence diversity is very low: 0–4% on the nt level, and 0–1% on the aa level. Full-length M segment sequences ($n = 23$) are 3651–3653 nt long, and they encode a glycoprotein precursor of 1133 aa. Sequence diversity is only 0–6% on the nt level, and 0–3% on the aa level. Full-length S segment sequences (33 with unique coding regions) are 1754–1785 nt long, encoding a 429 aa N protein with diversity of 0–5% (nt) and 0–3% (aa).

Gou virus (GOUV) is a tentative new hantavirus species discovered from the roof rat, captured in China (Wang *et al.,* 2000). GOUV was first identified as a variant of SEOV, although the clear distinction was noted already in the first report. Full-length S and M segment sequences are available for the first isolate, before and after cell culture adaptation. In addition, a recently deposited strain ZJ5, from *R. norvegicus* captured in China, appears to be more closely related to GOUV than to SEOV. The M segment sequences are 3651 nt long encoding a glycoprotein precursor of 1133 aa, the diversity from SEOV is 15–16% on the nt level, and 3–4% on the aa level. The S segment sequences are 1732–1759 nt long encoding a 429 aa N protein, and the diversity from SEOV is up to 13% on the nt level, and up to 3% on the aa level.

Thailand virus (THAIV) was discovered and isolated from bandicoot rats, *Bandicota indica,* captured in Thailand (Elwell *et al.,* 1985). Sequencing and phylogenetic analysis grouped THAIV with other Murinae-borne hantaviruses (Xiao *et al.,* 1994). Serological evidence suggests that THAIV is a human pathogen and one possible human case has been described (Pattamadilok *et al.,* 2006). The first isolate, Thai749, has been sequenced further and full S and M sequences are available (Pattamadilok *et al.,* 2006). In addition, four new strains, including one full S sequence, have been described (Hugot *et al.,* 2006). The full M sequence is 3613 nt long, encoding a 1132-aa glycoprotein precursor that differs from other Murinae-borne hantaviruses by 23–25% on the aa level. The two full S sequences are 1882 and 1884 nt long, encoding a 429-aa nucleocapsid protein with aa divergence of 1%, and more than 13% from other hantaviruses.

Serang virus (SERV) is a tentative hantavirus species recovered from the Asian house rat, *Rattus tanezumi,* captured in Indonesia (Plyusnina *et al.,* 2009). SERV seems to share most recent common ancestry with THAIV, and not with the other *Rattus*-carried hantaviruses such as SEOV. Closely related strains have been reported from Cambodia originating from *Rattus rattus* (Reynes *et al.,* 2003). These strains cluster together with SERV, and not GOUV that is carried by *R. rattus.*

Soricidae- and Talpidae-borne hantaviruses

The first hantavirus ever discovered was Thottapalayam virus (TPMV) isolated from shrews, although it was recognized as a hantavirus only later. For many years TPMV was taken as an exception to the rule that hantaviruses are rodent-borne. In the 1980s, hantavirus antibodies were reported from other Soricomorph insectivores as well, but no viruses or viral sequences could be detected. During the last couple of years, the recovery of these sequences has become successful, and several new hantaviruses have been characterized with rapid speed (Table 3.5). It has also become evident that the prevalence of hantaviruses in insectivore species is much higher than in rodents with up to 30% positivity. Isolation attempts have been mainly negative, which is maybe not a surprise, since many of the virus sequences have been recovered from archival tissue samples not ideal for virus isolation.

The insectivore-borne hantaviruses have a wide geographical distribution, spanning four continents. Genetically they are far more diverse than rodent-borne ones, and phylogenetically they appear to be ancestral to rodent-borne hantaviruses. Two or even three major lineages exist, one (or two) clearly ancestral to all other hantaviruses, and one lineage that shares a more recent ancestry with the Murinae-borne viruses. Insectivore-borne hantaviruses have so far been recovered in the families Soricidae and Talpidae of the order Soricomorpha. Yet unknown is whether other soricomorphs like hedgehogs and gymnures

Table 3.5 List of hantaviruses carried by rodents from the family Muridae, subfamily Murinae. In ***bold italics*** are the hantavirus species currently recognized by the International Committee on Taxonomy of Viruses; others are tentative species

				Full-length sequences		
Virus	Abbreviation	Host	Reference	S	M	L
Thottapalayam	TPMV	*Suncus murinus,* Asian house shrew	Carey *et al*. (1971)	NC_010704	NC_010708	NC_010707
Imjin	MJNV	*Crocidura lasiura,* Ussuri white-toothed shrew	Song *et al*. (2009)	EF641799	EF641805	EF641807
Nova	NVAV	*Talpa europaea,* European common mole	Kang *et al*. (2009)	FJ539168		FJ593498
Asama	ASAV	*Urotrichus talpoides,* Japanese shrew mole	Arai *et al*. (2008)	EU929072	EU929075	EU929078
Seewis	SWSV	*Sorex araneus,* Eurasian common shrew	Song *et al*. (2007)	EF636024		
Oxbow	OXBV	*Neurotrichus gibbsii,* American shrew mole	Kang *et al*. (2009)	FJ539166	FJ539167	FJ593497
Cao Bang	CBNV	*Anourosorex squamipes,* Chinese mole shrew	Song *et al*. (2007)	EF543524	EF543526	
Camp Ripley	RPLV	*Blarina brevicauda,* northern short-tailed shrew	Arai *et al*. (2007)			
Tanganya	TGNV	*Crodidura theresae,* Therese shrew	Klempa *et al*. (2007)			
Ash river	ARRV	*Sorex cinereus,* cinereus shrew	Arai *et al*. (2008)			
Jemez Springs	JMSV	*Sorex monticolus,* dusky shrew	Arai *et al*. (2008)			

(family Erinaceidae) also carry hantaviruses. An intriguing speculation is that since other bunyaviruses have insect or arthropod vectors, maybe a primordial hantavirus emerged through species jumping from an insect to soricomorph ancestor and then to rodents.

These viruses appear to be non-pathogenic to humans, but very little has been done owing to limitations in methodology, i.e. lack of isolated viruses. Seropositivity surveys have not been conducted yet for any of these viruses except for TPMV. Because rodent-borne hantaviruses are capable of causing clinical disease, the insectivore-borne viruses might also have a similar potential. On the other hand, human exposure to shrews and moles is rare, and thus it can be argued that the risk of human infection is negligible.

Thottapalayam virus (TPMV) was discovered in a survey for Japanese encephalitis virus in southern India in 1964 (Carey *et al.*, 1971). It was classified as a hantavirus based on its ultrastructural features (Zeller *et al.*, 1989) and genetic similarities (Xiao *et al.*, 1994). The virus was isolated from the Asian house shrew (*Suncus murinus*), but whether this animal is the natural reservoir of TPMV, remains to be confirmed. Only one small survey of sera from 14 wild shrews captured in Indonesia has been conducted, resulting in two positive samples (Okumura *et al.*, 2007). In small laboratory animals, TPMV seems to have a much broader experimental host range than rodent-borne hantaviruses.

The pathogenicity of TPMV to humans, or any other insectivore-borne hantaviruses, is unknown. Serum samples from residents of Mumbai, India have been studied for serological evidence of TPMV infection, and 3,3% of samples were determined positive. One possible human case of febrile illness has also been described in Laos, however, this patient lacked IgM antibodies against TPMV (Okumura *et al.*, 2007).

TPMV has been isolated in cell culture, and the full-length sequence has been recovered (Yadav *et al.*, 2007; Song *et al.*, 2007d). The full-length L segment sequence is 6581 nt long, encoding for a RNA-dependent RNA-polymerase of 2149/2150 aa (two variants of the same isolate have been deposited in GenBank). The L protein sequence differs from all other hantaviruses by 38–39% on the aa level. The full M segment sequence is 3621 nt long, and the encoded glycoprotein precursor is 1121 aa long, differing from other hantaviruses by 57–60%. The S segment sequence is the shortest of hantaviruses with the length of 1530 nt. The predicted N protein of 435 aa has a highly divergent amino-terminus, and it has a longer coiled coil structure than is found in rodent-borne hantaviruses. The N sequence differs by 30% from Imjin virus, and by 53–57% from all other hantaviruses.

Imjin virus (MJNV) was isolated from lung tissue samples of Ussuri white-toothed shrew (*Crocidura lasiura*) captured in Korea (Song *et al.*, 2009). Seven out of 115 Ussuri shrews were hantavirus positive in RT-PCR. MJNV was isolated in cell culture and full-length sequence was obtained. After the successful isolation of MJNV, the shrew sera were also screened for IgG antibodies against the new virus using IFA, and 18 of 115 samples were determined positive. Antisera against rodent-borne hantaviruses does not react with MJNV-infected cells, neither are antibodies cross-neutralizing.

Phylogenetically this virus shares common ancestry with TPMV. These two viruses are placed in a distinct group ancestral to all other hantaviruses except NVAV. The S segment sequences both from cell culture isolates and from tissues of Ussuri shrews are 1577 nt long, the predicted N protein is 436 aa long, and there is no ORF for NSs. The 5′-NCR is slightly longer (67 nt) and the 3′-NCR (199) shorter than most hantaviruses. The MJNV S segment sequence differs from that of TPMV by 30–31%, and by approximately 47% (nt) and 52–54% (aa) from other hantaviruses. The inter-strain diversity is 9% on the nt and

1% on the aa level. The M segment sequences are 3619 nt long, and the predicted glycoprotein precursor is 1120 aa long. Diversity from TPMV is 31% on the nt and 28% on the aa level; diversity from other hantaviruses is 47–48% (nt) and 56–58% (aa). Among the MJNV strains the variation is modest, sequences differ by 1–2%. The L segment sequences are 6570 nt long, and the predicted L protein is 2149 aa long with all five conserved motifs present. The L segment sequences are more conserved than S and M sequences: MJNV L differs from TPMV by 26% (nt) and 19% (aa), and from other hantaviruses by 36–38% (nt) and 39% (aa).

Nova virus (NVAV) was identified in archival tissue samples of European common mole (*Talpa europaea*) captured in Hungary (Kang *et al.*, 2009c). This virus is the first example of a hantavirus carried by a true mole species. Two other talpid species also harbour hantaviruses: ASAV in Japanese shrew mole and OXBV in American shrew mole. These two viruses are phylogenetically closer to Soricidae-borne hantaviruses than to NVAV, and host-switching to talpids has been proposed. In the phylogenetic tree, NVAV stands alone and appears to be ancestral to all other hantaviruses (Kang *et al.*, 2009c).

The full-length S- and L-segments reveal that NVAV is the most divergent hantavirus described. The S segment sequence is 1839 nt long, and it contains an ORF that encodes a 428 aa N protein, but not the NSs ORF found in Cricetidae-borne hantaviruses. Sequence diversity from all other hantaviruses is 47–57% on the aa level. Despite the low level of sequence similarity, the predicted secondary structure of the NVAV N protein resembles that of other hantaviruses. The L-segment is 6474 nt long, and it encodes a 2157-aa long RNA-dependent RNA-polymerase differing from other hantaviruses by 37–40%.

Asama virus (ASAV) was the first mole-borne hantavirus to be discovered, and it was found in Japanese shrew moles (family Talpidae) captured in Japan (Arai *et al.*, 2008b). Phylogenetically ASAV is related to hantaviruses harboured by Old World soricine shrews, and not to other Talpidae-borne ones, like NVAV, and thus a host-switching event has been suggested. ASAV belongs to a lineage of Soricidae-borne hantaviruses that is genetically closer to Muridae-borne hantaviruses than to crocidurine shrew-borne ones.

The ASAV S segment sequence is 1756–1801 nt long, encoding a predicted N protein of 434 aa. The inter-strain variation is very low, and diversity from other hantaviruses is 32–43% on the nt and 29–54% on the aa level. ASAV has the highest sequence similarity with Soricinae-borne hantaviruses, followed by Murinae-borne, Cricetidae-borne, and finally crocidurine shrew-borne hantaviruses. The M segment sequence is 3646 nt long, and it encodes a glycoprotein precursor of 1141 aa that includes the WAASA motif in the form WAVSA. The ASAV M sequence displays diversity of 25–44% (nt) and 20–57% (aa) from other hantaviruses. The nearly full-length L segment sequence encodes a 2041 aa L protein with a higher sequence similarity to other hantaviruses: ASAV L differs by 24–36% on the nt and 15–38% on the aa level.

Oxbow virus (OXBV) was detected in the tissues of American shrew mole, *Neurotrichus gibbsii*, captured in USA (Kang *et al.*, 2009b). OXBV has a widespread tissue distribution in the shrew mole, similarly to hantavirus-infected rodent hosts. Phylogenetically, OXBV is related to ASAV, and these two share common ancestry with Soricidae-borne hantaviruses, and not with the Talpidae-borne NVAV, suggesting again a host-switching event.

The full-length S segment sequence is 1705 nt long, encoding a 428-aa N protein. The overall structure is predicted to resemble that of other hantaviruses. Sequence diversity from other hantaviruses is 25–47% on the nt and 16–53% on the aa level; the closest one is JMSV.

OXBV has higher similarity to rodent-borne hantaviruses, especially the Murinae-borne ones, than to the crocidurine shrew-borne TPMV and MJNV. The M segment sequence of OXBV is 3645 nt long, encoding a glycoprotein precursor of 1141 aa. The OXBV M sequence differs by 27–49% (nt) and 22–58% (aa) from other hantaviruses. A partial L segment sequence is known covering the region with conserved motifs in the L protein, and differing by 18–40% from other hantaviruses.

Cao Bang virus (CBNV) was discovered from tissues of Chinese mole shrew, *Anourosorex squamipes*, captured in Vietnam (Song *et al.*, 2007b). CBNV-like sequences have also been reported from Chinese mole shrews captured in Taiwan (unpublished data). CBNV belongs to the same phylogenetic lineage as other Soricinae-borne hantaviruses. Full-length S and M segment sequences have been retrieved, as well as partial L segment. The CBNV S segment is 1833 nt long encoding for a 428 aa N protein that displays sequence diversity of 26–54% from all other hantaviruses. The CBNV M segment sequence is 3637 nt long, encoding a glycoprotein precursor of 1139 aa that differs from other hantaviruses by 29–59%.

Seewis virus (SWSV) is carried by the Eurasian common shrew, *Sorex araneus*, which is among the most widely dispersed small mammal species in Eurasia. As long ago as the 1980s, hantaviral antigens were reported in tissues of this species (Gavriloskaya *et al.*, 1983). SWSV was amplified finally from tissues of Eurasian common shrew captured in Switzerland in 2006 (Song *et al.*, 2007c). SWSV has since been recovered from at least Hungary and Finland (Kang *et al.*, 2009a), and recently from Russia (Yashina *et al.*, 2010). In Russia, extensive host sharing was detected for SWSV that was recovered not only from common shrews, but also *Sorex tundrensis* and *Sorex daphaenodon.*

The full-length SWSV S segment is 1641 nt long, encoding a 429-aa N protein that differs from other hantavirus N proteins by 31–57%. Only partial M and L segment sequences are available: these show a comparable level of diversity from other hantaviruses. Phylogenetically SWSV belongs to the same lineage as other Soricinae-borne hantaviruses.

Camp Ripley virus (RPLV) was the first shrew-borne hantavirus described in the Americas. It was found in archival tissue samples of northern short-tailed shrew (*Blarina brevicauda*) captured in Minnesota in 1998 (Arai *et al.*, 2007). More than 20 years earlier, serological evidence of hantavirus infection had been found in this shrew species, in a study in which Prospect Hill virus was discovered (Lee *et al.*, 1985). Only partial M and L segment sequences are available, and further sequencing is needed to confirm the phylogenetic position of RPLV.

Jemez Springs virus (JMSV) and *Ash River virus (ARRV)* were discovered in the same search for Soricinae-borne hantaviruses from archival tissue collections (Arai *et al.*, 2008a). JMSV was found in tissue samples of dusky shrew, *Sorex monticolus*, captured in New Mexico, USA; ARRV was found in masked shrew, *Sorex cinereus*, captured in Minnesota, USA. Partial S and L segment sequences are known, and phylogenetically these viruses belong to the same lineage as other Soricinae-borne hantaviruses.

Tanganya virus (TGNV) is one of the most intriguing insectivore-borne hantaviruses. It was discovered from Therese shrew (*Crocidura theresae*) captured in Tanganya, Africa (Klempa *et al.*, 2007). Partial S and L segments are available, and they have a different position in phylogenetic trees, yet in neither of the trees is TGNV clustering together with the other Crocidurinae-borne hantaviruses TPMV and MJNV.

In addition to the published insectivore-borne hantaviruses, there is a number of novel sequences discovered in different shrews that are not yet published. While this book has been in preparation, Kenkeme virus from *Sorex roboratus* was reported (Kang *et al.*, 2010). Sequences are also available for Artybash virus originating from the Laxmann's shrew (*Sorex caecutiens*) captured in Siberia, Russia. It has also been reported that, for example, *Blarina carolinensis* (Iamonia virus), *Neomys fodiens* (Laihia virus) and *Sorex minutus* (Asikkala virus) are potential hosts of novel hantaviruses. And based on the high positivity rate, more novel insectivore-borne hantaviruses will undoubtedly be discovered even as we are writing this chapter.

Concluding remarks

Recent developments in the field of hantavirus genetics have been fast and intriguing. The number of virus species has increased, their geographical distribution has been detailed, and basic knowledge about most of them has improved. It would be safe to assume swift progress in further inventory of hantaviruses. For instance, more genetic variants of Puumala virus (the current leader in the number of GeneBank entries, more than 700) from Europe are awaited. The same could be said about Andes and Sin Nombre viruses in the Americas, Hantaan and Seoul viruses in Asia and also Sangassou virus in Africa. It would be logical to expect also rapid accumulation of genetic information about insectivore-borne hantaviruses. New sequencing technologies should certainly help retrieving full-length hantaviral genomes, especially for novel virus species.

One of the most important tasks for the nearest future would be to convert sequence-related knowledge on hantaviral genes and proteins into the hard currency of functionality. It would be very important to find at least some virulence markers and understand what determines extremely high host- and tissue-specificity of hantaviruses. Reverse genetics of hantaviruses is still in infancy and no successful rescue of infectious viruses has been reported yet. The progress in this direction would be warmly welcomed. Animal models based on both rodents/insectivores and primates are needed as well and, by now, some progress in this direction was achieved (Hooper *et al.*, 2001; Klingström *et al.*, 2002; Milazzo *et al.*, 2002; Sironen *et al.*, 2008). In combination, these approaches would allow systematic search for the abovementioned markers. Useful lists of mutations to be evaluated exist already (see for example Lundkvist *et al.*, 1997; Nemirov *et al.*, 2003a,b) and more are coming from genetic studies.

References

Arai, S., Song, J.W., Sumibcay, L., Bennett, S.N., Nerurkar, V.R., Parmenter, C., Cook, J.A., Yates, T.L., and Yanagihara, R. (2007). Hantavirus in northern short-tailed shrew, United States. Emerg. Infect. Dis. *13*, 1420–1423.

Arai, S., Bennett, S.N., Sumibcay, L., Cook, J.A., Song, J.W., Hope, A., Parmenter, C., Nerurkar, V.R., Yates, T.L., and Yanagihara, R. (2008a). Phylogenetically distinct hantaviruses in the masked shrew (*Sorex cinereus*) and dusky shrew (*Sorex monticolus*) in the United States. Am. J. Trop. Med. Hyg. *78*, 348–351.

Arai, S., Ohdachi, S.D., Asakawa, M., Kang, H.J., Mocz, G., Arikawa, J., Okabe, N., and Yanagihara, R. (2008b). Molecular phylogeny of a newfound hantavirus in the Japanese shrew mole (*Urotrichus talpoides*). Proc. Natl. Acad. Sci. U.S.A. *105*, 16296–16301.

Avsic-Zupanc, T., Xiao, S.Y., Stojanovic, R., Gligic, A., van der Groen, G., and LeDuc, J.W. (1992). Characterization of Dobrava virus: a Hantavirus from Slovenia, Yugoslavia. J. Med. Virol. *38*, 132–137.

Avsic-Zupanc, T., Toney, A., Anderson, K., Chu, Y.K., and Schmaljohn, C. (1995). Genetic and antigenic properties of Dobrava virus: a unique member of the Hantavirus genus, family *Bunyaviridae*. J. Gen. Virol. *76*, 2801–2808.

Avsic-Zupanc, T., Nemirov, K., Petrovec, M., Trilar, T., Poljak, M., Vaheri, A., and Plyusnin, A. (2000). Genetic analysis of wild-type Dobrava hantavirus in Slovenia: co-existence of two distinct genetic lineages within the same natural focus. J. Gen. Virol. *81,* 1747–1755.

Baek, L.J., Kariwa, H., Lokugamage, K., Yoshimatsu, K., Arikawa, J., Takashima, I., Kang, J.I., Moon, S.S., Chung, S.Y., Kim, E.J., *et al.* (2006). Soochong virus: an antigenically and genetically distinct hantavirus isolated from *Apodemus peninsulae* in Korea. J. Med. Virol. *78,* 290–297.

Bharadwaj, M., Botten, J., Torrez-Martinez, N., and Hjelle, B. (1997). Rio Mamore virus: genetic characterization of a newly recognized hantavirus of the pygmy rice rat, *Oligoryzomys microtis,* from Bolivia. Am. J. Trop. Med. Hyg. *57,* 368–374.

Bohlman, M.C., Morzunov, S.P., Meissner, J., Taylor, M.B., Ishibashi, K., Rowe, J., Levis, S., Enria, D., and St Jeor, S.C. (2002). Analysis of hantavirus genetic diversity in Argentina: S segment-derived phylogeny. J. Virol. *76,* 3765–3773.

Brummer-Korvenkontio, M., Vaheri, A., Hovi, T., von Bonsdorff, C.H., Vuorimies, J., Manni, T., Penttinen, K., Oker-Blom, N., and Lähdevirta, J. (1980). Nephropathia epidemica: detection of antigen in bank voles and serologic diagnosis of human infection. J. Infect. Dis. *141,* 131–134.

Carey, D.E., Reuben, R., Panicker, K.N., Shope, R.E., and Myers, R.M. (1971). Thottapalayam virus: a presumptive arbovirus isolated from a shrew in India. Indian J. Med. Res. *59,* 1758–1760.

Carroll, D.S., Mills, J.N., Montgomery, J.M., Bausch, D.G., Blair, P.J., Burans, J.P., Felices, V., Gianella, A., Iihoshi, N., Nichol, S.T., *et al.* (2005). Hantavirus pulmonary syndrome in Central Bolivia: relationships between reservoir hosts, habitats, and viral genotypes. Am. J. Trop. Med. Hyg. *72,* 42–46.

Chare, E.R., Gould, E.A., and Holmes, E.C. (2003). Phylogenetic analysis reveals a low rate of homologous recombination in negative-sense RNA viruses. J. Gen. Virol. *84,* 2691–2703.

Chu, Y.K., Milligan, B., Owen, R.D., Goodin, D.G., and Jonsson, C.B. (2006). Phylogenetic and geographical relationships of hantavirus strains in eastern and western Paraguay. Am. J. Trop. Med. Hyg. *75,* 1127–1134.

Chu, Y.K., Owen, R.D., Sanchez-Hernandez, C., Romero-Almaraz Mde, L., and Jonsson, C.B. (2008). Genetic characterization and phylogeny of a hantavirus from Western Mexico. Virus Res. *131,* 180–188.

Chung, D.H., Sun, Y., Parker, W.B., Arterburn, J.B., Bartolucci, A., and Jonsson, C.B. (2007). Ribavirin reveals a lethal threshold of allowable mutation frequency for Hantaan virus. J. Virol. *81,* 11722–11729.

Dragoo, J.W., Lackey, J.A., Moore, K.E., Lessa, E.P., Cook, J.A., and Yates, T.L. (2006). Phylogeography of the deer mouse (*Peromyscus maniculatus*) provides a predictive framework for research on hantaviruses. J. Gen. Virol. *87,* 1997–2003.

Elwell, M.R., Ward, G.S., Tingpalapong, M., and LeDuc, J.W. (1985). Serologic evidence of Hantaan-like virus in rodents and man in Thailand. Southeast Asian J. Trop. Med. Public Health *16,* 349–354.

Eyzaguirre, E.J., Milazzo, M.L., Koster, F.T., and Fulhorst, C.F. (2008). Choclo virus infection in the Syrian golden hamster. Am. J. Trop. Med. Hyg. *78,* 669–674.

Feuer, R., Boone, J.D., Netski, D., Morzunov, S.P., and St Jeor, S.C. (1999). Temporal and spatial analysis of Sin Nombre virus quasispecies in naturally infected rodents. J. Virol. *73,* 9544–9554.

French, G.R., Foulke, R.S., Brand, O.A., Eddy, G.A., Lee, H.W., and Lee, P.W. (1981). Korean hemorrhagic fever: propagation of the etiologic agent in a cell line of human origin. Science *211,* 1046–1048.

Fulhorst, C.F., Monroe, M.C., Salas, R.A., Duno, G., Utrera, A., Ksiazek, T.G., Nichol, S.T., de Manzione, N.M., Tovar, D., and Tesh, R.B. (1997). Isolation, characterization and geographic distribution of Cano Delgadito virus, a newly discovered South American hantavirus (family *Bunyaviridae*). Virus Res. *51,* 159–171.

Fulhorst, C.F., Cajimat, M.N., Utrera, A., Milazzo, M.L., and Duno, G.M. (2004). Maporal virus, a hantavirus associated with the fulvous pygmy rice rat (*Oligoryzomys fulvescens*) in western Venezuela. Virus Res. *104,* 139–144.

Gavrilovskaya, I.N., Apekina, N.S., Myasnikov, Y., Bernshtein, A.D., Ryltseva, E.V., Gorbachkova, E.A., and Chumakov, M.P. (1983). Features of circulation of hemorrhagic fever with renal syndrome (HFRS) virus among small mammals in the European U.S.S.R. Arch. Virol. *75,* 313–316.

Gligic, A., Frusic, M., Obradovic, M., Stojanovic, R., Hlaca, D., Gibbs, C.J.,Jr, Yanagihara, R., Calisher, C.H., and Gajdusek, D.C. (1989a). Hemorrhagic fever with renal syndrome in Yugoslavia: antigenic characterization of hantaviruses isolated from Apodemus flavicollis and Clethrionomys glareolus. Am. J. Trop. Med. Hyg. *41,* 109–115.

Gligic, A., Obradovic, M., Stojanovic, R., Vujosevic, N., Ovcaric, A., Frusic, M., Gibbs, C.J.,Jr, Calisher, C.H., and Gajdusek, D.C. (1989b). Epidemic hemorrhagic fever with renal syndrome in Yugoslavia, 1986. Am. J. Trop. Med. Hyg. *41,* 102–108.

Gligic, A., Dimkovic, N., Xiao, S.Y., Buckle, G.J., Jovanovic, D., Velimirovic, D., Stojanovic, R., Obradovic, M., Diglisic, G., and Micic, J. (1992). Belgrade virus: a new hantavirus causing severe hemorrhagic fever with renal syndrome in Yugoslavia. J. Infect. Dis. *166,* 113–120.

Golovljova, I., Vasilenko, V., Mittzenkov, V., Prükk, T., Seppet, E., Vene, S., Settergren, B., Plyusnin, A., and Lundkvist, Å. (2007). Characterization of hemorrhagic fever with renal syndrome caused by hantaviruses, Estonia. Emerg. Infect. Dis. *13,* 1773–1775.

Hjelle, B., Chavez-Giles, F., Torrez-Martinez, N., Yates, T., Sarisky, J., Webb, J., and Ascher, M. (1994). Genetic identification of a novel hantavirus of the harvest mouse *Reithrodontomys megalotis.* J. Virol. *68,* 6751–6754.

Hjelle, B., Anderson, B., Torrez-Martinez, N., Song, W., Gannon, W.L., and Yates, T.L. (1995a). Prevalence and geographic genetic variation of hantaviruses of New World harvest mice (*Reithrodontomys*): identification of a divergent genotype from a Costa Rican *Reithrodontomys mexicanus.* Virology *207,* 452–459.

Hjelle, B., Krolikowski, J., Torrez-Martinez, N., Chavez-Giles, F., Vanner, C., and Laposata, E. (1995b). Phylogenetically distinct hantavirus implicated in a case of hantavirus pulmonary syndrome in the northeastern United States. J. Med. Virol. *46,* 21–27.

Hjelle, B., Lee, S.W., Song, W., Torrez-Martinez, N., Song, J.W., Yanagihara, R., Gavrilovskaya, I., and Mackow, E.R. (1995c). Molecular linkage of hantavirus pulmonary syndrome to the white-footed mouse, *Peromyscus leucopus*: genetic characterization of the M genome of New York virus. J. Virol. *69,* 8137–8141.

Hörling, J., Chizhikov, V., Lundkvist, Å., Jonsson, M., Ivanov, L., Dekonenko, A., Niklasson, B., Dzagurova, T., Peters, C.J., Tkachenko, E., and Nichol, S. (1996). Khabarovsk virus: a phylogenetically and serologically distinct hantavirus isolated from *Microtus fortis* trapped in far-east Russia. J. Gen. Virol. *77,* 687–694.

Hughes, A.L., and Friedman, R. (2000). Evolutionary diversification of protein-coding genes of hantaviruses. Mol. Biol. Evol. *17,* 1558–1568.

Hugot, J.P., Plyusnina, A., Herbreteau, V., Nemirov, K., Laakkonen, J., Lundkvist, Å., Supputamongkol, Y., Henttonen, H., and Plyusnin, A. (2006). Genetic analysis of Thailand hantavirus in *Bandicota indica* trapped in Thailand. Virol. J. *3,* 72.

Jackson, A.P., and Charleston, M.A. (2004). A cophylogenetic perspective of RNA-virus evolution. Mol. Biol. Evol. *21,* 45–57.

Jiang, J.F., Zhang, W.Y., Wu, X.M., Zhang, P.H., and Cao, W.C. (2007). Soochong virus and Amur virus might be the same entities of hantavirus. J. Med. Virol. *79,* 1792–1795.

Johnson, A.M., Bowen, M.D., Ksiazek, T.G., Williams, R.J., Bryan, R.T., Mills, J.N., Peters, C.J., and Nichol, S.T. (1997). Laguna Negra virus associated with HPS in western Paraguay and Bolivia. Virology *238,* 115–127.

Johnson, A.M., de Souza, L.T., Ferreira, I.B., Pereira, L.E., Ksiazek, T.G., Rollin, P.E., Peters, C.J., and Nichol, S.T. (1999). Genetic investigation of novel hantaviruses causing fatal HPS in Brazil. J. Med. Virol. *59,* 527–535.

Kang, H.J., Arai, S., Hope, A.G., Song, J.W., Cook, J.A., and Yanagihara, R. (2009a). Genetic diversity and phylogeography of Seewis virus in the Eurasian common shrew in Finland and Hungary. Virol. J. *6,* 208.

Kang, H.J., Bennett, S.N., Dizney, L., Sumibcay, L., Arai, S., Ruedas, L.A., Song, J.W., and Yanagihara, R. (2009b). Host switch during evolution of a genetically distinct hantavirus in the American shrew mole (*Neurotrichus gibbsii*). Virology *388,* 8–14.

Kang, H.J., Bennett, S.N., Sumibcay, L., Arai, S., Hope, A.G., Mocz, G., Song, J.W., Cook, J.A., and Yanagihara, R. (2009c). Evolutionary insights from a genetically divergent hantavirus harbored by the European common mole (*Talpa europaea*). PLoS One *4,* e6149.

Kang, H.J., Arai, S., Hope, A.G., Cook, J.A., and Yanagihara, R. (2010). Novel hantavirus in the flat-skulled shrew (*Sorex roboratus*). Vector Borne Zoonotic Dis. *10,* 593–597.

Kariwa, H., Yoshizumi, S., Arikawa, J., Yoshimatsu, K., Takahashi, K., Takashima, I., and Hashimoto, N. (1995). Evidence for the existence of Puumala-related virus among *Clethrionomys rufocanus* in Hokkaido, Japan. Am. J. Trop. Med. Hyg. *53,* 222–227.

Kariwa, H., Yoshimatsu, K., Sawabe, J., Yokota, E., Arikawa, J., Takashima, I., Fukushima, H., Lundkvist, A., Shubin, F.N., Isachkova, L.M., *et al.* (1999). Genetic diversities of hantaviruses among rodents in Hokkaido, Japan and Far East Russia. Virus Res. *59,* 219–228.

Kariwa, H., Yoshimatsu, K., Araki, K., Chayama, K., Kumada, H., Ogino, M., Ebihara, H., Murphy, M.E., Mizutani, T., Takashima, I., and Arikawa, J. (2000). Detection of hantaviral antibodies among patients with hepatitis of unknown etiology in Japan. Microbiol. Immunol. *44,* 357–362.

Kariwa, H., Yoshimatsu, K., and Arikawa, J. (2007). Hantavirus infection in East Asia. Comp. Immunol. Microbiol. Infect. Dis. *30,* 341–356.

Kaukinen, P., Vaheri, A., and Plyusnin, A. (2005). Hantavirus nucleocapsid protein: a multifunctional molecule with both housekeeping and ambassadorial duties. Arch. Virol. *150,* 1693–1713.

Khan, A.S., Gaviria, M., Rollin, P.E., Hlady, W.G., Ksiazek, T.G., Armstrong, L.R., Greenman, R., Ravkov, E., Kolber, M., Anapol, H., *et al.* (1996). Hantavirus pulmonary syndrome in Florida: association with the newly identified Black Creek Canal virus. Am. J. Med. *100,* 46–48.

Klempa, B., Meisel, H., Rath, S., Bartel, J., Ulrich, R., and Kruger, D.H. (2003). Occurrence of renal and pulmonary syndrome in a region of northeast Germany where Tula hantavirus circulates. J. Clin. Microbiol. *41,* 4894–4897.

Klempa, B., Fichet-Calvet, E., Lecompte, E., Auste, B., Aniskin, V., Meisel, H., Denys, C., Koivogui, L., ter Meulen, J., and Kruger, D.H. (2006). Hantavirus in African wood mouse, Guinea. Emerg. Infect. Dis. *12,* 838–840.

Klempa, B., Fichet-Calvet, E., Lecompte, E., Auste, B., Aniskin, V., Meisel, H., Barriere, P., Koivogui, L., ter Meulen, J., and Kruger, D.H. (2007). Novel hantavirus sequences in Shrew, Guinea. Emerg. Infect. Dis. *13,* 520–522.

Klempa, B., Tkachenko, E.A., Dzagurova, T.K., Yunicheva, Y.V., Morozov, V.G., Okulova, N.M., Slyusareva, G.P., Smirnov, A., and Kruger, D.H. (2008). Hemorrhagic fever with renal syndrome caused by 2 lineages of Dobrava hantavirus, Russia. Emerg. Infect. Dis. *14,* 617–625.

Kukkonen, S.K., Vaheri, A., and Plyusnin, A. (2005). L protein, the RNA-dependent RNA polymerase of hantaviruses. Arch. Virol. *150,* 533–556.

Lee, P.W., Amyx, H.L., Yanagihara, R., Gajdusek, D.C., Goldgaber, D., and Gibbs, C.J.,Jr. (1985). Partial characterization of Prospect Hill virus isolated from meadow voles in the United States. J. Infect. Dis. *152,* 826–829.

Lee, H.W., and Lee, P.W. (1976). Korean hemorrhagic fever. I. Demonstration of causative agent and antibodies. Kor J Intern Med *19,* 371–383.

Lee, H.W., Lee, P.W., and Johnson, K.M. (1978). Isolation of the etiologic agent of Korean Hemorrhagic fever. J. Infect. Dis. *137,* 298–308.

Lee, H.W., Baek, L.J., and Johnson, K.M. (1982). Isolation of Hantaan virus, the etiologic agent of Korean hemorrhagic fever, from wild urban rats. J. Infect. Dis. *146,* 638–644.

Levis, S., Rowe, J.E., Morzunov, S., Enria, D.A., and St Jeor, S. (1997). New hantaviruses causing hantavirus pulmonary syndrome in central Argentina. Lancet *349,* 998–999.

Levis, S., Morzunov, S.P., Rowe, J.E., Enria, D., Pini, N., Calderon, G., Sabattini, M., and St Jeor, S.C. (1998). Genetic diversity and epidemiology of hantaviruses in Argentina. J. Infect. Dis. *177,* 529–538.

Li, D., Schmaljohn, A.L., Anderson, K., and Schmaljohn, C.S. (1995). Complete nucleotide sequences of the M and S segments of two hantavirus isolates from California: evidence for reassortment in nature among viruses related to hantavirus pulmonary syndrome. Virology *206,* 973–983.

Liang, M., Li, D., Xiao, S.Y., Hang, C., Rossi, C.A., and Schmaljohn, C.S. (1994). Antigenic and molecular characterization of hantavirus isolates from China. Virus Res. *31,* 219–233.

Löber, C., Anheier, B., Lindow, S., Klenk, H.D., and Feldmann, H. (2001). The Hantaan virus glycoprotein precursor is cleaved at the conserved pentapeptide WAASA. Virology *289,* 224–229.

Lokugamage, K., Kariwa, H., Hayasaka, D., Cui, B.Z., Iwasaki, T., Lokugamage, N., Ivanov, L.I., Volkov, V.I., Demenev, V.A., Slonova, R., *et al.* (2002). Genetic characterization of hantaviruses transmitted by the Korean field mouse (*Apodemus peninsulae*), Far East Russia. Emerg. Infect. Dis. *8,* 768–776.

Lokugamage, K., Kariwa, H., Lokugamage, N., Miyamoto, H., Iwasa, M., Hagiya, T., Araki, K., Tachi, A., Mizutani, T., Yoshimatsu, K., Arikawa, J., and Takashima, I. (2004). Genetic and antigenic characterization of the Amur virus associated with hemorrhagic fever with renal syndrome. Virus Res. *101,* 127–134.

Lopez, N., Padula, P., Rossi, C., Lazaro, M.E., and Franze-Fernandez, M.T. (1996). Genetic identification of a new hantavirus causing severe pulmonary syndrome in Argentina. Virology *220,* 223–226.

Lundkvist, Å., and Niklasson, B. (1992). Bank vole monoclonal antibodies against Puumala virus envelope glycoproteins: identification of epitopes involved in neutralization. Arch. Virol. *126,* 93–105.

Lundkvist, Å., Cheng, Y., Sjölander, K.B., Niklasson, B., Vaheri, A., and Plyusnin, A. (1997). Cell culture adaptation of Puumala hantavirus changes the infectivity for its natural reservoir, *Clethrionomys glareolus,* and leads to accumulation of mutants with altered genomic RNA S segment. J. Virol. *71,* 9515–9523.

Lundkvist, Å., Wiger, D., Hörling, J., Sjölander, K.B., Plyusnina, A., Mehl, R., Vaheri, A., and Plyusnin, A. (1998). Isolation and characterization of Puumala hantavirus from Norway: evidence for a distinct phylogenetic sublineage. J. Gen. Virol. *79,* 2603–2614.

Medina, R.A., Torres-Perez, F., Galeno, H., Navarrete, M., Vial, P.A., Palma, R.E., Ferres, M., Cook, J.A., and Hjelle, B. (2009). Ecology, genetic diversity, and phylogeographic structure of andes virus in humans and rodents in Chile. J. Virol. *83*, 2446–2459.

Meissner, J.D., Rowe, J.E., Borucki, M.K., and St Jeor, S.C. (2002). Complete nucleotide sequence of a Chilean hantavirus. Virus Res. *89*, 131–143.

Melo-Silva, C.R., Maranhao, A.Q., Nagasse-Sugahara, T.K., Bisordi, I., Suzuki, A., and Brigido, M.M. (2009). Characterization of hantaviruses circulating in Central Brazil. Infect. Genet. Evol. *9*, 241–247.

Milazzo, M.L., Eyzaguirre, E.J., Molina, C.P., and Fulhorst, C.F. (2002). Maporal viral infection in the Syrian golden hamster: a model of hantavirus pulmonary syndrome. J. Infect. Dis. *186*, 1390–1395.

Milazzo, M.L., Cajimat, M.N., Hanson, J.D., Bradley, R.D., Quintana, M., Sherman, C., Velasquez, R.T., and Fulhorst, C.F. (2006). Catacamas virus, a hantaviral species naturally associated with *Oryzomys couesi* (Coues' oryzomys) in Honduras. Am. J. Trop. Med. Hyg. *75*, 1003–1010.

Monroe, M.C., Morzunov, S.P., Johnson, A.M., Bowen, M.D., Artsob, H., Yates, T., Peters, C.J., Rollin, P.E., Ksiazek, T.G., and Nichol, S.T. (1999). Genetic diversity and distribution of *Peromyscus*-borne hantaviruses in North America. Emerg. Infect. Dis. *5*, 75–86.

Morzunov, S.P., Feldmann, H., Spiropoulou, C.F., Semenova, V.A., Rollin, P.E., Ksiazek, T.G., Peters, C.J., and Nichol, S.T. (1995). A newly recognized virus associated with a fatal case of hantavirus pulmonary syndrome in Louisiana. J. Virol. *69*, 1980–1983.

Morzunov, S.P., Rowe, J.E., Ksiazek, T.G., Peters, C.J., St Jeor, S.C., and Nichol, S.T. (1998). Genetic analysis of the diversity and origin of hantaviruses in Peromyscus leucopus mice in North America. J. Virol. *72*, 57–64.

Nemirov, K., Vapalahti, O., Lundkvist, Å., Vasilenko, V., Golovljova, I., Plyusnina, A., Niemimaa, J., Laakkonen, J., Henttonen, H., Vaheri, A., and Plyusnin, A. (1999). Isolation and characterization of Dobrava hantavirus carried by the striped field mouse (*Apodemus agrarius*) in Estonia. J. Gen. Virol. *80*, 371–379.

Nemirov, K., Henttonen, H., Vaheri, A., and Plyusnin, A. (2002). Phylogenetic evidence for host switching in the evolution of hantaviruses carried by *Apodemus* mice. Virus Res. *90*, 207–215.

Nemirov, K., Vaheri, A., and Plyusnin, A. (2004). Hantaviruses: Co-evolution with natural hosts. Recent Res. Devel. Virol. *6*, 201–228.

Nichol, S., Beaty, B.J., Elliott, R.M., Goldbach, R., Plyusnin, A., Schmaljohn, C.S., and Tesh, R.B. (2005). *Bunyaviridae*. In Virus taxonomy. VIIIth report of the International Committee on Taxonomy of Viruses. Fauquet, C. M., Mayo, M. A., Maniloff, J., Desselberger, U., and Ball, L. A. eds., (Amsterdam: Elsevier Academic Press), pp. 695–716.

Nichol, S.T., Spiropoulou, C.F., Morzunov, S., Rollin, P.E., Ksiazek, T.G., Feldmann, H., Sanchez, A., Childs, J., Zaki, S., and Peters, C.J. (1993). Genetic identification of a hantavirus associated with an outbreak of acute respiratory illness. Science *262*, 914–917.

Okumura, M., Yoshimatsu, K., Kumperasart, S., Nakamura, I., Ogino, M., Taruishi, M., Sungdee, A., Pattamadilok, S., Ibrahim, I.N., Erlina, S., *et al.* (2007). Development of serological assays for Thottapalayam virus, an insectivore-borne Hantavirus. Clin. Vaccine Immunol. *14*, 173–181.

Pattamadilok, S., Lee, B.H., Kumperasart, S., Yoshimatsu, K., Okumura, M., Nakamura, I., Araki, K., Khoprasert, Y., Dangsupa, P., Panlar, P., *et al.* (2006). Geographical distribution of hantaviruses in Thailand and potential human health significance of Thailand virus. Am. J. Trop. Med. Hyg. *75*, 994–1002.

Plyusnin, A. (2002). Genetics of hantaviruses: implications to taxonomy. Arch. Virol. *147*, 665–682.

Plyusnin, A., and Morzunov, S.P. (2001). Virus evolution and genetic diversity of hantaviruses and their rodent hosts. Curr. Top. Microbiol. Immunol. *256*, 47–75.

Plyusnin, A., Vapalahti, O., Lankinen, H., Lehväslaiho, H., Apekina, N., Myasnikov, Y., Kallio-Kokko, H., Henttonen, H., Lundkvist, Å., and Brummer-Korvenkontio, M. (1994). Tula virus: a newly detected hantavirus carried by European common voles. J. Virol. *68*, 7833–7839.

Plyusnin, A., Vapalahti, O., Lehväslaiho, H., Apekina, N., Mikhailova, T., Gavrilovskaya, I., Laakkonen, J., Niemimaa, J., Henttonen, H., and Brummer-Korvenkontio, M. (1995). Genetic variation of wild Puumala viruses within the serotype, local rodent populations and individual animal. Virus Res. *38*, 25–41.

Plyusnin, A., Vapalahti, O., Lundkvist, A., Henttonen, H., and Vaheri, A. (1996a). Newly recognised hantavirus in Siberian lemmings. Lancet *347*, 1835.

Plyusnin, A., Cheng, Y., Lehväslaiho, H., and Vaheri, A. (1996b). Quasispecies in wild-type Tula hantavirus populations. J. Virol. *70*, 9060–9063.

Plyusnin, A., Vapalahti, O., Vasilenko, V., Henttonen, H., and Vaheri, A. (1997). Dobrava hantavirus in Estonia: does the virus exist throughout Europe? Lancet *349,* 1369–1370.
Plyusnin, A., Kukkonen, S.K., Plyusnina, A., Vapalahti, O., and Vaheri, A. (2002). Transfection-mediated generation of functionally competent Tula hantavirus with recombinant S RNA segment. EMBO J. *21,* 1497–1503.
Plyusnin, A., Beaty, B.J., Elliott, R.M., Goldbach, R., Kormelink, R., Lundkvist, Å, Schmaljohn, C.S., and Tesh, R.B. (2011). *Bunyaviridae.* In Virus Taxonomy. Ninth report of the International Committee on Taxonomy of Viruses, King, A.M.Q., Adams, M.J., Carstens, E.B., and Lefkowitz, E.J., eds (Elsevier, San Diego, CA).
Plyusnina, A., and Plyusnin, A. (2005). Recombinant Tula hantavirus shows reduced fitness but is able to survive in the presence of a parental virus: analysis of consecutive passages in a cell culture. Virol. J. *2,* 12.
Plyusnina, A., Laakkonen, J., Niemimaa, J., Nemirov, K., Muruyeva, G., Pohodiev, B., Lundkvist, A., Vaheri, A., Henttonen, H., Vapalahti, O., and Plyusnin, A. (2008). Genetic analysis of hantaviruses carried by *Myodes* and *Microtus* rodents in Buryatia. Virol. J. *5,* 4.
Plyusnina, A., Ibrahim, I.N., and Plyusnin, A. (2009). A newly recognized hantavirus in the Asian house rat (*Rattus tanezumi*) in Indonesia. J. Gen. Virol. *90,* 205–209.
Powers, A.M., Mercer, D.R., Watts, D.M., Guzman, H., Fulhorst, C.F., Popov, V.L., and Tesh, R.B. (1999). Isolation and genetic characterization of a hantavirus (*Bunyaviridae*: Hantavirus) from a rodent, *Oligoryzomys microtis* (Muridae), collected in northeastern Peru. Am. J. Trop. Med. Hyg. *61,* 92–98.
Raboni, S.M., Probst, C.M., Bordignon, J., Zeferino, A., and dos Santos, C.N. (2005). Hantaviruses in Central South America: phylogenetic analysis of the S segment from HPS cases in Parana, Brazil. J. Med. Virol. *76,* 553–562.
Ramsden, C., Holmes, E.C., and Charleston, M.A. (2009). Hantavirus evolution in relation to its rodent and insectivore hosts: no evidence for codivergence. Mol. Biol. Evol. *26,* 143–153.
Ravkov, E.V., Rollin, P.E., Ksiazek, T.G., Peters, C.J., and Nichol, S.T. (1995). Genetic and serologic analysis of Black Creek Canal virus and its association with human disease and *Sigmodon hispidus* infection. Virology *210,* 482–489.
Rawlings, J.A., Torrez-Martinez, N., Neill, S.U., Moore, G.M., Hicks, B.N., Pichuantes, S., Nguyen, A., Bharadwaj, M., and Hjelle, B. (1996). Cocirculation of multiple hantaviruses in Texas, with characterization of the small (S) genome of a previously undescribed virus of cotton rats (*Sigmodon hispidus*). Am. J. Trop. Med. Hyg. *55,* 672–679.
Razzauti, M., Plyusnina, A., Henttonen, H., and Plyusnin, A. (2008). Accumulation of point mutations and reassortment of genomic RNA segments are involved in the microevolution of Puumala hantavirus in a bank vole (Myodes glareolus) population. J. Gen. Virol. *89,* 1649–1660.
Razzauti, M., Plyusnina, A., Sironen, T., Henttonen, H., and Plyusnin, A. (2009). Analysis of Puumala hantavirus in a bank vole population in northern Finland: evidence for co-circulation of two genetic lineages and frequent reassortment between strains. J. Gen. Virol. *90,* 1923–1931.
Reynes, J.M., Soares, J.L., Hue, T., Bouloy, M., Sun, S., Kruy, S.L., Flye Sainte Marie, F., and Zeller, H. (2003). Evidence of the presence of Seoul virus in Cambodia. Microbes Infect. *5,* 769–773.
Richter, M.H., Hanson, J.D., Cajimat, M.N., Milazzo, M.L., and Fulhorst, C. 2010. Geographical range of Rio Mamoré virus (family *Bunyaviridae,* genus *Hantavirus*) in association with the small-eared pygmy rice rat (*Oligoryzomys microtis*). *10,* 613–619.
Rodriguez, L.L., Owens, J.H., Peters, C.J., and Nichol, S.T. (1998). Genetic reassortment among viruses causing hantavirus pulmonary syndrome. Virology *242,* 99–106.
Rollin, P.E., Ksiazek, T.G., Elliott, L.H., Ravkov, E.V., Martin, M.L., Morzunov, S., Livingstone, W., Monroe, M., Glass, G., and Ruo, S. (1995). Isolation of black creek canal virus, a new hantavirus from *Sigmodon hispidus* in Florida. J. Med. Virol. *46,* 35–39.
Rosa, E.S., Mills, J.N., Padula, P.J., Elkhoury, M.R., Ksiazek, T.G., Mendes, W.S., Santos, E.D., Araujo, G.C., Martinez, V.P., Rosa, J.F., Edelstein, A., and Vasconcelos, P.F. (2005). Newly recognized hantaviruses associated with hantavirus pulmonary syndrome in northern Brazil: partial genetic characterization of viruses and serologic implication of likely reservoirs. Vector Borne Zoonotic Dis. *5,* 11–19.
Rowe, J.E., St Jeor, S.C., Riolo, J., Otteson, E.W., Monroe, M.C., Henderson, W.W., Ksiazek, T.G., Rollin, P.E., and Nichol, S.T. (1995). Coexistence of several novel hantaviruses in rodents indigenous to North America. Virology *213,* 122–130.
Sanchez, A.J., Abbott, K.D., and Nichol, S.T. (2001). Genetic identification and characterization of limestone canyon virus, a unique *Peromyscus*-borne hantavirus. Virology *286,* 345–353.

Scharninghausen, J.J., Pfeffer, M., Meyer, H., Davis, D.S., Honeycutt, R.L., and Faulde, M. (2002). Genetic evidence for tula virus in *Microtus arvalis* and *Microtus agrestis* populations in Croatia. Vector Borne Zoonotic Dis. *2,* 19–27.

Severson, W.E., Schmaljohn, C.S., Javadian, A., and Jonsson, C.B. (2003). Ribavirin causes error catastrophe during Hantaan virus replication. J. Virol. *77,* 481–488.

Shi, X., McCaughey, C., and Elliott, R.M. (2003). Genetic characterisation of a Hantavirus isolated from a laboratory-acquired infection. J. Med. Virol. *71,* 105–109.

Sibold, C., Meisel, H., Kruger, D.H., Labuda, M., Lysy, J., Kozuch, O., Pejcoch, M., Vaheri, A., and Plyusnin, A. (1999). Recombination in Tula hantavirus evolution: analysis of genetic lineages from Slovakia. J. Virol. *73,* 667–675.

Sironen, T., Kallio, E.R., Vaheri, A., Lundkvist, Å., and Plyusnin, A. (2008). Quasispecies dynamics and fixation of a synonymous mutation in hantavirus transmission. J. Gen. Virol. *89,* 1309–1313.

Sironen, T., Vaheri, A., and Plyusnin, A. (2001). Molecular evolution of Puumala hantavirus. J. Virol. *75,* 11803–11810.

Sjölander, K.B., Golovljova, I., Vasilenko, V., Plyusnin, A., and Lundkvist, Å. (2002). Serological divergence of Dobrava and Saaremaa hantaviruses: evidence for two distinct serotypes. Epidemiol. Infect. *128,* 99–103.

Song, W., Torrez-Martinez, N., Irwin, W., Harrison, F.J., Davis, R., Ascher, M., Jay, M., and Hjelle, B. (1995). Isla Vista virus: a genetically novel hantavirus of the California vole Microtus californicus. J. Gen. Virol. *76,* 3195–3199.

Song, J.W., Baek, L.J., Nagle, J.W., Schlitter, D., and Yanagihara, R. (1996). Genetic and phylogenetic analyses of hantaviral sequences amplified from archival tissues of deer mice (*Peromyscus maniculatus nubiterrae*) captured in the eastern United States. Arch. Virol. *141,* 959–967.

Song, J.W., Gligic, A., and Yanagihara, R. (2002). Identification of Tula hantavirus in *Pitymys subterraneus* captured in the Cacak region of Serbia-Yugoslavia. Int. J. Infect. Dis. *6,* 31–36.

Song, K.J., Baek, L.J., Moon, S., Ha, S.J., Kim, S.H., Park, K.S., Klein, T.A., Sames, W., Kim, H.C., Lee, J.S., Yanagihara, R., and Song, J.W. (2007a). Muju virus, a novel hantavirus harboured by the arvicolid rodent *Myodes regulus* in Korea. J. Gen. Virol. *88,* 3121–3129.

Song, J.W., Kang, H.J., Song, K.J., Truong, T.T., Bennett, S.N., Arai, S., Truong, N.U., and Yanagihara, R. (2007b). Newfound hantavirus in Chinese mole shrew, Vietnam. Emerg. Infect. Dis. *13,* 1784–1787.

Song, J.W., Gu, S.H., Bennett, S.N., Arai, S., Puorger, M., Hilbe, M., and Yanagihara, R. (2007c). Seewis virus, a genetically distinct hantavirus in the Eurasian common shrew (*Sorex araneus*). Virol. J. *4,* 114.

Song, J.W., Baek, L.J., Schmaljohn, C.S., and Yanagihara, R. (2007d). Thottapalayam virus, a prototype shrewborne hantavirus. Emerg. Infect. Dis. *13,* 980–985.

Song, J.W., Kang, H.J., Gu, S.H., Moon, S.S., Bennett, S.N., Song, K.J., Baek, L.J., Kim, H.C., O'Guinn, M.L., Chong, S.T., Klein, T.A., and Yanagihara, R. (2009). Characterization of Imjin virus, a newly isolated hantavirus from the Ussuri white-toothed shrew (*Crocidura lasiura*). J. Virol. *83,* 6184–6191.

Suzuki, A., Bisordi, I., Levis, S., Garcia, J., Pereira, L.E., Souza, R.P., Sugahara, T.K., Pini, N., Enria, D., and Souza, L.T. (2004). Identifying rodent hantavirus reservoirs, Brazil. Emerg. Infect. Dis. *10,* 2127–2134.

Taller, A.M., Xiao, S.Y., Godec, M.S., Gligic, A., Avsic-Zupanc, T., Goldfarb, L.G., Yanagihara, R., and Asher, D.M. (1993). Belgrade virus, a cause of hemorrhagic fever with renal syndrome in the Balkans, is closely related to Dobrava virus of field mice. J. Infect. Dis. *168,* 750–753.

Torrez-Martinez, N., Bharadwaj, M., Goade, D., Delury, J., Moran, P., Hicks, B., Nix, B., Davis, J.L., and Hjelle, B. (1998). Bayou virus-associated hantavirus pulmonary syndrome in Eastern Texas: identification of the rice rat, *Oryzomys palustris,* as reservoir host. Emerg. Infect. Dis. *4,* 105–111.

Vapalahti, O., Lundkvist, A., Kukkonen, S.K., Cheng, Y., Gilljam, M., Kanerva, M., Manni, T., Pejcoch, M., Niemimaa, J., Kaikusalo, A., *et al.* (1996). Isolation and characterization of Tula virus, a distinct serotype in the genus Hantavirus, family *Bunyaviridae.* J. Gen. Virol. *77 (Pt 12),* 3063–3067.

Vapalahti, O., Lundkvist, Å., Fedorov, V., Conroy, C.J., Hirvonen, S., Plyusnina, A., Nemirov, K., Fredga, K., Cook, J.A., Niemimaa, J., *et al.* (1999). Isolation and characterization of a hantavirus from *Lemmus sibiricus*: evidence for host switch during hantavirus evolution. J. Virol. *73,* 5586–5592.

Vincent, M.J., Quiroz, E., Gracia, F., Sanchez, A.J., Ksiazek, T.G., Kitsutani, P.T., Ruedas, L.A., Tinnin, D.S., Caceres, L., Garcia, A., *et al.* (2000). Hantavirus pulmonary syndrome in Panama: identification of novel hantaviruses and their likely reservoirs. Virology *277,* 14–19.

Wang, H., Yoshimatsu, K., Ebihara, H., Ogino, M., Araki, K., Kariwa, H., Wang, Z., Luo, Z., Li, D., Hang, C., and Arikawa, J. (2000). Genetic diversity of hantaviruses isolated in china and characterization of novel hantaviruses isolated from *Niviventer confucianus* and *Rattus rattus.* Virology *278,* 332–345.

Xiao, S.Y., Leduc, J.W., Chu, Y.K., and Schmaljohn, C.S. (1994). Phylogenetic analyses of virus isolates in the genus Hantavirus, family *Bunyaviridae*. Virology *198*, 205–217.
Yadav, P.D., Vincent, M.J., and Nichol, S.T. (2007). Thottapalayam virus is genetically distant to the rodent-borne hantaviruses, consistent with its isolation from the Asian house shrew (Suncus murinus). Virol. J. *4*, 80.
Yashina, L.N., Patrushev, N.A., Ivanov, L.I., Slonova, R.A., Mishin, V.P., Kompanez, G.G., Zdanovskaya, N.I., Kuzina, I.I., Safronov, P.F., Chizhikov, V.E., *et al.* (2000). Genetic diversity of hantaviruses associated with hemorrhagic fever with renal syndrome in the far east of Russia. Virus Res. *70*, 31–44.
Yashina, L., Mishin, V., Zdanovskaya, N., Schmaljohn, C., and Ivanov, L. (2001). A newly discovered variant of a hantavirus in Apodemus peninsulae, far Eastern Russia. Emerg. Infect. Dis. *7*, 912–913.
Yashina, L., Abramov, S., Gutorov, V., Dupal, T., Krivopalov, A., Panov, V., Danchinova, G., Vinogradov, V., Luchnikova, E., Hay, J., *et al.* (2010). Seewis virus: phylogeography of a shrew-borne hantavirus in Siberia, Russia. Vector Borne Zoonotic Dis. *10*, 585–591.
Zeller, H.G., Karabatsos, N., Calisher, C.H., Digoutte, J.P., Cropp, C.B., Murphy, F.A., and Shope, R.E. (1989). Electron microscopic and antigenic studies of uncharacterized viruses. II. Evidence suggesting the placement of viruses in the family *Bunyaviridae*. Arch. Virol. *108*, 211–227.
Zhang, Y.Z., Zou, Y., Yan, Y.Z., Hu, G.W., Yao, L.S., Du, Z.S., Jin, L.Z., Liu, Y.Y., Li, M.H., Chen, H.X., and Fu, Z.F. (2007). Detection of phylogenetically distinct Puumala-like viruses from red-grey vole *Clethrionomys rufocanus* in China. J. Med. Virol. *79*, 1208–1218.
Zou, Y., Hu, J., Wang, Z.X., Wang, D.M., Li, M.H., Ren, G.D., Duan, Z.X., Fu, Z.F., Plyusnin, A., and Zhang, Y.Z. (2008a). Molecular diversity and phylogeny of Hantaan virus in Guizhou, China: evidence for Guizhou as a radiation center of the present Hantaan virus. J. Gen. Virol. *89*, 1987–1997.
Zou, Y., Hu, J., Wang, Z.X., Wang, D.M., Yu, C., Zhou, J.Z., Fu, Z.F., and Zhang, Y.Z. (2008b). Genetic characterization of hantaviruses isolated from Guizhou, China: evidence for spillover and reassortment in nature. J. Med. Virol. *80*, 1033–1041.
Zou, Y., Wang, J.B., Gaowa, H.S., Yao, L.S., Hu, G.W., Li, M.H., Chen, H.X., Plyusnin, A., Shao, R., and Zhang, Y.Z. (2008c). Isolation and genetic characterization of hantaviruses carried by Microtus voles in China. J. Med. Virol. *80*, 680–688.
Zou, Y., Xiao, Q.Y., Dong, X., Lv, W., Zhang, S.P., Li, M.H., Plyusnin, A., and Zhang, Y.Z. (2008d). Genetic analysis of hantaviruses carried by reed voles *Microtus fortis* in China. Virus Res. *137*, 122–128.

Molecular Biology of Phleboviruses

4

Michèle Bouloy

Abstract

Phleboviruses are emerging pathogens in many countries of the world. Rift Valley fever virus, which in the past was responsible for dramatic zoonosis only in Africa, spread to Saudi Arabia and Yemen in 2000, and other phleboviruses such as Toscana virus or sandfly fever viruses Sicilian or Naples are now widely circulating in the Mediterranean basin. Like all bunyaviruses, phleboviruses are trisegmented negative-stranded RNA viruses but differ from others in having an ambisense S segment to encode the nucleoprotein and the non-structural protein. In this review, we will present the latest advances in the understanding of structure and pathogenesis and in the recent development of reverse genetics.

Introduction

This review summarizes current knowledge on the phleboviruses, with a particular focus on the new data published on the molecular biology, such as the morphology and structure of the particles, the interactions of the virus with the host cell and the innate immunity and the development of the reverse genetics, a potent tool to produce recombinant viruses. For more extensive information on specific viruses such as Rift Valley fever virus (RVFV) and sandfly fever viruses, the reader is referred to comprehensive reviews (Tesh, 1988a,b; Beaty and Calisher, 1991; Verani and Nicoletti, 1995; Swanepoel and Coetzer, 1994; Gerdes, 2002, 2004; Valassina *et al.*, 2003; Swanepoel and Coetzer, 2004; Flick and Bouloy, 2005; Soldan and Gonzalez-Scarano, 2005; Bouloy and Weber, 2010; Cusi *et al.*, 2010; Pépin *et al.*, 2010). The virus detection and diagnosis will not be covered here but will be addressed in a separate chapter (see Chapter 7).

Classification

Originally created in 1973, the *Bunyaviridae* family was composed of only one defined genus, *Bunyavirus* (Porterfield *et al.*, 1975). In 1980, after additional studies on viruses considered as bunyavirus-like, the ICTV committee recognized that other genera should be included (Bishop *et al.*, 1980), leading to the current division of the *Bunyaviridae* family into five genera *Orthobunyavirus, Hantavirus, Nairovirus, Phlebovirus* and *Tospovirus* (Nichol *et al.*, 2005; Plyusnin *et al.*, 2010). Until 1991, Uukuniemi virus (UUKV) and the related viruses were classified in a separate genus, the *UUKuvirus* but these viruses were reclassified and incorporated as members of the *Phlebovirus* genus owing to their strong biochemical and molecular similarities to phleboviruses (Calisher *et al.*, 1990).

The ninth report of the International Committee on Taxonomy of viruses described the presence of nine species (or antigenic complexes) within the *Phlebovirus* genus, RVFV being the type species (Plyusnin *et al.*, 2010). The nine species included 37 viruses while 33 are considered as tentative species in the genus (Table 4.1). The recently isolated Punique virus (Zhioua *et al.*, 2010) should be added to the list of tentative phleboviruses. Also, a novel phlebovirus causing fatality rate of approximately 30% has been isolated in Central and North-east China (Zhang *et al.* 2011; Yu *et al.* 2011). This virus designated under two names, Huaiyangshan virus (HYSV) or severe fever with thrombocytopenia syndrome virus (SFTSV), depending on the laboratory which isolated it, is most closely related to UUKV and appears to be transmitted by *Haemaphysalis* ticks.

History and geographic distribution

Phleboviruses are geographically distributed in North and South America, Europe, Asia and Africa (reviewed in Tesh, 1988a,b; Beaty and Calisher, 1991; Verani and Nicoletti, 1995). To our knowledge, no sandfly fever virus has been isolated so far in South-East Asia and Australia. Historically, sandfly fevers have been traced back during Napoleonic War,s when the Austrian troops stationed in the Adriatic. These diseases had also an important impact on the British colonial troops in India and Pakistan or during the Second World War in 1943–44, when the American, British and German soldiers in North Africa and the Mediterranean basin were affected by the disease. Epidemics have also occurred in civilian populations, such as in Serbia in 1948 with more than one million cases. More recently, tourists visiting sandfly fever endemic area became a target population.

Although many phleboviruses are transmitted by sandflies, other arthropods are also recognized vectors. This is the case for UUKV and RVFV which are respectively transmitted by ticks and mosquitoes (Daubney *et al.*, 1931; Oker-Blom *et al.*, 1964). The diversity of vectors can explain the differences in the biology of the phleboviruses. Therefore, the three subgroups of viruses transmitted respectively by *Phlebotomus* (sandfly fever viruses), mosquitoes (RVFV) and ticks (UUKV) will be addressed individually in the course of this review.

Phlebovirus vectors are members of the *Phlebotomus and Lutzomyia* genera. A few words on the phlebotomines will help to understand the ecology of these vectors. More than 500 species are registered; they are abundant in the tropical areas as well as in temperate regions of Europe (Tesh, 1988b). Their life cycle involves four stages: eggs, larvae, pupa and adults. The larvae develop in microhabitats feeding on organic matter. During cold, prolonged dry periods or winter in temperate regions, some species diapause as fourth instar larvae and emerge as adults at spring. Most sandflies are silent nocturnal fliers with short flight range, biting at night and showing some host preference: some species feed only on cold blood animals while others prefer birds or mammals.

In Europe and Middle East, the principal vectors are *Phlebotomus* spp. Phleboviruses have been isolated from various species such as *P. papatasi, P. perniciosus or P. perfileiwi* (Tesh, 1988b; Beaty and Calisher, 1991). These insects live in dry peri-urban and rural environments close to domestic animals and humans. In the New World, the principal vectors are *Lutzomyia* spp.; they are associated with tropical environment. Virus isolation from male

Table 4.1 Members of the phlebovirus genus

Species	
Bajaru virus	*Salehebad virus*
Bajaru virus - BeAn47693	Arbia virus - ISS.Phl.18
Munguba virus - BeAr389707	Salehebad virus -I-81
Candiru virus	*Sandfly fever Naples virus*
Alenquer virus - BeH301101	Karimadad virus - I-58
Candiru virus - BeH 22511	Massila virus
Itaituba virus - BeAn213452	Sandfly fever Naples virus-Sabin
Nique virus - Nique-9C	Tehran virus - I-47
Oriximina virus - BeAr385309	Toscana virus - ISS.Phl.3
Turuna virus - BeAr352492	*Uukuniemi virus*
Chilibre virus	EgAN 1825-61 virus
Cacao virus - VP-137R	Fin V 707 virus
Chilibre virus-VP118D	Grand Arbaud virus - Argas2
Frijoles virus	Manawa virus - Argas T461
Frijoles virus VP-161A	Murre virus
Joa virus -BeAr371637	Oceanside virus
Punta Toro virus	Ponteves virus - Larves 6
Buenaventura virus - CoAr3319	Precarious Point virus - MI 19334
Punta Toro virus - D4021A	RML 105355 virus
Rift Valley fever virus	St Abbs Head virus
Belterra virus - BeAn356637	Tunis virus
Icoaraci virus - BeAn24262	Uukuniemi virus-S23
Rift Valley fever virus	Zaliv Terpenyia virus
Tentative species in the genus	
Aguacate virus - VP-175A	Jacunda virus – BeAn428329
Ambe virus – BeAr407981	Leticia virus – CoAr171616
Anhanga virus – BeAn46852	Mariquita virus – Mariquita A
Arboledas virus – CoAr170152	Morolillo virus – HTN351
Ariquemes virus – BeAr485687	Morumbi virus – BeH475236
Armero virus – CoAr171096	Mucura virus – BeAr455230
Arumowot virus – SudAr1284-64	Odrenisrou virus – ArA1131/80
Caimito virus – VP-488A	Pacui virus – BeAn27326
Chagres virus – JW-10	Rio Grande virus – TBM3-24
Corfou virus – PaAr814	Salobo virus – BeAn578142
Durania virus – CoAr171162	Sandfly fever Sicilian virus
Escharte virus – OBS-6528	Saint-Floris virus – DakAnBR512d
Gabek forest virus – SudAn754-61	Serra Norte virus – BeH505240
Gordil virus – DakAnBR496d	Tapara virus – BeAr413570
Itaporanga virus	Uriurana virus – BeAr479776
Ixcanal virus – CA Ar170897	Uricuri virus – BeAn100049

sandflies brought some evidence that the virus can be transovarially transmitted since only females are hematophagous. This mode of transmission, confirmed experimentally for several phleboviruses, can explain the persistence of virus in nature. The question of the vertebrate reservoir remains unanswered. It is suggested that vertebrates (wildlife or rodents) play little role in virus transmission and ecology but instead provide the blood for egg development. However, the question of a putative animal reservoir may be pertinent as, for example, Toscana virus (TOSV) was isolated from the brain of a bat captured in endemic area in Italy.

RVFV is known to cause recurrent epizootics and epidemics in East and West Africa but since 2000, it expanded its geographic range as illustrated by the two simultaneous outbreaks in the Arabic Peninsula (Anonymous, 1998, 2000a,b) (for recent reviews see Bouloy and Flick, 2009; Pepin *et al.*, 2010). During the last two decades, the frequency of RVF outbreaks in East Africa increased significantly, the main ones occurring in Kenya, Somalia, and Tanzania in 1996–97 and 2006–7 (Anonymous, 2007; CDC, 2007) and in South Africa in 2010 (Anonymous, 2010). During these outbreaks the virus spread to the Comoros in 2006–7 (Sissoko *et al.*, 2009) and to Madagascar in 2008 (Anonymous, 2008) provoking a serious epidemic/epizootic. Historically, the virus was isolated for the first time in 1930 during an epizootic involving sheep, lambs and ewes on a farm near the lake Naivasha in the Great Rift Valley in Kenya (Daubney *et al.*, 1931). Since then, many RVFV variants have been isolated in various regions of Africa and from different hosts during endemic or epidemic periods. The virus is transmitted by many species of mosquitoes particularly the *Aedes* and *Culex* species but also a few other hematophagous arthropods (Swanepoel and Coetzer, 2004). Most of the major epizootics/epidemics are associated with periods of excessive rains as it happened during the outbreak in 1997–98 in the Horn of Africa or with alteration of the ecological conditions as during the construction of the dam in Aswan or on the Senegal River which was at the origin of the epidemics in Egypt in 1977 and Mauritania in 1987, respectively (Meegan and Bailey, 1988; Digoutte and Peters, 1989). In addition, humidity and flood favour hatching of mosquito eggs and contribute to amplify virus circulation. Earth observation by satellite remote sensing over the last 30 years has enabled systematic mapping of indicators of climate variability including sea-surface temperature patterns, rainfalls, and ecological indicators (essentially vegetation) on a global scale at high-temporal and moderate spatial resolutions. In a prospective way, vegetation measurements linked to an anomalous increase in sea surface temperatures in Central and Eastern Pacific Ocean and western Indian Ocean can be used to forecast RVF before it reaches epidemic proportion (Linthicum *et al.*, 1999; Anyamba *et al.*, 2006, 2009; Martin *et al.*, 2008). Because of its impact on human and animal health, the economic consequences of epizootics and the mode of transmission by arthropods and aerosols, RVFV is classified as a select agent in many countries in the world (Peters, 2000).

Viruses of the Uukuniemi group are all transmitted by ticks. UUKV was isolated from an engorged *Ixodes ricinus* tick collected from a cow in Finland (Oker-Blom *et al.*, 1964). Other viruses have been isolated from argasid ticks, birds and rodents.

Disease and clinical features

Of the many phleboviruses known to cause human diseases, RVFV has the most dramatic impact. Among the sandfly fever viruses, the most important ones are Sicilian and Naples sandfly fever viruses, TOSV, Karimabad, Punta Toro, Alenquer, Chagres and Candiru viruses (Tesh, 1988a,b; Verani and Nicoletti, 1995). During the history of sandfly fevers, two very early studies were performed by inoculating serum from acute cases to non-immune human volunteers. The first was performed in 1909 by a medical team investigating the cause of epidemics in Austrian troops along the Adriatic coast (Doerr *et al.*, 1909). They demonstrated that the disease was caused by a filterable agent present in the blood of patients and that *P. papatasi* was the vector. The second study was carried out in 1951 by Sabin, who isolated two different viruses from human volunteers inoculated with serum from patients, the Sicilian and Naples viruses (Sabin, 1951).

Sandfly fever is an acute, self limited influenza-like illness of 2–5 days' duration. It begins with fever, severe headache, generalized myalgia, conjunctiva injection, photophobia and malaise. In some cases, the patient suffers from nausea, dizziness and stiff neck. It is usually marked with a severe leucopenia. Recovery is complete with no death associated with sandfly fevers. However, weakness and depression may persist for one to several weeks after the acute phase. In the sandfly fever group, only TOSV exhibits neurotropism and may lead to aseptic meningitis and meningoencephalitis. Typically, TOSV infections occur in spring with a peak in August which correlates with the life cycle of *P. perniciosus* and *P. perfiliewi* (Valassina *et al.*, 2000; Verani *et al.*, 1984a). TOSV virus circulates in central Italy as well as in other Mediterranean countries such as Spain, the south of France, Portugal, Cyprus and Turkey, where sandflies are widely distributed (Niklasson and Eitrem, 1985; Valassina *et al.*, 1998; Sanchez-Seco *et al.*, 2003; Valassina *et al.*, 2003; Peyrefitte *et al.*, 2005; Sanbonmatsu-Gamez *et al.*, 2005; Papa *et al.*, 2006; Charrel *et al.*, 2007). In contrast to RVFV, sandfly fever viruses are not known to cause diseases in domestic animals or wildlife. Nevertheless, antibodies against Sicilian virus have been detected in different species including sheep and cattle in Europe and the Middle East (Tesh, 1988a,b).

RVFV is a pathogen of concern for public and animal health with disastrous economic consequences (Meegan and Bailey, 1988; Swanepoel and Coetzer, 2004). Infection in humans leads to a wide spectrum of symptoms. In a significant proportion of cases, which may vary from 30% to 60%, the infection is asymptomatic. The most frequent course is a benign febrile illness, but the disease can be more severe manifesting in encephalitis, retinitis and hepatitis often associated with fatal haemorrhagic fever. People with jaundice, neurological disease or haemorrhagic complications are at increased risk. Although lethality has been estimated to be between 0.5% and 2% during outbreaks, in recent outbreaks the case fatality ratio appeared to be substantially higher (LaBeaud *et al.*, 2008). This raises the question of whether the virus has acquired an increased virulence or whether the surveillance has been improved. Human infection occurs frequently in groups having contact with livestock such as herdsmen, abattoir workers and veterinarians. Although transmission by mosquitoes is the preferred mode of infection by this arbovirus at the beginning of an outbreak, aerosols of infected blood and contacts with infected tissues probably play an important role in the course of the epizootic/epidemic. Transmission by direct contact with the virus is largely due to the high viraemia developed in many vertebrates and especially in domestic animals such as sheep, cattle and goats which are considered as the principal amplifying host of the virus (Gerdes, 2004; Swanepoel and Coetzer, 2004). Aborted materials are also infectious, being a cause of virus transmission through direct contact with fetal envelopes, placenta and the fetus. Within these tissues, virus may stay infectious over a few days as RVFV particles are rather resistant to inactivation in a protein-rich environment. Virus has been detected in nasal and lacrimal secretions and in milk during the viraemic phase but these data need further confirmation. The shedding of RVFV into milk would have potentially large consequences for public health.

In susceptible animals and particularly in ruminants, RVF disease has been well described since the first identification of the virus in the Rift Valley of Kenya during an epizootic in 1930 (Daubney *et al.*, 1931). Several well-documented descriptions of the symptoms in naturally infected animals have been recorded (Coetzer, 1977, 1982; Coetzer and Ishak, 1982; Swanepoel and Coetzer, 2004). In experimental infections in ruminants, the incubation period lasts for two to four days and is followed by clinical manifestations like fever, hepatitis and/or abortions. During an epizootic, sheep are susceptible to the disease and the mortality rate in newborn lambs may reach 90–100% but the animals become more resistant with age. The classical hallmark of RVF epizootics is the large number of abortions among pregnant ruminants regardless of the stage of pregnancy. These massive abortion events are referred to as 'abortion storms'. They are characteristic of RVF infection in adult animals and should be distinguished from the other most common infectious causes of abortion in ruminants such as chlamydiosis, salmonellosis, listeriosis or toxoplasmosis. In non-endemic countries, surveillance relying on the detection and rapid reporting of significant abortion events to national authorities (for example, more than 20% of pregnant animals in a herd suddenly abort with accompanying signs of jaundice among survivors) could provide effective means to detect the emergence of this significant veterinary and human health threat.

The viruses of the Uukuniemi group have been isolated from ticks in Europe, Asia and Australia. Antibodies against UUKV have been detected in humans and other vertebrates indicating that they can be infected but no clinical signs have been reported (Saikku, 1973).

Strain evolution

Owing to the paucity in genomic information, the taxonomy of phleboviruses is based on the degree of cross-reactivity of serological tests (haemagglutination-inhibition and complement-fixation), the species being distinguishable by 4-fold differences in two-way neutralization tests. A preliminary genetic characterization of different phleboviruses of the sandfly fever Naples, sandfly fever Sicilian and Punta Toro complexes was undertaken by sequencing and analysing a region of the M segment (Liu *et al.*, 2003). The three genetic lineages were consistent with the serological data. Moreover, a high genetic divergence was observed among viruses of Sicilian, Naples and Punta Toro serocomplexes, the values being 48.2% and 57.6% at the nucleotide and amino acid levels, respectively. A more extensive analysis including Icoaraci and Frijoles serogroup viruses and involving the nucleoprotein and the NSs sequence in the S segment showed also a good correlation between genetic lineages and serological data (Xu *et al.*, 2007). The genetic divergence is lower in the sequences of the S segments than in the M segments. Another study performed by the same laboratory on Salobo virus, a newly isolated phlebovirus pointed out a discordance between serological data and phylogenies of Icoaraci, Belterra and RVFV viruses which exhibit serological cross-reactions (Travassos da Rosa *et al.*, 1983) while the genetic relationships showed that Icoaraci and Belterra viruses did not cluster together with RVFV (Xu *et al.*, 2007) (Fig. 4.1). There are presently nine species (serocomplexes) identified in the ninth ICTV report, the RVFV species including also Icoaraci and Belterra viruses. The authors of the phylogenetic analysis suggested addition of a tenth species corresponding to a new Icoaraci serocomplex comprising Belterra, Icoaraci and the newly described Salobo virus while the RVFV species remains alone.

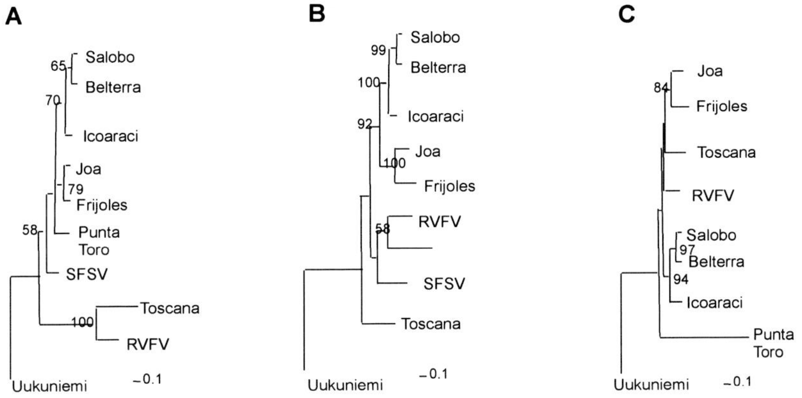

Figure 4.1 Phylogenetic relationship between phleboviruses based on nucleic acid sequences of the S segments (A), M segments (B) and L segments (C). Analysis were performed using neighbour joining, maximum parsimony maximum likelihood and Bayesian methods. Numbers represent the percentage bootstrap support calculated for 2000 replicates. Uukuniemi virus was set as the outgroup to root the trees (adapted from Xu *et al.*, 2007).

Within the last decade, TOSV, a major cause of aseptic meningitis and encephalitis, became a prevalent phlebovirus in Europe in the Mediterranean basin as revealed by multiple isolations from patients and from sandflies in Italy, Spain, Portugal, France, Greece and Cyprus (reviewed by Charrel *et al.*, 2005). Through the phylogenetic analysis performed on partial L and S sequences of various isolates from patients and *P. perniciosus*, it appears that two genotypes of TOSV are circulating, the Italian and Spanish genotypes (genotypes A and B, respectively) by reference to the origin of the isolates (Sanbonmatsu-Gamez *et al.*, 2005; Charrel *et al.*, 2007). Both genotypes were identified in France and probably in Portugal as well (Charrel *et al.*, 2005; Peyrefitte *et al.*, 2005). A relatively low diversity was reported within the sequence of the nucleoprotein (Valassina *et al.*, 1998). In contrast, among the sequences of the Gn amplicon performed on various isolates of TOSV collected over a period of 23 years, the divergence ranged from 0% to 5.7% at the nucleotide level and phylogenetic analysis revealed four clusters (Venturi *et al.*, 2007). A more detailed study (including TOSV isolates collected over 35 years) based on complete sequence of the L, M and S segments came also to the conclusion that the M segment is the most divergent and S the most conserved (Collao *et al.*, 2009). The authors noted that the variability in the M sequences among TOSV isolates is higher than among the RVFV strains and that these viruses have evolved from a most recent common ancestor more ancient for TOSV than for RVFV. The modifications in the distribution of phlebotomines owing to climate changes might have a significant impact on the epidemiology of TOSV.

Phylogenetic studies on RVFV isolates collected in various geographic areas from ruminants, mosquitoes and humans over 50 years showed that the genetic variability is relatively low, with a maximum of 4–5% at the nucleotide and 1–2% at the amino acid level (Bird *et al.*, 2007b; Sall *et al.*, 1997, 1999). This probably reflects the existence of only one serotype. Furthermore, genetic analyses place the most recent common ancestor of these viruses in the late 1880s, suggesting that RVFV strains have spread recently after introduction of 'exotic' breeds into endemic regions (Bird *et al.*, 2007b). It is speculated that at some point between

1850 and 1910 an unknown arbovirus ancestor which has evolved into RVFV exploited a newly formed ecological niche created by the presence of highly susceptible cattle and sheep breeds imported into eastern and southern Africa (Bird *et al.*, 2007b). RVFV isolates cluster into several distinct lineages generally in correlation with the geographic origin. However, some isolates, although geographically distant, are found within the same lineage, indicating a widespread dispersal and movement of the RVFV genotype across Africa (Bird *et al.*, 2008b). In some cases, the phylogenetic trees based on the L, M and S sequences exhibit limited incongruence, suggesting that these variants were generated by reassortment (Sall *et al.*, 1999; Bird *et al.*, 2007b). While the genetic diversity is relatively low among strains from an outbreak, strains from an enzootic period can be of relatively higher diversity. Isolates with low diversity have been observed during the Egyptian or the Mauritanian outbreak as well as among the Kenyan strains that remained similar for more than 20 years in East Africa and during the outbreaks in Saudi Arabia and Yemen (Sall *et al.*, 1998; Bird *et al.*, 2008b). This genetic variability, even if low, allowed the tracking of viruses isolated in Yemen and Saudi Arabia and suggested that they had been imported from East Africa through livestock infected during the Kenyan outbreak of 1996–7 (Shoemaker *et al.*, 2002). In spite of low diversity, it should be noted that during the 2006–7 outbreak evidence of recent increase in genomic diversity was observed, 2–4 years prior to the outbreak, indicating ongoing RVFV activity during a non-epidemic period (Bird *et al.*, 2008b).

Virus structure

Bunyaviruses are all enveloped spherical viruses approximately 80–120 nm in diameter, with the viral ribonucleoproteins (RNP) packaged into the particle. Early work on UUKV and RVFV reported that the particle possesses a particular arrangement of the glycoprotein spikes resembling hollow cylinders regularly arranged at the surface of the particle (von Bonsdorff and Pettersson, 1975; Pettersson, 1975; Ellis *et al.*, 1979). Indeed, when compared with other members of the family like La Crosse and Hantaan viruses, phleboviruses appear to be the most regular, having a constant size and being only mildly pleiomorphic. More recent studies by cryo-electron tomography and single particle averaging on UUKV and RVFV (Freiberg *et al.*, 2008; Overby *et al.*, 2008; Huiskonen *et al.*, 2009; Sherman *et al.*, 2009) have confirmed these observations and changed our view of phleboviruses and probably of other bunyaviruses. These studies indicated that the virion is likely to have icosahedral symmetry: the structure is highly ordered and the surface covered by a shell of 120–122 glycoprotein capsomers arranged in an icosahedral lattice with $T = 12$. Three-dimensional reconstructions at 22 or 27Å resolution revealed that the capsomers resemble hollow cylinders composed of pentamers and hexamers of Gn–Gc heterodimers. Inside the envelope, the viral ribonucleoproteins formed by each of the three genomic segments associated with numerous copies of the nucleoprotein (N) and a few copies of the RNA-dependent RNA polymerase (L), are packaged into the virion. The three-dimensional reconstructions showed that a layer of RNP is located proximal to the inner leaflet of the membrane, strongly suggesting an interaction between the cytosolic tail of the glycoproteins and the RNPs. This interaction would compensate for the absence of matrix protein in viruses of this family. Therefore, the structure of the phlebovirus, with the internally packaged and not well-organized RNPs and the highly ordered surface of glycoprotein capsomers, appears to represent an intermediate between strictly icosahedral and pleiomorphic viruses. In the same vein, new data on

Tula hantavirus also described novel features of virions with an ordered organization of the glycoprotein spikes composed of tetramers of Gn–Gc dimers (Huiskonen *et al.*, 2010).

Genome organization

Like all bunyaviruses, phleboviruses are characterized by a genome composed of three segments of negative polarity RNA, designated L, M and S. However, the S segment of phleboviruses presents a variation, compared with other bunyaviruses, in utilizing an ambisense strategy for N and NSs expression. In tospoviruses, the ambisense strategy is also used for both S and M segments. All the replication steps occur in the cytoplasm of infected cells and virions mature by budding in the Golgi compartment (Plyusnin *et al.*, 2010). Maturation at Golgi membranes differentiates bunyaviruses from other enveloped viruses which bud at the plasma membrane and/or at intracellular membranes of the endoplasmic reticulum (ER) as in the case of flaviviruses or rotaviruses. To date, RVFV-infected hepatocytes in primary cultures seem to represent the only known exception in which budding of particles was observed at the plasma membrane in addition to the Golgi vesicles (Anderson and Smith, 1987).

The virion is composed of four proteins, the two envelope glycoproteins Gn and Gc present at the surface of the particle and two internal proteins, the viral RNA-dependent RNA polymerase (L) and the nucleoprotein (N) that are tightly associated with the viral genome in the form of ribonucleoproteins (RNP). As illustrated for UUKV, purified RNPs are circular and possess a pseudo-helical structure (Pettersson and von Bonsdorff, 1975), a feature reported for orthobunyaviruses (Samso *et al.*, 1975; Obijeski *et al.*, 1976) and characteristic probably for other bunyaviruses as well. The structure of the RVFV nucleoprotein was recently reported (Raymond *et al.*, 2010). It has a compact helical fold consisting of N and C terminal lobes of approximately equal size. No structure similar to either lobe was identified in search with the server Dali and VAST (Holm and Sander, 1995; Gibrat *et al.*, 1996). Cross-linked N species in the viral or recombinant RNPs appeared as multiples of two, consistent with the dimeric association of N (Le May *et al.*, 2005). The N lobe of RVFV N protein was assumed to be a potential RNA interaction site because it is more positively charged than the C terminus and to engage the phosphate backbone of the RNA because the multimer is highly resistant to ribonuclease. Unlike other negative sense RNA viruses, RNFV N protein has no positively charged surface cleft for RNA binding. The RNPs purified from the virions or reconstituted from recombinant nucleoprotein and a defined RNA have an extended structure without helical symmetry, unlike the RNPs from other negative stranded RNA viruses like influenza virus, rabies virus or Borna virus which have RNP with helical symmetry forming high-order ring structures with specific contacts of loops or chain termini of neighbouring N subunits. Thus, the N protein of RVFV and probably of all phleboviruses appears unrelated to any nucleoprotein of other negative stranded RNA viruses.

The L and M segments are of negative polarity, coding respectively for the L protein, which is the viral RNA-dependent RNA polymerase (Elliott *et al.*, 1992; Accardi *et al.*, 1993; Muller *et al.*, 1994), and for a polypeptide precursor to the glycoproteins (Collett *et al.*, 1985; Ihara *et al.*, 1985b; Ronnholm and Pettersson, 1987). The S segment utilizes an ambisense strategy and codes for two proteins, the nucleoprotein N and a non-structural protein called NSs (Ihara *et al.*, 1984; Marriott *et al.*, 1989; Simons *et al.*, 1990; Giorgi *et al.*, 1991). The coding capacity of the genome is depicted in Fig. 4.2. Of note, the 3′ and

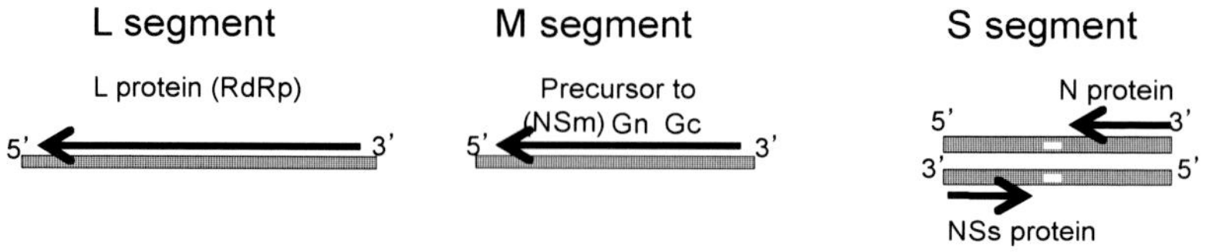

Figure 4.2 Genome organization of phleboviruses. The L, M and S segments are represented in the genomic sense (and antigenomic sense for the S ambisense segment) by the grey line and the open reading frame are orientated as indicated by the arrow.

5′ terminal sequences of each segment possess the consensus sequence 5′-ACACAAAG... CUUUGUGU-3′ with the first eight nucleotides strictly conserved within the three segments and in all the phleboviruses. The extremities are complementary to each other, allowing self-annealing and formation of panhandle structures. This explains why the RNPs appear circular when observed by electron microscopy.

The general view that only the viral genome is incorporated into the mature particle has been revisited since a small but significant fraction of antigenomes, i.e. replicative intermediates, have been detected in purified RVFV particles (Ikegami *et al.*, 2005a). These data are in agreement with previous studies on UUKV showing that the S segment of genomic and antigenomic polarities could be detected in purified virions (Simons *et al.*, 1990).

Viral cycle

Over the two last decade, the development of tools such as the reverse genetics or the production of virus-like particles (VLP) has helped to make progress in the understanding of the particular sequences or proteins involved in transcription and replication as well as in virus assembly.

Transcription and replication

General features

The general features of phlebovirus transcription and replication are similar to those of other negative stranded RNA viruses (Elliott, 1996; Giorgi, 1996; Schmaljohn and Nichol, 2007). During the replication cycle, each genome-sense RNA molecule is transcribed into mRNA and is replicated through a process involving a replicative intermediate which is the exact copy of the genome, called complementary RNA (cRNA) or antigenome. In infected cells, neither genomes nor antigenomes are found as naked RNA; they are associated with N and L proteins in the form of RNPs. Owing to its ambisense organization, the S segment antigenome serves as a template for the synthesis of the NSs mRNA. It is worth recalling that the S antigenome is present in the input virus together with the S vRNA; thus both the N and NSs proteins are expressed at early stages of the viral cycle (Simons *et al.*, 1990; Ikegami *et al.*, 2005b).

Similar to other bunyaviruses, mRNA synthesis is initiated through a cap-snatching mechanism while the synthesis of cRNA is initiated with 5′ nucleoside triphosphates (Simons and Pettersson, 1991; Gro *et al.*, 1992). In fact, cRNAs and mRNAs can be distinguished by the nature of their 5′ end as well as by their 3′ end since mRNAs terminate before the 5′ end of the template while cRNAs are the complete copies of the vRNA. During mRNA synthesis, transcription termination occurs some 100–150 nucleotides before the end of the M template (Collett, 1986) or in the intergenic region of the S segment (Ihara *et al.*, 1985a; Simons and Pettersson, 1991; Gro *et al.*, 1992). Since the L protein has two functions, there should be a switch between transcriptase and replicase activities but the mechanism remains unknown. Several polymerase consensus motifs were found in the L protein of RVFV and other bunyaviruses (Muller *et al.*, 1994). An N-terminal endonuclease domain responsible for the cap-snatching activities has been identified in La Crosse virus (Reguera *et al.*, 2011). This domain appears to be conserved in other bunyaviruses as well as in arenaviruses, tenuiviruses and orthomyxoviruses.

Reverse genetics systems

After the pioneering work of R Elliott's group to rescue infectious Bunyamwera virus from cDNA plasmids (Bridgen and Elliott, 1996; Bridgen *et al.*, 2001; Lowen *et al.*, 2004), different systems allowing manipulation of the genomes of other bunyaviruses have now been developed (Blakqori and Weber, 2005; Ogawa *et al.*, 2007). Among phleboviruses, RVFV remains the only one amenable for infectious virus rescue (Ikegami *et al.*, 2006; Gerrard *et al.*, 2007; Billecocq *et al.*, 2008; Habjan *et al.*, 2008b) but minigenomes mimicking L, M or S genome segment in which the viral ORF is replaced by a reporter gene, have been developed for UUKV (Flick and Pettersson, 2001; Flick *et al.*, 2002) and TOSV virus (Accardi *et al.*, 2001). Minigenome expression and infectious virus rescue utilize one of the classical systems: the T7 or Pol-I promoter-based plasmids for synthesis of the viral RNAs (Neumann *et al.*, 1999). Both systems have advantages and disadvantages: expression under the control of the T7 promoter requires transfection in cells such as the BSR-T7 or BHK-T7–9 cells, that constitutively express T7 RNA polymerase, while RNA synthesis from the Pol-I promoter is species-specific so that the cells used for the study must correspond to the species of the promoter sequence.

Besides being an essential step in the development of reverse genetics and infectious virus rescue, minigenomes have been helpful in analysing various steps in RNA synthesis, i.e. transcription (including transcription termination), replication, and packaging. Studies with minigenomes clearly established that transcription and replication of the viral-like RNA requires expression of both N and L proteins (Lopez *et al.*, 1995; Flick and Pettersson, 2001; Ikegami *et al.*, 2005a; Gauliard *et al.*, 2006). These proteins are necessary to reconstitute RNPs confirming the concept, valid for all the negative stranded genomes, that naked RNA cannot be transcribed. Interestingly, both N and L of RVFV have the ability to form homo-dimers or -oligomers, which were detected in purified RNPs (Le May *et al.*, 2005; Zamoto-Niikura *et al.*, 2009; Raymond *et al.*, 2010). The role of the oligomers is yet unknown but proteins which cannot oligomerize are not functional in the minigenome assay, a property shared by Bunyamwera virus (Leonard *et al.*, 2005). In some studies, mini-antigenomes were used instead of mini-genomes to assess the role of L in replication since a reporter activity can be observed only if the mini-antigenome has been replicated into a minigenome (Flick *et al.*, 2004; Gauliard *et al.*, 2006). These experiments confirmed that the L protein is able to perform both transcription and replication, excluding the possibility that another viral factor like NSs modifies the transcriptase into a replicase. One should note, however, that the RVFV NSs was reported to promote viral RNA replication and transcription in a minigenome system (Ikegami *et al.*, 2005a) but this particular characteristic was not confirmed by our group (unpublished data) which found it inhibitory, like the NSs of Bunyamwera virus (Weber *et al.*, 2001). This discrepancy may be due to the differences between T7- and Pol I-based systems as well as to the cell lines (IFN-competent 293 cells vs. IFN-incompetent BHK cells) used to express the minigenomes (Ikegami *et al.*, 2005a; Gauliard *et al.*, 2006). This effect observed in IFN competent cells might be linked to IFN production which is highly activated by the T7 transcripts and has antiviral effects at multiple levels. In contrast, such effects were not observed in the presence of NSs, owing to the lack of IFN production since the NSs protein is a potent inhibitor of the IFN expression (see later).

When compared with each other, the L, M and S segment-based minigenomes do not express identical levels of reporter gene, indicating differential promoter activities associated with the non-coding regions (Gauliard *et al.*, 2006). For RVFV, the L segment-based minigenome was 1.7 times more efficient than the S or M segment-based minigenomes whereas for UUKV, the M segment has a stronger promoter activity than the L or S segments (Flick *et al.*, 2002, 2004). Extensive mutational analysis in the 3′ and 5′ terminal sequences of the UUKV M segment-based minigenome indicated that both ends which are able to form a panhandle owing to base pairing, build up the promoter region (Flick and Pettersson, 2001). An earlier study on the RVFV non-coding regions showed also the importance of the conserved eight terminal nucleotides in promoter activity and regulation of gene expression as well as a particular role of the purine at position 13, which appear to be specific for RVFV L polymerase (Prehaud *et al.*, 1997).

Transcription termination

In contrast to most viral or cellular mRNAs, bunyavirus messengers are not polyadenylated (Schmaljohn and Nichol, 2007). The M segment mRNA of Sin Nombre hantavirus seems to be the only exception reported to date (Hutchinson *et al.*, 1996). Bunyavirus mRNAs terminate prematurely on their template, strongly suggesting that the transcriptase can recognize a transcription termination signal. This question was addressed in the studies of ambisense S segment: the 3′ end of the N and NSs mRNAs are overlapping and terminate either within or near the intergenic region (Ihara *et al.*, 1985a; Simons and Pettersson, 1991; Gro *et al.*, 1992). Preliminary analysis showed that some phleboviruses like TOSV or RVFV have a C-rich intergenic sequence while others, like Punta Toro virus (PTV) or UUKV have the ability to form hairpin structures in their intergenic region, the latter being similar to the situation observed in arenaviruses (Iapalucci *et al.*, 1991; Lopez and Franze-Fernandez, 2007). It was suggested that there are two types of transcription termination signals, depending on the virus and the characteristics of its intergenic region (Giorgi *et al.*, 1991; Gro *et al.*, 1992): a hairpin structure for viruses like PTV or UUKV and a sequence motif for viruses with the C-rich intergenic region. To further understand the termination process in RVFV transcription, Albarino *et al.* (2007) mapped precisely the 3′ end of the four RVFV mRNAs, i.e. N mRNA, NSs mRNA and also M- and L-derived mRNA, and identified a conserved sequence motif "-C_{1-3}GUCG/A′ on both M and ambisense S segment sequences of several phleboviruses of the sandfly serogroup confirming previous data on the 3′ end of RVFV M mRNA (Collett, 1986). To demonstrate the role of this sequence, they created by means of reverse genetics, a recombinant RVFV lacking the termination motif in the S segment. Analysis of the mRNAs synthesized in the cells infected with this mutated virus showed that the transcriptase failed to terminate correctly the S mRNAs. Using the RNase protection method to map the 3′ end, Ikegami *et al.* (2007) came to similar conclusion on the transcription termination on the S and M segments. However, the two reports did not agree on the termination in the L segment which lacks the pentanucleotide motif in the 5′ non-coding region. Albarino *et al.* (2007) reported that L mRNA represents the complete copy of its template whereas, according to Ikegami and colleagues, the L mRNA lacks the last 16–41 nucleotides when compared with the full-length cRNA copy and the termination signal corresponds to two complementary sequences, each 13 nucleotides long, present in the 5′ non-coding region of the L genomic segment.

The role of the glycoproteins and virus assembly

Maturation and trafficking

During the viral cycle, the glycoproteins play an essential role in attachment and entry of the virus, and their proper processing is crucial for the maturation and budding of the virion. During infection, the glycoproteins are the most exposed components of the virus, they are recognized by the immune system, and induce the production of neutralizing antibodies which play a predominant role in protection.

Present at the surface of the particle, the glycoproteins mediate virus attachment and entry into many cell types through specific receptors which, except for the hantaviruses, remain to be identified. Entry is predicted to employ a class II fusion mechanism that is activated by low pH following endocytosis of the virion (Filone *et al.*, 2006). Little is known yet regarding the early phases of infection that precede the release of virus RNPs into the cytosol. A recent study showed that UUKV is rapidly (within 10 min) endocytosed into cells, mainly but not exclusively, via non-coated vesicles (Lozach *et al.*, 2010). The virus traverses via early endosomes to late endosomes where acid-activated penetration (pH~5.4) occurs between 20 and 40 min after internalization. This study provided strong evidence that UUKV penetrates host cells from late endosomal compartments.

Among phleboviruses, UUKV, RVFV and to a lesser extend PTV have been used as models to analyse the glycoproteins processing and trafficking (for reviews see Matsuoka *et al.*, 1991; Pettersson and Melin, 1996). The phlebovirus M segment codes for a polyprotein precursor which is cleaved co-translationally, generating Gn (encoded by amino-terminal sequences of the precursor), Gc (encoded by carboxy-terminal sequences) and eventually a non-structural protein called NSm. In the precursor sequence, the NSm protein is located at the N terminal region of the precursor and corresponds to the PreGn region but the M segment of UUKV does not encode such a non-structural protein. PTV and TOSV express a unique NSm protein of 30 kDa (Ihara *et al.*, 1985b; Di Bonito *et al.*, 1999) whereas the RVFV precursor can generate two non-structural proteins, the 78-kDa (also called NSm1) and the 14-kDa (NSm2) proteins (Kakach *et al.*, 1989, 1988; Suzich *et al.*, 1990). These proteins are produced by alternative use of the first or the second of five in-frame AUG codons present at the 5′ end of the M mRNA and located upstream of the Gn sequence (Fig. 4.3). The role of RVFV NSm proteins has been addressed recently and will be discussed in the following paragraph. The first AUG initiates the synthesis of the 78-kDa NSm1 product while the second AUG is utilized for the synthesis of the 14-kDa NSm2. The first and fourth AUGs precede a sequence typical for signal peptides, suggesting that signal peptidases are

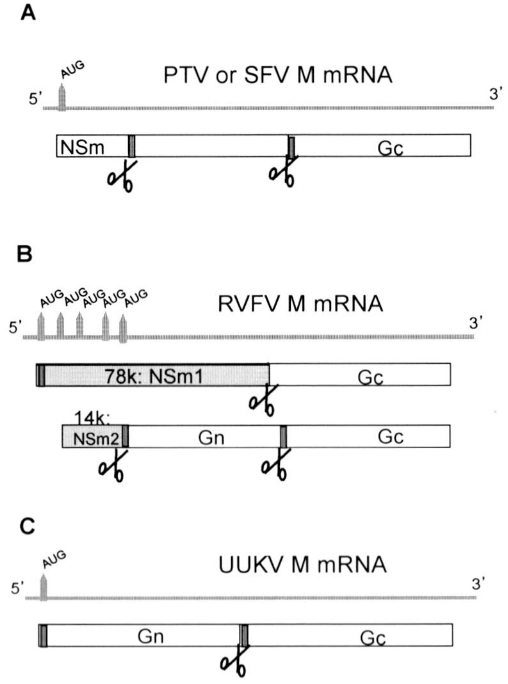

Figure 4.3 Expression of the glycoprotein and non-structural protein NSm. Schematic representation of the processing of the polyprotein precursor coded by the M segment mRNA of Punta Toro (PTV) or sandfly fever viruses (SFV) (A), Rift Valley fever virus (RVFV) (B) and Uukuniemi virus (UUKV) (C). The cleavage site is indicated as well as the hydrophobic region corresponding to the signal peptide (in dark grey).

involved in the cleavage of the precursor and that NSm1 and Gn, but not NSm2, are directed to the endoplasmic reticulum. Interestingly, the 78-kDa protein comprises the PreG and Gn sequences but the N-terminus of Gn is not cleaved at its signal peptide (Suzich *et al.*, 1990). Immunofluorescence studies with sequence-specific antibodies revealed Golgi localization of the 78-kDa protein, similar to Gn and Gc proteins, while the 14-kDa protein exhibited a slightly different distribution, with predominantly Golgi but also some reticular staining (Wasmoen *et al.*, 1988). Previous studies addressed the question of whether any of the NSm proteins plays a role in Golgi retention of the two glycoproteins. By expressing the glycoproteins in a set of recombinant vaccinia viruses possessing progressive deletions of the N-terminal ORF, Wasmoen *et al.* (1988) showed that the absence of NSm1 and NSm2 did not alter the Golgi localization of Gn and Gc, suggesting that the signals for Golgi localization reside in Gn and/or Gc themselves and that the PreG region itself does not contain signals for transport or targeting of the glycoproteins. Thus, the question must be raised of how Gn and Gc reach the Golgi complex. Indeed, localization of the glycoproteins in the Golgi complex together with the viral RNPs, which accumulate in the cytosolic side underneath the Golgi membranes, is essential for proper maturation of the bunyavirus particles that bud into the lumen of the Golgi. Studies on PTV, UUKV and RVFV converged to indicate that Gc, when expressed in the absence of Gn, was not targeted to Golgi; Gerrard and Nichol, 2002) but localized to the endoplasmic reticulum (ER) and therefore only Gn seems to contain the Golgi retention signal. These data strongly suggest that Gc associates with Gn in the ER and moves to the Golgi apparatus owing to its physical association with Gn. Heterodimeric association could well occur through disulfide bridges formed between the numerous cysteine residues present in both glycoproteins. Following translocation and cleavage, the two glycoproteins undergo core glycosylation at asparagines residues of the consensus motif Asn-X-Ser/Thr. During intracellular transport to the Golgi, the glycans are trimmed to various degrees and acquire partial or complete endo H-resistant glycans (reviewed in Pettersson and Melin, 1996). In Gn, the Golgi retention signal was localized within a region comprising the Gn transmembrane domain and the cytoplasmic tail for PTV (Chen and Compans, 1991; Chen *et al.*, 1991a,b; Matsuoka *et al.*, 1994) and within the cytoplasmic tail for UUKV and RVFV (Andersson *et al.*, 1997a,b; Gerrard and Nichol, 2002). For RVFV, it was shown that Gc is retained in the ER owing to the presence of a lysine-based ER retrieval signal at its C-terminus (Gerrard and Nichol, 2002).

Once budding has occurred, virions are transported within vacuoles to the cell surface where fusion of the vacuole and plasma membrane allows releasing the virus into the extracellular medium. This has been described in detail for cells infected by PTV and Karimabad virus (Smith and Pifat, 1982) and UUKV (Pettersson and Melin, 1996). As the bunyavirus cycle progresses, the Golgi complex undergoes a drastic morphologic change with a progressive vacuolization and expansion. These morphological changes are probably due to the accumulation of the glycoproteins and the N protein, which are found to line up on the membrane of the distended Golgi vesicles, or are caused by the budding process itself.

Expression of VLPs

Expression of the glycoproteins together with the minigenome and the transcription machinery allows formation of VLPs which are then released into the medium resembling infectious virus (Overby *et al.*, 2006; Habjan *et al.*, 2009a). Such a methodology was developed for UUKV and RVFV. For UUKV, VLPs were utilized to analyse the signals needed for particle packaging and release (Overby *et al.*, 2006). For RVFV, it was used to produce one-step infectious particles and to analyse their immunogenic potential (see below). Expression of VLPs was also exploited to analyse how the three genomic segments are copackaged to generate infectious particles (Terasaki *et al.*, 2011). The long 5′ noncoding region of the M segment was shown to play a critical role in copackaging of the three viral segments. Intermolecular interactions among the three RNAs are also involved. These data suggest a low particle-to-pfu ratio during virus production.

When studied by electron microscopy and negative staining, VLPs resemble phlebovirus particles both in size and morphology. The VLPs were able to infect naïve cells, undergo the first step of the replication cycle and express reporter protein thus allowing to monitor the capacity of the VLPs to package minigenomes. If both L and N proteins are provided *in trans* to VLP-infected cells, the signal of the reporter gene is amplified (Habjan *et al.*, 2009a). Extensive mutagenesis of the cytosolic tail of Gn and Gc of UUKV helped to determine the functional motifs within the cytosolic domains of the two glycoproteins (Overby *et al.*, 2007a,b). By alanine scanning mutagenesis of the 98 amino acid long Gn C-terminal tail, two short regions (amino acid residues 21–25, 46–50) were found important for VLP release in the medium. Amino acid residues 46–50 are involved in the proper targeting of Gn to the Golgi whereas amino acid residues 21–25 are associated with budding into Golgi vesicles. Because of the lack of a matrix protein in bunyaviruses, it has been assumed that the packaging of RNPs into the viral envelope is based on close contacts between the RNPs and the cytosolic tail(s) of the glycoproteins at the budding sites. Mutational analysis of UUKV Gn protein showed that four amino acid residues (76–81) are essential for the binding of RNPs to the cytosolic tail of Gn and hence for RNP packaging. Regarding the Gc protein, its 5-amino acid long C-terminal domain was found to play an important role in intracellular targeting since mutations in this region led to some staining at the plasma membrane (Overby *et al.*, 2007b). In the case of RVFV, it was shown that efficient release of VLPs requires Gn, which is sufficient for packaging the viral RNPs according to a model in which the cytosolic tail of this glycoprotein binds to N and L (Piper *et al.*, 2011).

Effects of virus infection on host cells

In most cases, vertebrate cells infected with phleboviruses die from the infection. In rare cases, a few cells survive and establish a persistent infection. Persistent infection of Vero cells infected by RVFV and TOSV was associated with the production of viruses with defective L segments (Verani *et al.*, 1984b; Billecocq *et al.*, 1996; Marchi *et al.*, 1998). Only a few studies have been performed to characterize the cell death pathways. When hepatocytes were infected with PTV, they underwent an apoptotic process (Ding *et al.*, 2005; Xu *et al.*, 2008) during which the N and the NSs proteins triggered the extrinsic and intrinsic pathways involving respectively caspase-8 and caspase 9 (Li *et al.*, 2010). This mimics the situation observed in hamster hepatocytes infected with the virus where apoptosis contributes to the severe liver disease (Fisher *et al.*, 2003). Similarly, the neuroadapted strain of TOSV virus was shown to induce neuronal apoptosis (Cusi *et al.*, 2005).

In insect cells, phlebovirus infection is usually chronic or persistent but at present this is documented rather poorly (Nicoletti and Verani, 1985). For other bunyaviruses, persistence is correlated to the production of siRNAs, a phenomenon known as RNA interference (RNAi) and described for several arboviruses and notably for La Crosse orthobunyavirus (Blakqori *et al.*, 2007). However some viruses have developed countermeasures against RNAi by expressing a protein acting as an RNAi inhibitor. Conflicting results were reported for the NSs of La Crosse virus (Soldan *et al.*, 2005; Blakqori *et al.*, 2007) but the activity was clearly identified in the NSs protein of tomato spotted wilt tospovirus, which, like phleboviruses, utilizes an ambisense strategy for its expression (Takeda *et al.*, 2002). However, such an activity was not observed for the NSs protein of RVFV, when tested in arthropod cells using a Semliki Forest virus replicon as an inducer of RNAi (Garcia *et al.*, 2006).

Host cell response and viral suppression of the innate immune response

Being an obligate internal parasite, a virus must propagate to survive. Not only has the virus to usurp the cellular machinery for its replication and protein expression, but also, as the host reacts by deploying defence mechanisms, the virus has to evade this antiviral cell response to win the war against its host. The more serious the disease, the more efficacious the mechanisms should be in counteracting the host defence. Both innate and adaptive immune responses play essential roles. During the last decade, research has been developed to characterize the strategies deployed by bunyaviruses to counteract the innate immune response.

Interferons constitute the first line of defence mechanisms activated in virus-infected cells. In most cell types IFN-β is secreted first and in turn activates the production of IFN-αs in an autocrine and paracrine manner. For recent data on the innate immune response, the reader is referred to several recent excellent reviews (Goodbourn *et al.*, 2000; Weber *et al.*, 2003, 2004; Haller *et al.*, 2006; Borden *et al.*, 2007; Katze *et al.*, 2008; Randall and Goodbourn, 2008). In summary, IFN induction occurs mainly by an intracellular pathway involving two RNA helicases, RIG-I and MDA5, which are the intracellular receptors for viral RNA (Onomoto *et al.*, 2007; Yoneyama and Fujita, 2007). The dsRNA-binding protein kinase PKR contributes to the activation as well. RIG-I binds to short dsRNA as well as to the triphosphate groups at the 5′ ends of uncapped viral genomes characteristic for the negative strand RNA viruses. In contrast to some bunyaviruses like Hantaan virus or Crimean–Congo haemorrhagic fever virus that process their 5′ end to avoid RIG-I dependent IFN induction, RVFV was shown to strongly trigger the IFN pathway through the RIG-I- but not the MDA-dependent induction (Habjan *et al.*, 2008a). The binding of a viral RNA to RIG-I and MDA5 induces a signalling pathway that eventually results in phosphorylation of the transcription factor IRF3. Phosphorylated IRF3 forms homodimers which are relocated to the nucleus. IFN-β is induced through the coordinate activation of NF-κB, IRF3 and AP1 factors which are recruited at the IFN-β promoter. Once expressed and secreted, IFN-β binds to a cell surface receptor common to all type I IFNs and activates the signal transducers and transcriptional activators STAT1 and STAT2, leading to expression of IFN-stimulated genes which code for a large panel of proteins with antiviral activities including Mx GTPase, PKR and 2′–5′ oligoadenylate synthetase OAS/RNaseL. Expression of Mx proteins in cells or in transgenic mice infected with bunyaviruses including RVFV was shown to have potent antiviral effects (Frese *et al.*, 1996; Stertz *et al.*, 2007).

The NSs protein

Although its function remained undetermined for many years, the NSs protein of bunyaviruses has now been identified as a major factor of virulence, primarily characterized as an interferon antagonist (reviewed in Elliott and Weber, 2009). In the case of phleboviruses, the IFN antagonism by NSs was clearly established (Bouloy *et al.*, 2001; Billecocq *et al.*, 2004), with the first evidence coming from a comparative analysis of reassortants between the virulent RVFV strain ZH548 and the avirulent Clone 13, a natural variant isolated serendipitously, which has a large deletion in the NSs ORF (Muller *et al.*, 1995). This analysis demonstrated that all the reassortants possessing the S segment of Clone 13 were avirulent whereas the reassortants possessing the S segment of strain ZH548 were virulent in a murine model (Vialat *et al.*, 2000). A similar genetic approach was carried out for PTV using two different isolates (Baillet and Ames) which induce a lethal or non-lethal infection in the Syrian hamster model (Perrone *et al.*, 2007). Like RVFV NSs, the PTV counterpart was identified as a potent inhibitor of IFN-β mRNA expression. Knowledge on the NSs proteins of other phleboviruses is scarce. In the case of sandfly fever Sicilian virus, the NSs protein was expressed within the context of a recombinant RVFV in which its own NSs ORF was replaced by that from sandfly fever virus (Habjan *et al.*, 2009b); IFN antagonistic activity was also observed. Interestingly, it has been shown that, compared with wild type mice, animals defective for TLR3 demonstrated an increased resistance to lethal infection with PTV and have reduced liver damage (Gowen *et al.*, 2006, 2007). TLR3$^{-/-}$ mice were able to clear systemic virus at a slightly faster rate. In addition, TLR3 appeared to play an important role in the pathogenesis of PTV infection through the overproduction of inflammatory mediators, and interleukin 6 in particular.

For RVFV, the molecular mechanisms sustaining the IFN-β antagonism involve several cellular proteins interacting with NSs. One of them, SAP30 belongs to the Sin3A/NCoR/HDAC repressor complexes intervening in gene transcription regulation. Moreover, it was shown that SAP30 interacts directly with YY1, a transcription factor involved in the regulation of expression of numerous genes, including IFN-β (Le May *et al.*, 2008). Through a series of co-immunoprecipitation, confocal microscopy and chromatin immunoprecipitation experiments, it was demonstrated that NSs, SAP30, YY1, HDAC3 and Sin3A-associated co-repressor factors are recruited on the IFN-β promoter together with IRF3 and that this multiprotein complex excludes CBP (a co-activator known as CREB-binding protein) loading and prevents histone acetylation and transcriptional activation. To ascertain the role of NSs interaction with SAP30 in this mechanism, a recombinant ZH548, was produced by reverse genetics in which the specific domain of NSs required for the interaction with SAP30 was deleted. In contrast to the virulent ZH548 RVFV, this mutant, called ZH548-NSsΔ210–230, was not able to antagonize IFN-β expression and was avirulent in the mouse model.

Investigations on the role of NSs have clearly established that the protein has multiple functions, all aimed at evading the host antiviral response (Ikegami *et al.*, 2009a). In addition to the SAP30-dependent mechanism which specifically inhibits IFN-β gene transcription, NSs is also able to inhibit general transcription of the infected cell (Le May *et al.*, 2004). The two phenomena are timely orchestrated and complementary to each other: IFN-β transcription is an early event occurring at 3–4 h p.i. whereas the general transcription occurs relatively late (8–9 h p.i.) during the viral cycle. The strategy to circumvent host cell transcription relies on the interaction with the p44 subunit of the TFIIH basal transcription factor, which is sequestered by the NSs filamentous structure characteristic of RVFV infection. As a consequence, TFIIH cannot assemble and its concentration drops rapidly. This explains the drastically reduced transcriptional activity of cells infected with RVFV expressing NSs.

While the NSs protein was characterized as an inhibitor of cellular transcription, other studies indicated that it is also able to affect translational activity of the infected cell. This novel function of NSs was described simultaneously by two groups (Habjan *et al.*, 2009b; Ikegami *et al.*, 2009b): NSs promotes post-transcriptional down-regulation of dsRNA-dependent protein kinase (PKR) through the proteasomal pathway and prevents phosphorylation of the eukaryotic initiation factor 2 alpha (eIF2a). It was also observed that this activity is specific to RVFV NSs and is not shared by other phleboviruses like sandfly fever Sicilian virus or La Crosse orthobunyavirus. This conclusion was based on the production of a recombinant RVFV/ZH548 in which the NSs sequence was replaced by the heterologous NSs from either sandfly fever Sicilian phlebovirus or La Crosse orthobunyavirus. It is worth noting that while the PKR degradation activity seems specific for RVFV, the IFN antagonistic activity of NSs is conserved among bunyaviruses (for a review, see Elliott and Weber, 2009).

The cellular localization of the phleboviral NSs protein seems to depend on the virus type. For PTV, the protein is mainly (if not exclusively) present in the cytoplasm and possibly associated with nucleoproteins (Overton *et al.*, 1987). In the case of UUKV and TOSV it was found associated with ribosomes in the cytoplasm (Simons *et al.*, 1992; Di Bonito *et al.*, 1999). The significance of this interaction and localization remains obscure. In contrast, RVFV NSs protein is unique among bunyaviruses as it forms a filamentous structure in the nucleus, which was not expected for a virus replicating in the cytoplasm (Swanepoel and Blackburn, 1977; Struthers and Swanepoel, 1982). Investigations on the formation of the NSs filament showed that cellular DNA is predominantly excluded from the NSs filaments except for some specific DNA regions of the host genome such as clusters of a pericentromeric gamma-satellite sequence to which the NSs is strongly attached (Mansuroglu *et al.*, 2010). Targeting of these sequences by NSs was correlated with the induction of chromosome cohesion and segregation defects in RVFV-infected cells of murine as well as ovine origin. Such phenomenon may contribute to the pathology associated to foetal deformities observed in infected animals.

Altogether, the NSs proteins of RVFV and Punta Toro virus studied so far appear to antagonize IFN-β expression like their counterparts expressed by orthobunyaviruses and some hantaviruses (Elliott and Weber, 2009). No biological role has been assigned to UUKV NSs but it can be speculated that similar functions are lacking in this mildly pathogenic virus. In the case of RVFV, which has been studied in more details, NSs has multiple functions to counteract the actions of the host cell interferon response, either at the transcriptional or at the translational levels by degrading PKR which in turn facilitates translation of the viral products.

The NSm protein: a suppressor of virus-induced apoptosis

Except for UUKV and probably all UUKV-related viruses, phleboviruses express a polyprotein precursor generating the glycoproteins Gn and Gc as well as a pre-glycoprotein which so far is considered as a non-structural protein. Little is known on this non-structural protein. In the case of RVFV, the role of the two non-structural proteins was addressed by producing a recombinant RVFV unable to synthesize the two NSm proteins due to deletion or point mutation in the sequence of the first three initiating AUGs of the M segment (Won *et al.*, 2006; Bird *et al.*, 2007a). Although viral growth was similar to that of the parental virus, the mutated virus caused extensive cell death and produced larger plaques than the parent (Won *et al.*, 2007). Further analyses indicated that the deletion mutant triggered apoptosis through the caspase 3, 8 and 9 pathways, thus revealing that NSm has an anti-apoptotic function contributing to pathogenesis. While a number of studies have shown that NSs and, to a lesser extent, NSm play an important role in virulence, little is known on the involvment of other proteins like Gn and Gc or L and N in pathogenicity. Interestingly, a single nucleotide substitution in Gn affected RVFV mouse virulence (Morill *et al.*, 2010).

Pathogenesis

Studies on virulence and pathogenicity are greatly facilitated by the use animal models which exist for a number of phleboviruses. PTV usually induces a disease in Syrian hamsters and laboratory mice (Fisher *et al.*, 2003; Perrone *et al.*, 2007; Mendenhall *et al.*, 2009). A mouse model has been developed to study pathogenesis of TOSV infection TOSV (Cusi *et al.*, 2005). For RVFV studies, mice, rats and hamsters are good models (Peters and Anderson,

1981). Ruminants are also used for experimental infections with RVFV (Swanepoel and Coetzer, 2004). Nonhuman primates are susceptible to sandfly fever virus (McClain *et al.*, 1997) and to RVFV (Peters *et al.*, 1981; Smith *et al.*, 2010).

The widespread hepatic necrosis is a classic hallmark of severe RVFV disease both in experimental and natural infections (Coetzer, 1982; Coetzer and Ishak, 1982). In animal and human RVFV infections, the primary site of tissue pathology is the liver in which the virus causes a marked cytopathic effect leading to multifocal and coalescing zones of hepatic necrosis with inflammatory infiltrates of immune effector cells. These features have been frequently observed by histopathological examination; they are consistent with elevations of liver enzymes and decrease in total leucocyte counts associated with severe disease (Coetzer, 1982; Coetzer and Ishak, 1982). The rapid onset of severe hepatic damage, particularly in ruminants, may explain many of the early clinical signs associated with severe disease caused by RVFV. Although RVFV is hepatotropic, during severe infections, the virus can be found in many tissues and cell types. The primary tropism to the liver may suffer a few exceptions particularly when the virus is transmitted by aerosols, a mode of contamination probably exceptional but already described for slaughterhouse or for laboratory workers (Smithburn *et al.*, 1949; Chambers and Swanepoel, 1980; Abu-Elyazeed *et al.*, 1996). In mice experimentally infected by inhalation of infectious aerosols, lungs become the primary site of multiplication of the fully virulent viral strain ZH-501, without any clear signs of pneumonia. Infected mice die from a fulminant hepatitis associated with massive hepatic necrosis, as already described when the virus is inoculated by other routes (Brown *et al.*, 1981).

Among the animal models, inbred strains of rats have shown a striking difference in susceptibility to RVFV infection: Wistar-Furth (WF) and Lewis rats were respectively found susceptible and resistant to experimental RVFV infection (Peters and Anderson, 1981; Anderson and Peters, 1988; Anderson *et al.*, 1991). However, the differences are dependent on the commercial breeding source, since in another study that used rats from European breeding colony (mol), the observed picture was reverse (Ritter *et al.*, 2000). In spite of this difference, these results suggested that there is a genetic basis for the resistance or susceptibility to RVFV in these animals. Surprisingly, an influence of genetic factors was not observed in a large survey of 34 classical inbred laboratory mouse strains (Peters and Anderson, 1981). More recent studies involved mouse strains derived from wild catches of different *Mus* subspecies (do Valle *et al.*, 2010). This collection of wild-derived inbred strains encompasses genetic variation accumulated over one million years (Guenet and Bonhomme, 2003) offering a much larger genetic variation than classical laboratory strains. The latter originated from a small number of founder animals and showed a remarkably high level of shared ancestry largely contributed by the *M. m. domesticus* subspecies (Beck *et al.*, 2000; Yang *et al.*, 2007). The MBT/Pas inbred strain was derived from *M. m. musculus* animals trapped in Bulgaria in 1980; the mouse colony was later propagated by sib-mating at the Pasteur Institute (Bonhomme and Guenet, 1996). MBT/Pas mice exhibited an extreme susceptibility to experimental infection with the virulent RVFV ZH548 strain compared with BALB/cByJ mice, demonstrating a natural variation to RVF severity in inbred strains of mice. Investigations are in progress to determine the underlying mechanism and to identify the genes responsible for this difference. These results suggest that, if applicable to ruminants, it may be possible to select resistant domestic animals through controlled breeding programs. This will be of interest since differences in susceptibility were reported among different breeds of sheep in field conditions (Fagbami *et al.*, 1975), although these results were not reproduced during controlled experimental infections (Tomori, 1979; Olaleye *et al.*, 1996). Similarly, genetic determinants may play an important role in modelling RVF outcomes in humans in which infection may lead to a great variety of clinical manifestations that range from a febrile influenza-like illness to retinitis, encephalitis and hepatitis with fatal haemorrhagic fever (Meegan and Bailey, 1988; Gerdes, 2004).

Prevention and vaccines

As for many diseases caused by arboviruses, there is no treatment or vaccine and only supportive therapy can be provided to patients. However, simple measures to avoid mosquito or phlebotomine bites can be relatively efficient for disease prevention. As an example, both Sicilian and Naples sandfly fevers decreased or disappeared after 1940 in countries where the malaria eradication campaigns using insecticide spraying were implemented. Similarly, during RVFV outbreaks, when mosquitoes are abundant it is recommended to relocate animals to altitude where mosquitoes are absent. Application of residual insecticides to animals and to their pens and barns was also suggested.

Sandfly fevers

Although there would be a limited need for vaccines against sandfly fevers for populations living in endemic area, as well as for tourists or military troops stationing in these regions, none is available, probably due to the relatively modest pathogenicity of the viruses. There is however, a study on TOSV showing that inoculation of Gc in combination with N was able to protect all the challenged mice whereas partial protection was observed in animals vaccinated with the glycoproteins or the nucleoprotein alone (Savellini *et al.*, 2008). This and other studies on TOSV and RVFV pointed out the role of the N protein which is the major immunogen during infection, in inducing a strong cell-mediated immune response (Cusi *et al.*, 2001; Wallace *et al.*, 2006).

As an alternative to vaccines, attempts have been made to develop antiviral agents, such as favipiravir T-705 and the related pyravine derivative T-1106 both in PTV- infected cells or in animal models (Gowen *et al.*, 2010). Also, recombinant proteins consisting of IFN-alpha or beta fused to human albumin with prolonged half life compared with unmodified IFNs was found efficient against PTV infection (Subramanian *et al.*, 2008). Treatment with IFN or ribavirin was also found protective against RVF (Peters *et al.*, 1986). However, ribavirin might have undesired effects as it was observed in Saudi Arabia, where administration of this drug was effective against haemorrhagic icterus forms but did not prevent the development of severe meningo-encephalitic complications. This failure might be due the fact that ribavirin is a large molecule which does not pass the meningeal barrier (P. Formenty, personal communication).

Existing RVFV vaccines

A RVFV vaccine, still in use, was developed by Smithburn (1949), who adapted the Entebbe isolate by serial intracerebral inoculation into mice to attenuate the virulence, a procedure already used by Theiler to produce the yellow fever 17D vaccine (Theiler and Smith, 1937). However, the passaged virus, called the Smithburn strain, had lost its virulence only partially: it induced abortions and teratogenesis in ewes, cows and goats (Swanepoel and Coetzer,

2004; Botros *et al.*, 2006; Kamal, 2009), so that its use is restricted mainly to outbreaks and non-pregnant females only (Barnard, 1979; Swanepoel and Coetzer, 2004). To circumvent these difficulties, an inactivated vaccine has been produced, allowing vaccination of animals safely, but this vaccine is not as efficacious as live-attenuated vaccines and needs booster inoculations (Barnard and Botha, 1977). In contrast with veterinary vaccines, there is no commercially available vaccine for humans. However, formalin-inactivated vaccines NDBR103, and later also TSI-GSD200, have been manufactured in limited amounts and used to vaccinate persons at risk to RVFV exposition such as laboratory workers or veterinarians (Pittman *et al.*, 1999). After three subcutaneous injections in human volunteers, a good level of neutralizing antibodies was elicited. However, a regular booster administration is necessary to maintain protective immunity. Unfortunately, this vaccine is difficult to produce and is in short supply.

After the Egyptian outbreak in 1977, USARMRIID initiated research to produce a new live-attenuated RVFV vaccine. The attenuated MP12 variant was obtained from the virulent Egyptian strain (ZH548) after random mutagenesis with 5-fluorouracil during 12 successive passages in the presence of the mutagen (Caplen *et al.*, 1985). This virus acquired mutations in all three RNA segments, had lost its virulence for mice (Saluzzo and Smith, 1990; Vialat *et al.*, 1997) and, after experimental inoculation in ruminants and sheep, induced a good immunity (Morrill *et al.*, 1987, 1991, 1997a,b). However, other trials carried out in South Africa using pregnant ewes revealed abortions and teratogenic effects during the early stages of pregnancy (Hunter *et al.*, 2002). This might corroborate data reporting that MP12 still showed some neurovirulence when inoculated to hamsters, which are considered as a very sensitive animal model (Rossi and Turell, 1988).

Appearing as an alternative candidate, Clone 13, a plaque purified isolate obtained from a mild case, was found to be avirulent owing to a large deletion in the NSs protein (Muller *et al.*, 1995). This virus grows to high titres in cell cultures and the deletion in the sequence coding for the virulence factor made it attractive owing to its inability to revert to the pathogenic wild-type. Vaccination trials in sheep and ewes as well as in bovines were carried out. The virus did not provoke any adverse effects and the animals elicited a high antibody response and were protected against a virulent challenge. After the challenge, pregnant ewes were also protected from abortion which was observed in the unvaccinated animals (Dungu *et al.*, 2010).

Designing new vaccines

Because of the socio-economical impact of RVF in Africa, it appeared that vaccines, at least for veterinarian use, were urgently needed. Beside the naturally avirulent virus Clone 13 (Muller *et al.*, 1995) or R566, a reassortant with the L and M segments of MP12 and the S segment of Clone 13 (Bouloy *et al*, unpublished data), several other candidates lacking the complete NSs gene were created by reverse genetics (reviewed in Bouloy and Flick, 2009; Ikegami and Makino, 2009). A recombinant MP12 equivalent to R566 was produced in which the S segment carries a mutation identical to the one in Clone 13 (Ikegami *et al.*, 2006). Other recombinant viruses were obtained by removing the complete NSs sequence from MP12 or from the otherwise virulent backbone of the strain ZH501 (Bird *et al.*, 2007a, 2008a). In addition, correlating with its role as an anti-apoptotic factor, NSm was found to contribute to pathogenesis (Bird *et al.*, 2007a). Thus, a double mutant ΔNSs/NSm abrogating expression of both NSs and NSm proteins was created and shown to be avirulent in the rat model (Bird *et al.*, 2008a).

Sub-unit recombinant vaccine candidates expressing the RVFV glycoproteins have been described: they used different vectors such as the lumpy skin disease virus (Wallace *et al.*, 2006), an alphavirus (Filone *et al.*, 2006; Gorchakov *et al.*, 2007; Heise *et al.*, 2009), a non-replicating adenovirus (Holman *et al.*, 2009), or Newcastle disease virus (Kortekaas *et al.*, 2010a). Other groups have developed sub-unit vaccine by expressing the RVFV glycoproteins which can assemble into 'empty' particles or VLPs (Naslund *et al.*, 2009; de Boer *et al.*, 2010). For a more efficient production of VLPs, recombinant baculoviruses were constructed expressing the glycoproteins alone or in association with N (Liu *et al.*, 2008). DNA-based vaccine administrated by gene gun has been described but it required several immunizations (Spik *et al.*, 2006; Wallace *et al.*, 2006; Lagerqvist *et al.*, 2009).

Interestingly, most of the vaccine candidates, live-attenuated virus or subunit vaccine, have some genes missing when compared with natural strains circulating during endemic or epidemic periods. This is also the case of Clone 13 or R566, the truncated NSs of which is rapidly degraded by the proteasome (Vialat *et al.*, 2000). This raised the theoretical possibility to distinguish infected and vaccinated animals (DIVA) (Bird *et al.*, 2008a). Recently an ELISA was developed, based on the detection or absence of detection of antibodies against the NSs protein. The test was evaluated using sera from experimentally infected rats and from a few number of ruminants (McElroy *et al.*, 2009). These sera contained antibodies against the protein, although to a low titre. Considering that the NSs protein is poorly immunogenic, a survey on a large number of animals should be done to determine if a high proportion of naturally infected livestock elicits detectable and long lasting immune response against this protein.

Conclusions

During the last decade, many efforts have been made to better understand the biology of phleboviruses and their interactions with the host cell. However, little is elucidated on the molecular mechanisms developed by the arthropods which apparently mount an efficient response to limit the virus spread, opening possible avenues to develop intervention strategies against diseases caused by these viruses. In the field of RVFV, which can be considered among the most pathogenic phleboviruses, good progress has been made both at the academic research level and the development of vaccines. A large gap still exists in the field of antiviral agents which will remain a challenging issue for future research.

References

Abu-Elyazeed, R., el-Sharkawy, S., Olson, J., Botros, B., Soliman, A., Salib, A., Cummings, C., and Arthur, R. (1996). Prevalence of anti-Rift-Valley-fever IgM antibody in abattoir workers in the Nile delta during the 1993 outbreak in Egypt. Bull. World Health Organ. *74*, 155–158.

Accardi, L., Gro, M.C., Di Bonito, P., and Giorgi, C. (1993). Toscana virus genomic L segment: molecular cloning, coding strategy and amino acid sequence in comparison with other negative strand RNA viruses. Virus Res. *27*, 119–131.

Accardi, L., Prehaud, C., Di Bonito, P., Mochi, S., Bouloy, M., and Giorgi, C. (2001). Activity of Toscana and Rift Valley fever virus transcription complexes on heterologous templates. J. Gen. Virol. *82*, 781–785.

Albarino, C.G., Bird, B.H., and Nichol, S.T. (2007). A shared transcription termination signal on negative and ambisense RNA genome segments of Rift Valley fever, sandfly fever Sicilian, and Toscana viruses. J. Virol. *81*, 5246–5256.

Anderson, G.W., Jr., and Smith, J.F. (1987). Immunoelectron microscopy of Rift Valley fever viral morphogenesis in primary rat hepatocytes. Virology *161*, 91–100.
Anderson, G.W., Jr., and Peters, C.J. (1988). Viral determinants of virulence for Rift Valley fever (RVF) in rats. Microb. Pathog. *5*, 241–250.
Anderson, G.W., Jr., Rosebrock, J.A., Johnson, A.J., Jennings, G.B., and Peters, C.J. (1991). Infection of inbred rat strains with Rift Valley fever virus: development of a congenic resistant strain and observations on age-dependence of resistance. Am. J. Trop. Med. Hyg. *44*, 475–480.
Andersson, A.M., Melin, L., Bean, A., and Pettersson, R.F. (1997a). A retention signal necessary and sufficient for Golgi localization maps to the cytoplasmic tail of a *Bunyaviridae* (Uukuniemi virus) membrane glycoprotein. J. Virol. *71*, 4717–4727.
Andersson, A.M., Melin, L., Persson, R., Raschperger, E., Wikstrom, L., and Pettersson, R.F. (1997b). Processing and membrane topology of the spike proteins G1 and G2 of Uukuniemi virus. J. Virol. *71*, 218–225.
Anonymous (1998). An outbreak of Rift valley fever, Eastern Africa, 1997–1998. Wkly Epidemiol. Rec. *73*, 105–109.
Anonymous (2000a). Outbreak of Rift Valley fever, Yemen, August–October 2000. Wkly Epidemiol. Rec. *75*, 392–395.
Anonymous (2000b). Rift Valley fever, Saudi Arabia, August–October 2000. Wkly Epidemiol. Rec. *75*, 370–371.
Anonymous (2007). Outbreaks of Rift Valley fever in Kenya, Somalia and United Republic of Tanzania, December 2006–April 2007. Wkly Epidemiol. Rec. *82*, 169–178.
Anonymous (2008). Outbreak news. Rift Valley fever, Madagascar. Wkly Epidemiol. Rec. *83*, 157.
Anonymous (2010). Rift Valley fever, South Africa – update. Wkly Epidemiol. Rec. *85*, 185–186.
Anyamba, A., Chretien, J.P., Formenty, P.B., Small, J., Tucker, C.J., Malone, J.L., El Bushra, H., Martin, V., and Linthicum, K.J. (2006). Rift Valley Fever potential, Arabian Peninsula. Emerg. Infect. Dis. *12*, 518–520.
Anyamba, A., Chretien, J.P., Small, J., Tucker, C.J., Formenty, P.B., Richardson, J.H., Britch, S.C., Schnabel, D.C., Erickson, R.L., and Linthicum, K.J. (2009). Prediction of a Rift Valley fever outbreak. Proc. Natl. Acad. Sci. U.S.A. *106*, 955–959.
Barnard, B.J. (1979). Rift Valley fever vaccine--antibody and immune response in cattle to a live and an inactivated vaccine. J. S. Afr. Vet. Assoc. *50*, 155–157.
Barnard, B.J., and Botha, M.J. (1977). An inactivated rift valley fever vaccine. J. S. Afr. Vet. Assoc. *48*, 45–48.
Beaty, B.J., and Calisher, C.H. (1991). *Bunyaviridae* – natural history. In Current Topics in Microbiology and Immunology, Kolakoksky, D.M ed. (Springer-Verlag, Berlin-Heidelberg).
Beck, J.A., Lloyd, S., Hafezparast, M., Lennon-Pierce, M., Eppig, J.T., Festing, M.F., and Fisher, E.M. (2000). Genealogies of mouse inbred strains. Nat. Genet. *24*, 23–25.
Billecocq, A., Vialat, P., and Bouloy, M. (1996). Persistent infection of mammalian cells by Rift Valley fever virus. J. Gen. Virol. *77*, 3053–3062.
Billecocq, A., Spiegel, M., Vialat, P., Kohl, A., Weber, F., Bouloy, M., and Haller, O. (2004). NSs protein of Rift Valley fever virus blocks interferon production by inhibiting host gene transcription. J. Virol. *78*, 9798–9806.
Billecocq, A., Gauliard, N., Le May, N., Elliott, R.M., Flick, R., and Bouloy, M. (2008). RNA polymerase I-mediated expression of viral RNA for the rescue of infectious virulent and avirulent Rift Valley fever viruses. Virology *378*, 377–384.
Bird, B.H., Albarino, C.G., and Nichol, S.T. (2007a). Rift Valley fever virus lacking NSm proteins retains high virulence *in vivo* and may provide a model of human delayed onset neurologic disease. Virology *362*, 10–15.
Bird, B.H., Khristova, M.L., Rollin, P.E., Ksiazek, T.G., and Nichol, S.T. (2007b). Complete genome analysis of 33 ecologically and biologically diverse Rift Valley fever virus strains reveals widespread virus movement and low genetic diversity due to recent common ancestry. J. Virol. *81*, 2805–2816.
Bird, B.H., Albarino, C.G., Hartman, A.L., Erickson, B.R., Ksiazek, T.G., and Nichol, S.T. (2008a). Rift valley fever virus lacking the NSs and NSm genes is highly attenuated, confers protective immunity from virulent virus challenge, and allows for differential identification of infected and vaccinated animals. J. Virol. *82*, 2681–2691.
Bird, B.H., Githinji, J.W., Macharia, J.M., Kasiiti, J.L., Muriithi, R.M., Gacheru, S.G., Musaa, J.O., Towner, J.S., Reeder, S.A., Oliver, J.B., *et al.* (2008b). Multiple virus lineages sharing recent common ancestry were associated with a Large Rift Valley fever outbreak among livestock in Kenya during 2006–2007. J. Virol. *82*, 11152–11166.

Bishop, D.H., Calisher, C.H., Casals, J., Chumakov, M.P., Gaidamovich, S.Y., Hannoun, C., Lvov, D.K., Marshall, I.D., Oker-Blom, N., Pettersson, R.F., *et al.* (1980). *Bunyaviridae*. Intervirology *14*, 125–143.

Blakqori, G., Delhaye, S., Habjan, M., Blair, C.D., Sanchez-Vargas, I., Olson, K.E., Attarzadeh-Yazdi, G., Fragkoudis, R., Kohl, A., Kalinke, U., *et al.* (2007). La Crosse bunyavirus nonstructural protein NSs serves to suppress the type I interferon system of mammalian hosts. J. Virol. *81*, 4991–4999.

Blakqori, G., and Weber, F. (2005). Efficient cDNA-based rescue of La Crosse bunyaviruses expressing or lacking the nonstructural protein NSs. J. Virol. *79*, 10420–10428.

de Boer, S.M., Kortekaas, J., Antonis, A.F., Kant, J., van Oploo, J.L., Rottier, P.J., Moormann, R.J., and Bosch, B.J. (2010). Rift Valley fever virus subunit vaccines confer complete protection against a lethal virus challenge. Vaccine *28*, 2330–2339.

Bonhomme, F., and Guenet, J. (1996). The Laboratory Mouse and its Wild Relatives. (Oxford University Press, Avon).

von Bonsdorff, C.H., and Pettersson, R. (1975). Surface structure of Uukuniemi virus. J. Virol. *16*, 1296–1307.

Borden, E.C., Sen, G.C., Uze, G., Silverman, R.H., Ransohoff, R.M., Foster, G.R., and Stark, G.R. (2007). Interferons at age 50: past, current and future impact on biomedicine. Nat. Rev. Drug Discov. *6*, 975–990.

Botros, B., Omar, A., Elian, K., .Mohamed, G., Soliman, A., Salib, A., Salman, D., Saad, M., and Earhart, K. (2006). Adverse response of non-indigenous cattle of European breeds to live attenuated Smithburn Rift Valley fever vaccine. J. Med. Virol. *78*, 787–791.

Bouloy, M., and Flick, R. (2009). Reverse genetics technology for Rift Valley fever virus: current and future applications for the development of therapeutics and vaccines. Antiviral Res. *84*, 101–118.

Bouloy, M., and Weber, F. (2010). Molecular Biology of Rift Valley fever virus. Open Virol. J. *4*, 8–14.

Bouloy, M., Janzen, C., Vialat, P., Khun, H., Pavlovic, J., Huerre, M., and Haller, O. (2001). Genetic evidence for an interferon-antagonistic function of rift valley fever virus nonstructural protein NSs. J. Virol. *75*, 1371–1377.

Bridgen, A., and Elliott, R.M. (1996). Rescue of a segmented negative-strand RNA virus entirely from cloned complementary DNAs. Proc. Natl. Acad. Sci. U.S.A. *93*, 15400–15404.

Bridgen, A., Weber, F., Fazakerley, J.K., and Elliott, R.M. (2001). Bunyamwera bunyavirus nonstructural protein NSs is a nonessential gene product that contributes to viral pathogenesis. Proc. Natl. Acad. Sci. U.S.A. *98*, 664–669.

Brown, J., Dominik, J.W., and Morissey, R.L. (1981). Respiratory infectivity of a reecntly isolated egyptian strain of Rift Valley fever virus. Infect. Immun. *33*, 848–853.

Calisher, C.H., Beaty, B.J., Dalrymple, J.M., Elliott, R.M., Karabatsos, N., Lee, H.W., Lvov, D.K., Nuttall, P.A., Peters, D., Pettersson, R.F., *et al.* (1990). Report of the *Bunyaviridae* Study group. Arch. Virol. Suppl.

Caplen, H., Peters, C.J., and Bishop, D.H. (1985). Mutagen-directed attenuation of Rift Valley fever virus as a method for vaccine development. J. Gen. Virol. *66 (Pt 10)*, 2271–2277.

CDC. (2007). Rift Valley fever outbreak – Kenya, November 2006–January 2007. MMWR Morb. Mortal Wkly Rep. *56*, 73–76.

Chambers, P.G., and Swanepoel, R. (1980). Rift valley fever in abattoir workers. Cent. Afr. J. Med. *26*, 122–126.

Charrel, R.N., Gallian, P., Navarro-Mari, J.M., Nicoletti, L., Papa, A., Sanchez-Seco, M.P., Tenorio, A., and de Lamballerie, X. (2005). Emergence of Toscana virus in Europe. Emerg. Infect. Dis. *11*, 1657–1663.

Charrel, R.N., Izri, A., Temmam, S., Delaunay, P., Toga, I., Dumon, H., Marty, P., de Lamballerie, X., and Parola, P. (2007). Cocirculation of 2 genotypes of Toscana virus, southeastern France. Emerg. Infect. Dis. *13*, 465–468.

Charrel, R.N., Moureau, G., Temmam, S., Izri, A., Marty, P., Parola, P., da Rosa, A.T., Tesh, R.B., and de Lamballerie, X. (2009). Massila virus, a novel Phlebovirus (*Bunyaviridae*) isolated from sandflies in the Mediterranean. Vector Borne Zoonotic Dis. *9*, 519–530.

Chen, S.Y., and Compans, R.W. (1991). Oligomerization, transport, and Golgi retention of Punta Toro virus glycoproteins. J. Virol. *65*, 5902–5909.

Chen, S.Y., Matsuoka, Y., and Compans, R.W. (1991a). Assembly and polarized release of Punta Toro virus and effects of brefeldin A. J. Virol. *65*, 1427–1439.

Chen, S.Y., Matsuoka, Y., and Compans, R.W. (1991b). Golgi complex localization of the Punta Toro virus G2 protein requires its association with the G1 protein. Virology *183*, 351–365.

Coetzer, J.A. (1977). The pathology of Rift Valley fever. I. Lesions occurring in natural cases in new-born lambs. Onderstepoort J. Vet. Res. *44*, 205–211.

Coetzer, J.A. (1982). The pathology of Rift Valley fever. II. Lesions occurring in field cases in adult cattle, calves and aborted foetuses. Onderstepoort J. Vet. Res. *49*, 11–17.
Coetzer, J.A., and Ishak, K.G. (1982). Sequential development of the liver lesions in new-born lambs infected with Rift Valley fever virus. I. Macroscopic and microscopic pathology. Onderstepoort J. Vet. Res. *49*, 103–108.
Collao, X., Palacios, G., Sanbonmatsu-Gamez, S., Perez-Ruiz, M., Negredo, A.I., Navarro-Mari, J.M., Grandadam, M., Aransay, A.M., Lipkin, W.I., Tenorio, A., *et al.* (2009). Genetic diversity of Toscana virus. Emerg. Infect. Dis. *15*, 574–577.
Collett, M.S. (1986). Messenger RNA of the M segment RNA of Rift Valley fever virus. Virology *151*, 151–156.
Collett, M.S., Purchio, A.F., Keegan, K., Frazier, S., Hays, W., Anderson, D.K., Parker, M.D., Schmaljohn, C., Schmidt, J., and Dalrymple, J.M. (1985). Complete nucleotide sequence of the M RNA segment of Rift Valley fever virus. Virology *144*, 228–245.
Cusi, M.G., Gori Savellini, G., Terrosi, C., Di Genova, G., Valassina, M., Valentini, M., Bartolommei, S., and Miracco, C. (2005). Development of a mouse model for the study of Toscana virus pathogenesis. Virology *333*, 66–73.
Cusi, M.G., Savellini, G.G., and Zanelli, G. (2010). Toscana virus epidemiology: from Italy to beyond. Open Virol. J. *4*, 22–28.
Cusi, M.G., Valensin, P.E., Donati, M., and Valassina, M. (2001). Neutralization of Toscana virus is partially mediated by antibodies to the nucleocapsid protein. J. Med. Virol. *63*, 72–75.
Daubney, R.J., Hudson, J.R., and Garnham, P.C. (1931). Enzootic hepatitis of Rift Valley fever. An underdescribed virus disease of sheep, cattle and man from East Africa. J. Pathol. Bacteriol. *34*, 545–579.
Di Bonito, P., Nicoletti, L., Mochi, S., Accardi, L., Marchi, A., and Giorgi, C. (1999). Immunological characterization of Toscana virus proteins. Arch. Virol. *144*, 1947–1960.
Digoutte, J.P., and Peters, C.J. (1989). General aspects of the 1987 Rift Valley fever epidemic in Mauritania. Res. Virol. *140*, 27–30.
Ding, X., Xu, F., Chen, H., Tesh, R.B., and Xiao, S.Y. (2005). Apoptosis of hepatocytes caused by Punta Toro virus (*Bunyaviridae*: Phlebovirus) and its implication for Phlebovirus pathogenesis. Am. J. Pathol. *167*, 1043–1049.
Doerr, R., Franz, K., and Taussig, S. (1909). Das Papatacifieber (Deuticke, Leipzig).
Dungu, B., Louw, I., Lubisi, A., Hunter, P., von Teichman, B.F., and Bouloy, M. (2010). Evaluation of the efficacy and safety of the Rift Valley Fever Clone 13 vaccine in sheep. Vaccine *28*, 4581–4587.
Elliott, R.M. (1996). The *Bunyaviridae* (Plenum Press, New York and London).
Elliott, R.M., and Weber, F. (2009). Bunyaviruses and the type I Interferon system. Viruses *1*, 1003–1020.
Elliott, R.M., Dunn, E., Simons, J.F., and Pettersson, R.F. (1992). Nucleotide sequence and coding strategy of the Uukuniemi virus L RNA segment. J. Gen. Virol. *73 (Pt 7)*, 1745–1752.
Ellis, D.S., Simpson, D.I., Stamford, S., and Abdel Wahab, K.S. (1979). Rift Valley fever virus: some ultrastructural observations on material from the outbreak in Egypt 1977. J. Gen. Virol. *42*, 329–337.
Fagbami, A.H., Tomori, O., Fabiyi, A., and Isoun, T.T. (1975). Experimental Rift Valley fever in West African Dwarf sheep. Res. Vet. Sci. *18*, 334–335.
Filone, C.M., Heise, M., Doms, R.W., and Bertolotti-Ciarlet, A. (2006). Development and characterization of a Rift Valley fever virus cell–cell fusion assay using alphavirus replicon vectors. Virology *356*, 155–164.
Fisher, A.F., Tesh, R.B., Tonry, J., Guzman, H., Liu, D., and Xiao, S.Y. (2003). Induction of severe disease in hamsters by two sandfly fever group viruses, Punta toro and Gabek Forest (Phlebovirus, *Bunyaviridae*), similar to that caused by Rift Valley fever virus. Am. J. Trop. Med. Hyg. *69*, 269–276.
Flick, K., Katz, A., Overby, A., Feldmann, H., Pettersson, R.F., and Flick, R. (2004). Functional analysis of the noncoding regions of the Uukuniemi virus (*Bunyaviridae*) RNA segments. J. Virol. *78*, 11726–11738.
Flick, R., and Pettersson, R.F. (2001). Reverse genetics system for Uukuniemi virus (*Bunyaviridae*): RNA polymerase I-catalyzed expression of chimeric viral RNAs. J. Virol. *75*, 1643–1655.
Flick, R., and Bouloy, M. (2005). Rift Valley fever virus. Curr. Mol. Med. *5*, 827–834.
Flick, R., Elgh, F., and Pettersson, R.F. (2002). Mutational analysis of the Uukuniemi virus (*Bunyaviridae* family) promoter reveals two elements of functional importance. J. Virol. *76*, 10849–10860.
Freiberg, A.N., Sherman, M.B., Morais, M.C., Holbrook, M.R., and Watowich, S.J. (2008). Three-dimensional organization of Rift Valley fever virus revealed by cryoelectron tomography. J. Virol. *82*, 10341–10348.
Frese, M., Kochs, G., Feldmann, H., Hertkorn, C., and Haller, O. (1996). Inhibition of bunyaviruses, phleboviruses, and hantaviruses by human MxA protein. J. Virol. *70*, 915–923.

Garcia, S., Billecocq, A., Crance, J.M., Prins, M., Garin, D., and Bouloy, M. (2006). Viral suppressors of RNA interference impair RNA silencing induced by a Semliki Forest virus replicon in tick cells. J. Gen. Virol. *87*, 1985–1989.

Gauliard, N., Billecocq, A., Flick, R., and Bouloy, M. (2006). Rift Valley fever virus noncoding regions of L, M and S segments regulate RNA synthesis. Virology *351*, 170–179.

Gerdes, G.H. (2002). Rift Valley fever. Vet. Clin. North Am. Food Anim. Pract. *18*, 549–555.

Gerdes, G.H. (2004). Rift Valley fever. Rev. Sci. Tech. *23*, 613–623.

Gerrard, S.R., Bird, B.H., Albarino, C.G., and Nichol, S.T. (2007). The NSm proteins of Rift Valley fever virus are dispensable for maturation, replication and infection. Virology *359*, 459–465.

Gerrard, S.R., and Nichol, S.T. (2002). Characterization of the Golgi retention motif of Rift Valley fever virus G(N) glycoprotein. J. Virol. *76*, 12200–12210.

Gibrat, J.F., Madej, T., and Bryant, S.H. (1996). Surprising similarities in structure comparison. Curr. Opin. Struct. Biol. *6*, 377–385.

Giorgi, C. (1996). Molecular biology of phlebovirus. In The *Bunyaviridae*, Elliot, R.M., ed. (Plenum Press, New York, NY), pp. 105–128.

Giorgi, C., Accardi, L., Nicoletti, L., Gro, M.C., Takehara, K., Hilditch, C., Morikawa, S., and Bishop, D.H. (1991). Sequences and coding strategies of the S RNAs of Toscana and Rift Valley fever viruses compared with those of Punta Toro, Sicilian Sandfly fever, and Uukuniemi viruses. Virology *180*, 738–753.

Goodbourn, S., Didcock, L., and Randall, R.E. (2000). Interferons: cell signalling, immune modulation, antiviral response and virus countermeasures. J. Gen. Virol. *81*, 2341–2364.

Gorchakov, R., Volkova, E., Yun, N., Petrakova, O., Linde, N.S., Paessler, S., Frolova, E., and Frolov, I. (2007). Comparative analysis of the alphavirus-based vectors expressing Rift Valley fever virus glycoproteins. Virology *366*, 212–225.

Gowen, B.B., Hoopes, J.D., Wong, M.H., Jung, K.H., Isakson, K.C., Alexopoulou, L., Flavell, R.A., and Sidwell, R.W. (2006). TLR3 deletion limits mortality and disease severity due to Phlebovirus infection. J. Immunol. *177*, 6301–6307.

Gowen, B.B., Wong, M.H., Jung, K.H., Sanders, A.B., Mitchell, W.M., Alexopoulou, L., Flavell, R.A., and Sidwell, R.W. (2007). TLR3 is essential for the induction of protective immunity against Punta Toro Virus infection by the double-stranded RNA (dsRNA), poly(I:C12U), but not Poly(I:C): differential recognition of synthetic dsRNA molecules. J. Immunol. *178*, 5200–5208.

Gowen, B.B., Wong, M.H., Jung, K.H., Smee, D.F., Morrey, J.D., and Furuta, Y. (2010). Efficacy of favipiravir (T-705) and T-1106 pyrazine derivatives in phlebovirus disease models. Antiviral Res. *86*, 121–127.

Gro, M.C., Di Bonito, P., Accardi, L., and Giorgi, C. (1992). Analysis of 3′ and 5′ ends of N and NSs messenger RNAs of Toscana Phlebovirus. Virology *191*, 435–438.

Guenet, J.L., and Bonhomme, F. (2003). Wild mice: an ever-increasing contribution to a popular mammalian model. Trends Genet. *19*, 24–31.

Habjan, M., Andersson, I., Klingstrom, J., Schumann, M., Martin, A., Zimmermann, P., Wagner, V., Pichlmair, A., Schneider, U., Muhlberger, E., *et al.* (2008a). Processing of genome 5′ termini as a strategy of negative-strand RNA viruses to avoid RIG-I-dependent interferon induction. PLoS One 3, e2032.

Habjan, M., Penski, N., Spiegel, M., and Weber, F. (2008b). T7 RNA polymerase-dependent and -independent systems for cDNA-based rescue of Rift Valley fever virus. J. Gen. Virol. *89*, 2157–2166.

Habjan, M., Penski, N., Wagner, V., Spiegel, M., Overby, A.K., Kochs, G., Huiskonen, J.T., and Weber, F. (2009a). Efficient production of Rift Valley fever virus-like particles: The antiviral protein MxA can inhibit primary transcription of bunyaviruses. Virology *385*, 400–408.

Habjan, M., Pichlmair, A., Elliott, R.M., Overby, A.K., Glatter, T., Gstaiger, M., Superti-Furga, G., Unger, H., and Weber, F. (2009b). NSs protein of rift valley fever virus induces the specific degradation of the double-stranded RNA-dependent protein kinase. J. Virol. *83*, 4365–4375.

Haller, O., Kochs, G., and Weber, F. (2006). The interferon response circuit: Induction and suppression by pathogenic viruses. Virology *344*, 119–130.

Heise, M.T., Whitmore, A., Thompson, J., Parsons, M., Grobbelaar, A.A., Kemp, A., Paweska, J.T., Madric, K., White, L.J., Swanepoel, R., *et al.* (2009). An alphavirus replicon-derived candidate vaccine against Rift Valley fever virus. Epidemiol. Infect. *137*, 1309–1318.

Holm, L., and Sander, C. (1995). Dali: a network tool for protein structure comparison. Trends Biochem. Sci. *20*, 478–480.

Holman, D.H., Penn-Nicholson, A., Wang, D., Woraratanadharm, J., Harr, M.K., Luo, M., Maher, E.M., Holbrook, M.R., and Dong, J.Y. (2009). A complex adenovirus-vectored vaccine against Rift Valley

fever virus protects mice against lethal infection in the presence of preexisting vector immunity. Clin. Vaccine Immunol. *16*, 1624–1632.
Huiskonen, J.T., Overby, A.K., Weber, F., and Grunewald, K. (2009). Electron cryo-microscopy and single-particle averaging of Rift Valley fever virus: evidence for GN-GC glycoprotein heterodimers. J. Virol. *83*, 3762–3769.
Huiskonen, J.T., Hepojoki, J., Laurinmaki, P., Vaheri, A., Lankinen, H., Butcher, S.J., and Grunewald, K. (2010). Electron cryotomography of Tula hantavirus suggests a unique assembly paradigm for enveloped viruses. J. Virol. *84*, 4889–4897.
Hunter, P., Erasmus, B.J., and Vorster, J.H. (2002). Teratogenicity of a mutagenised Rift Valley fever virus (MVP 12) in sheep. Onderstepoort J. Vet. Res. *69*, 95–98.
Hutchinson, K.L., Peters, C.J., and Nichol, S.T. (1996). Sin Nombre virus mRNA synthesis. Virology *224*, 139–149.
Iapalucci, S., Lopez, N., and Franze-Fernandez, M.T. (1991). The 3′ end termini of the Tacaribe arenavirus subgenomic RNAs. Virology *182*, 269–278.
Ihara, T., Akashi, H., and Bishop, D.H. (1984). Novel coding strategy (ambisense genomic RNA) revealed by sequence analyses of Punta Toro Phlebovirus S RNA. Virology *136*, 293–306.
Ihara, T., Matsuura, Y., and Bishop, D.H. (1985a). Analyses of the mRNA transcription processes of Punta Toro phlebovirus (*Bunyaviridae*). Virology *147*, 317–325.
Ihara, T., Smith, J., Dalrymple, J.M., and Bishop, D.H. (1985b). Complete sequences of the glycoproteins and M RNA of Punta Toro phlebovirus compared with those of Rift Valley fever virus. Virology *144*, 246–259.
Ikegami, T., and Makino, S. (2009). Rift valley fever vaccines. Vaccine *27 (Suppl 4)*, D69–72.
Ikegami, T., Peters, C.J., and Makino, S. (2005a). Rift valley fever virus nonstructural protein NSs promotes viral RNA replication and transcription in a minigenome system. J. Virol. *79*, 5606–5615.
Ikegami, T., Won, S., Peters, C.J., and Makino, S. (2005b). Rift Valley fever virus NSs mRNA is transcribed from an incoming anti-viral-sense S RNA segment. J. Virol. *79*, 12106–12111.
Ikegami, T., Won, S., Peters, C.J., and Makino, S. (2006). Rescue of infectious rift valley fever virus entirely from cDNA, analysis of virus lacking the NSs gene, and expression of a foreign gene. J. Virol. *80*, 2933–2940.
Ikegami, T., Won, S., Peters, C.J., and Makino, S. (2007). Characterization of Rift Valley fever virus transcriptional terminations. J. Virol. *81*, 8421–8438.
Ikegami, T., Narayanan, K., Won, S., Kamitani, W., Peters, C.J., and Makino, S. (2009a). Dual functions of Rift Valley fever virus NSs protein: inhibition of host mRNA transcription and post-transcriptional downregulation of protein kinase PKR. Ann. N. Y. Acad. Sci. *1171 (Suppl 1)*, E75–85.
Ikegami, T., Narayanan, K., Won, S., Kamitani, W., Peters, C.J., and Makino, S. (2009b). Rift Valley fever virus NSs protein promotes post-transcriptional downregulation of protein kinase PKR and inhibits eIF2alpha phosphorylation. PLoS Pathog. *5*, e1000287.
Kakach, L.T., Wasmoen, T.L., and Collett, M.S. (1988). Rift Valley fever virus M segment: use of recombinant vaccinia viruses to study Phlebovirus gene expression. J. Virol. *62*, 826–833.
Kakach, L.T., Suzich, J.A., and Collett, M.S. (1989). Rift Valley fever virus M segment: phlebovirus expression strategy and protein glycosylation. Virology *170*, 505–510.
Kamal, S.A. (2009). Pathological studies on postvaccinal reactions of Rift Valley fever in goats. Virol. J. *6*, 94.
Katze, M.G., Fornek, J.L., Palermo, R.E., Walters, K.A., and Korth, M.J. (2008). Innate immune modulation by RNA viruses: emerging insights from functional genomics. Nat. Rev. Immunol. *8*, 644–654.
Kortekaas, J., de Boer, S.M., Kant, J., Vloet, R.P., Antonis, A.F., and Moormann, R.J. (2010a). Rift Valley fever virus immunity provided by a paramyxovirus vaccine vector. Vaccine *28*, 4394–4401.
Kortekaas, J., Dekker, A., de Boer, S.M., Weerdmeester, K., Vloet, R.P., de Wit, A.A., Peeters, B.P., and Moormann, R.J. (2010b). Intramuscular inoculation of calves with an experimental Newcastle disease virus-based vector vaccine elicits neutralizing antibodies against Rift Valley fever virus. Vaccine *28*, 2271–2276.
LaBeaud, A.D., Muchiri, E.M., Ndzovu, M., Mwanje, M.T., Muiruri, S., Peters, C.J., and King, C.H. (2008). Interepidemic Rift Valley fever virus seropositivity, northeastern Kenya. Emerg. Infect. Dis. *14*, 1240–1246.
Lagerqvist, N., Naslund, J., Lundkvist, A., Bouloy, M., Ahlm, C., and Bucht, G. (2009). Characterisation of immune responses and protective efficacy in mice after immunisation with Rift Valley Fever virus cDNA constructs. Virol. J. *6*, 6.

Le May, N., Dubaele, S., Proietti De Santis, L., Billecocq, A., Bouloy, M., and and Egly, J. (2004). TFIIH transcription factor, a target for the Rift Valley hemorrhagic Fever Virus. Cell *116*, 1–20.
Le May, N., Gauliard, N., Billecocq, A., and Bouloy, M. (2005). The N terminus of Rift Valley fever virus nucleoprotein is essential for dimerization. J. Virol. *79*, 11974–11980.
Le May, N., Mansuroglu, Z., Leger, P., Josse, T., Blot, G., Billecocq, A., Flick, R., Jacob, Y., Bonnefoy, E., and Bouloy, M. (2008). A SAP30 complex inhibits IFN-beta expression in Rift Valley fever virus infected cells. PLoS Pathog. *4*, e13.
Leonard, V.H., Kohl, A., Osborne, J.C., McLees, A., and Elliott, R.M. (2005). Homotypic interaction of Bunyamwera virus nucleocapsid protein. J. Virol. *79*, 13166–13172.
Li, G., Ren, J., Xu, F., and Fergusson, M. (2010). Non-structural and nucleocapsid proteins of Punta Toro virus induce apoptosis of hepatocuytes through both intrinsic and extrinsic pathways. Microbiol. Immunol. *54*, 20–30.
Linthicum, K.J., Anyamba, A., Tucker, C.J., Kelley, P.W., Myers, M.F., and Peters, C.J. (1999). Climate and satellite indicators to forecast Rift Valley fever epidemics in Kenya. Science *285*, 397–400.
Liu, D.Y., Tesh, R.B., Travassos Da Rosa, A.P., Peters, C.J., Yang, Z., Guzman, H., and Xiao, S.Y. (2003). Phylogenetic relationships among members of the genus Phlebovirus (*Bunyaviridae*) based on partial M segment sequence analyses. J. Gen. Virol. *84*, 465–473.
Liu, L., Celma, C.C., and Roy, P. (2008). Rift Valley fever virus structural proteins: expression, characterization and assembly of recombinant proteins. Virol. J. *5*, 82.
Lopez, N., and Franze-Fernandez, M.T. (2007). A single stem–loop structure in Tacaribe arenavirus intergenic region is essential for transcription termination but is not required for a correct initiation of transcription and replication. Virus Res. *124*, 237–244.
Lopez, N., Muller, R., Prehaud, C., and Bouloy, M. (1995). The L protein of Rift Valley fever virus can rescue viral ribonucleoproteins and transcribe synthetic genome-like RNA molecules. J. Virol. *69*, 3972–3979.
Lowen, A.C., Noonan, C., McLees, A., and Elliott, R.M. (2004). Efficient bunyavirus rescue from cloned cDNA. Virology *330*, 493–500.
Lozach, P.Y., Mancini, R., Bitto, D., Meier, R., Oestereich, L., Overby, A.K., Pettersson, R.F., and Helenius, A. (2010). Entry of bunyaviruses into mammalian cells. Cell Host Microbe *7*, 488–499.
McClain, D.J., Summers, P.L., Pratt, W.D., Davis, K.J., and Jennings, G.B. (1997). Experimental infection of nonhuman primates with sandfly fever virus. Am. J. Trop. Med. Hyg. *56*, 554–560.
McElroy, A.K., Albarino, C.G., and Nichol, S.T. (2009). Development of a RVFV ELISA that can distinguish infected from vaccinated animals. Virol. J. *6*, 125.
Mansuroglu, Z., Josse, T., Gilleron, J., Billecocq, A., Leger, P., Bouloy, M., and Bonnefoy, E. (2010). Nonstructural NSs protein of Rift valley fever virus interacts with pericentromeric DNA sequences of the host cell, inducing chromosome cohesion and segregation defects. J. Virol. *84*, 928–939.
Marchi, A., Nicoletti, L., Accardi, L., Di Bonito, P., and Giorgi, C. (1998). Characterization of Toscana virus-defective interfering particles generated *in vivo*. Virology *246*, 125–133.
Marriott, A.C., Ward, V.K., and Nuttall, P.A. (1989). The S RNA segment of Sandfly Fever Sicilian virus: evidence for an ambisense genome. Virology *169*, 341–345.
Martin, V., Chevalier, V., Ceccato, P., Anyamba, A., De Simone, L., Lubroth, J., de La Rocque, S., and Domenech, J. (2008). The impact of climate change on the epidemiology and control of Rift Valley fever. Rev. Sci. Tech. *27*, 413–426.
Matsuoka, Y., Chen, S.Y., and Compans, R.W. (1991). Bunyavirus protein transport and assembly. Curr. Top. Microbiol. Immunol. *169*, 161–179.
Matsuoka, Y., Chen, S.Y., and Compans, R.W. (1994). A signal for Golgi retention in the bunyavirus G1 glycoprotein. J. Biol. Chem. *269*, 22565–22573.
Meegan, J.M., and Bailey, C.J. (1988). Rift Valley Fever, vol. IV (CRC Press Inc, Boca Raton, FL).
Mendenhall, M., Wong, M.H., Skirpstunas, R., Morrey, J.D., and Gowen, B.B. (2009). Punta Toro virus (*Bunyaviridae*, Phlebovirus) infection in mice: strain differences in pathogenesis and host interferon response. Virology *395*, 143–151.
Morrill, J.C., Jennings, G.B., Caplen, H., Turell, M.J., Johnson, A.J., and Peters, C.J. (1987). Pathogenicity and immunogenicity of a mutagen-attenuated Rift Valley fever virus immunogen in pregnant ewes. Am J. Vet. Res. *48*, 1042–1047.
Morrill, J.C., Carpenter, L., Taylor, D., Ramsburg, H.H., Quance, J., and Peters, C.J. (1991). Further evaluation of a mutagen-attenuated Rift Valley fever vaccine in sheep. Vaccine *9*, 35–41.
Morrill, J.C., Mebus, C.A., and Peters, C.J. (1997a). Safety and efficacy of a mutagen-attenuated Rift Valley fever virus vaccine in cattle. Am. J. Vet. Res. *58*, 1104–1109.

Morrill, J.C., Mebus, C.A., and Peters, C.J. (1997b). Safety of a mutagen-attenuated Rift Valley fever virus vaccine in fetal and neonatal bovids. Am. J. Vet. Res. *58*, 1110–1114.

Morrill, J.C., Ikegami, T., Yoshikawa-Iwata, N., Lokugamage, N., Won, S., Terasaki, K., Zamoto-Niikura, A., Peters, C.J., and Makino, S. (2010). Rapid accumulation of virulent Rift Valley fever virus in mice from an attenuated virus carrying a single nucleotide substitution in the m RNA. PLoS One *5*, e9986.

Muller, R., Poch, O., Delarue, M., Bishop, D.H., and Bouloy, M. (1994). Rift Valley fever virus L segment: correction of the sequence and possible functional role of newly identified regions conserved in RNA-dependent polymerases. J. Gen. Virol. *75 (Pt 6)*, 1345–1352.

Muller, R., Saluzzo, J.F., Lopez, N., Dreier, T., Turell, M., Smith, J., and Bouloy, M. (1995). Characterization of clone 13, a naturally attenuated avirulent isolate of Rift Valley fever virus, which is altered in the small segment. Am. J. Trop. Med. Hyg. *53*, 405–411.

Naslund, J., Lagerqvist, N., Habjan, M., Lundkvist, A., Evander, M., Ahlm, C., Weber, F., and Bucht, G. (2009). Vaccination with virus-like particles protects mice from lethal infection of Rift Valley Fever Virus. Virology *385*, 409–415.

Neumann, G., Watanabe, T., Ito, H., Watanabe, S., Goto, H., Gao, P., Hughes, M., Perez, D.R., Donis, R., Hoffmann, E., *et al.* (1999). Generation of influenza A viruses entirely from cloned cDNAs. Proc. Natl. Acad. Sci. U.S.A. *96*, 9345–9350.

Nichol, S.T., Beaty, B.J., Elliott, R.M., Goldbach, R.W., Plyusnin, A., and Tesh, R.B. (2005). *The Bunyaviridae.* In Virus Taxonomy Classification and Nomenclature of Viruses VIII: The report of the International Committee on Taxonomy of Viruses, Fauquet, C. , Mayo, M.A., Maniloff, L.S.M., Desselberger, J.U., and Ball, L.A., eds. (Elsevier Academic Press, London), pp. 695–716.

Nicoletti, L., and Verani, P. (1985). Growth of the Phlebovirus Toscana in a mosquito (*Aedes pseudoscutellaris*) cell line (AP-61): establishment of a persistent infection. Arch. Virol. *85*, 35–45.

Niklasson, B., and Eitrem, R. (1985). Sandfly fever among Swedish UN troops in Cyprus. Lancet *1*, 1212.

Obijeski, J.F., Bishop, D.H., Palmer, E.L., and Murphy, F.A. (1976). Segmented genome and nucleocapsid of La Crosse virus. J. Virol. *20*, 664–675.

Ogawa, Y., Sugiura, K., Kato, K., Tohya, Y., and Akashi, H. (2007). Rescue of Akabane virus (family *Bunyaviridae*) entirely from cloned cDNAs by using RNA polymerase I. J. Gen. Virol. *88*, 3385–3390.

Oker-Blom, N., Salminen, A., Brummer-Korvenkontio, M., Kaeaeriaeinen, L., and Weckstroem, P. (1964). Isolation of some viruses other than typical tick-borne encephalitis viruses from *Ixodes ricinus* ticks in Finland. Ann. Med. Exp. Biol. Fenn. *42*, 109–112.

Olaleye, O.D., Tomori, O., Fajimi, J.L., and Schmitz, H. (1996). Experimental infection of three Nigerian breeds of sheep with the Zinga strain of the Rift Valley Fever virus. Rev. Elev. Med. Vet. Pays Trop. *49*, 6–16.

Onomoto, K., Yoneyama, M., and Fujita, T. (2007). Regulation of antiviral innate immune responses by RIG-I family of RNA helicases. Curr. Top. Microbiol. Immunol. *316*, 193–205.

Overby, A.K., Popov, V., Neve, E.P., and Pettersson, R.F. (2006). Generation and analysis of infectious virus-like particles of uukuniemi virus (bunyaviridae): a useful system for studying bunyaviral packaging and budding. J. Virol. *80*, 10428–10435.

Overby, A.K., Pettersson, R.F., and Neve, E.P. (2007a). The glycoprotein cytoplasmic tail of Uukuniemi virus (*Bunyaviridae*) interacts with ribonucleoproteins and is critical for genome packaging. J. Virol. *81*, 3198–3205.

Overby, A.K., Popov, V.L., Pettersson, R.F., and Neve, E.P. (2007b). The cytoplasmic tails of Uukuniemi Virus (*Bunyaviridae*) G(N) and G(C) glycoproteins are important for intracellular targeting and the budding of virus-like particles. J. Virol. *81*, 11381–11391.

Overby, A.K., Pettersson, R.F., Grunewald, K., and Huiskonen, J.T. (2008). Insights into bunyavirus architecture from electron cryotomography of Uukuniemi virus. Proc. Natl. Acad. Sci. U.S.A. *105*, 2375–2379.

Overton, H.A., Ihara, T., and Bishop, D.H. (1987). Identification of the N and NSS proteins coded by the ambisense S RNA of Punta Toro phlebovirus using monospecific antisera raised to baculovirus expressed N and NSS proteins. Virology *157*, 338–350.

Papa, A., Konstantinou, G., Pavlidou, V., and Antoniadis, A. (2006). Sandfly fever virus outbreak in Cyprus. Clin. Microbiol. Infect. *12*, 192–194.

Pepin, M., Bouloy, M., Bird, B.H., Kemp, A., and Paweska, J. (2010). Rift Valley fever virus (*Bunyaviridae*: Phlebovirus): an update on pathogenesis, molecular epidemiology, vectors, diagnostics and prevention. Vet. Res. *41*, 61.

Perrone, L.A., Narayanan, K., Worthy, M., and Peters, C.J. (2007). The S segment of Punta Toro virus (*Bunyaviridae*, Phlebovirus) is a major determinant of lethality in the Syrian hamster and codes for a type I interferon antagonist. J. Virol. *81*, 884–892.
Peters, C.J. (2000). Are hemorrhagic viruses practical agents for biological terrorism? (ASM Press, Washington DC).
Peters, C.J., and Anderson, G.W. (1981). Pathogenesis of Rift Valley fever. Contr. Epidemiol. Biostat. *3*, 21–41.
Peters, C.J., Reynolds, J.A., Slone, T.W., Jones, D.E., and Stephen, E.L. (1986). Prophylaxis of Rift Valley fever with antiviral drugs, immune serum, an interferon inducer, and a macrophage activator. Antiviral Res. *6*, 285–297.
Peters, C.J., Jones, D., Trotter, R., Donaldson, J., White, J., Stephen, E., and Slone, T.W., Jr. (1988). Experimental Rift Valley fever in rhesus macaques. Arch. Virol. *99*, 31–44.
Pettersson, R.F. (1975). The structure of Uukuniemi virus, a proposed member of the bunyaviruses. Med. Biol. *53*, 418–424.
Pettersson, R.F., and Melin, L. (1996). Synthesis, ssembly and intracellular transport of bunyaviridae membrane proteins. In The *Bunyaviridae*, Elliott, R.M., ed. (Plenum Press, New York), pp. 159–188.
Pettersson, R.F., and von Bonsdorff, C.H. (1975). Ribonucleoproteins of Uukuniemi virus are circular. J. Virol. *15*, 386–392.
Peyrefitte, C.N., Devetakov, I., Pastorino, B., Villeneuve, L., Bessaud, M., Stolidi, P., Depaquit, J., Segura, L., Gravier, P., Tock, F., *et al.* (2005). Toscana virus and acute meningitis, France. Emerg. Infect. Dis. *11*, 778–780.
Piper, M.E., Sorenson, D.R., and Gerrard, S.R. (2001) Efficient cellular release of Rift Valley fever virus requires genomic RNA. PLoS One *6*, e18070.
Pittman, P.R., Liu, C.T., Cannon, T.L., Makuch, R.S., Mangiafico, J.A., Gibbs, P.H., and Peters, C.J. (1999). Immunogenicity of an inactivated Rift Valley fever vaccine in humans: a 12-year experience. Vaccine *18*, 181–189.
Plyusnin, A., Beaty, B.J., Elliott, R. M., Goldbach, R., Kormelink, R., Lundkvist, Å, Schmaljohn, C. S., and Tesh, R.B. (2010). *Bunyaviridae*. In Virus Taxonomy. IXth Report of the International Committee on Taxonomy of Viruses, Ball, A., eds (in press).
Porterfield, J.S., Casals, J., Chumakov, M.P., Gaidamovich, S.Y., Hannoun, C., Holmes, I.H., Horzinek, M.C., Mussgay, M., Oker-Blom, N., and Russell, P.K. (1975). Bunyaviruses and *Bunyaviridae*. Intervirology *6*, 13–24.
Prehaud, C., Lopez, N., Blok, M.J., Obry, V., and Bouloy, M. (1997). Analysis of the 3′ terminal sequence recognized by the Rift Valley fever virus transcription complex in its ambisense S segment. Virology *227*, 189–197.
Randall, R.E., and Goodbourn, S. (2008). Interferons and viruses: an interplay between induction, signalling, antiviral responses and virus countermeasures. J. Gen. Virol. *89*, 1–47.
Raymond, D.D., Piper, M.E., Gerrard, S.R., and Smith, J.L. (2010). Structure of the Rift Valley fever virus nucleocapsid protein reveals another architecture for RNA encapsidation. Proc. Natl. Acad. Sci. U.S.A. *107*, 11769–11774.
Reguera, J., Weber, F., and Cusack, S. (2011). Bunyaviridae RNA polymerases (L-protein) have an N-terminal, influenza-like endonuclease domain, essential for viral cap-dependent transcription. PLoS Pathog. *6*, e1001101.
Ritter, M., Bouloy, M., Vialat, P., Janzen, C., Haller, O., and Frese, M. (2000). Resistance to Rift Valley fever virus in Rattus norvegicus: genetic variability within certain 'inbred' strains. J. Gen. Virol. *81*, 2683–2688.
Ronnholm, R. (1992). Localization to the Golgi complex of Uukuniemi virus glycoproteins G1 and G2 expressed from cloned cDNAs. J. Virol. *66*, 4525–4531.
Ronnholm, R., and Pettersson, R.F. (1987). Complete nucleotide sequence of the M RNA segment of Uukuniemi virus encoding the membrane glycoproteins G1 and G2. Virology *160*, 191–202.
Rossi, C.A., and Turell, M.J. (1988). Characterization of attenuated strains of Rift Valley fever virus. J. Gen. Virol. *69*, 817–823.
Sabin, A.B. (1951). Experimental studies on Phlebotomus (pappataci, sandfly) fever during World War II. Arch. Gesamte Virusforsch. *4*, 367–410.
Saikku, P. (1973). Arboviruses in Finland. 3. Uukuniemi virus antibodies in human, cattle, and reindeer sera. Am. J. Trop. Med. Hyg. *22*, 400–403.
Sall, A.A., de Zanotto, A.P.M., Zeller, H.G., Digoutte, J.P., Thiongane, Y., and Bouloy, M. (1997). Variability of the NS(S) protein among Rift Valley fever virus isolates. J. Gen. Virol. *78 (Pt 11)*, 2853–2858.

Sall, A.A., de Zanotto, P.M., Vialat, P., Sene, O.K., and Bouloy, M. (1998). Origin of 1997–98 Rift Valley fever outbreak in East Africa. Lancet 352, 1596–1597.

Sall, A.A., de Zanotto, P.M., Sene, O.K., Zeller, H.G., Digoutte, J.P., Thiongane, Y., and Bouloy, M. (1999). Genetic reassortment of Rift Valley fever virus in nature. J. Virol. 73, 8196–8200.

Saluzzo, J.F., and Smith, J.F. (1990). Use of reassortant viruses to map attenuating and temperature-sensitive mutations of the Rift Valley fever virus MP-12 vaccine. Vaccine *8*, 369–375.

Samso, A., Bouloy, M., and Hannoun, C. (1975). [Circular ribonucleoproteins in the virus Lumbo (Bunyavirus)]. C. R. Acad. Sci. Hebd. Seances Acad. Sci. D. *280*, 779–782.

Sanbonmatsu-Gamez, S., Perez-Ruiz, M., Collao, X., Sanchez-Seco, M.P., Morillas-Marquez, F., de la Rosa-Fraile, M., Navarro-Mari, J.M., and Tenorio, A. (2005). Toscana virus in Spain. Emerg. Infect. Dis. *11*, 1701–1707.

Sanchez-Seco, M.P., Echevarria, J.M., Hernandez, L., Estevez, D., Navarro-Mari, J.M., and Tenorio, A. (2003). Detection and identification of Toscana and other phleboviruses by RT-nested-PCR assays with degenerated primers. J. Med. Virol. *71*, 140–149.

Savellini, G., Di Genova, G., Terrosi, C., Di Bonito, P., Giorgi, C., Valentini, M., Docquier, J.D., and Cusi, M.G. (2008). Immunization with Toscana virus N-Gc proteins protects mice against virus challenge. Virology *375*, 521–528.

Schmaljohn, C.S., and Nichol, S.T. (2007). *Bunyaviridae*. In Fields Virology, Knipe, D.M., and Howley, P.M., eds. (Lippincott Williams & Wilkins, Philadelphia), pp. 1741–1789.

Sherman, M.B., Freiberg, A.N., Holbrook, M.R., and Watowich, S.J. (2009). Single-particle cryo-electron microscopy of Rift Valley fever virus. Virology *387*, 11–15.

Shoemaker, T., Boulianne, C., Vincent, M.J., Pezzanite, L., Al-Qahtani, M.M., Al-Mazrou, Y., Khan, A.S., Rollin, P.E., Swanepoel, R., Ksiazek, T.G., *et al.* (2002). Genetic analysis of viruses associated with emergence of Rift Valley fever in Saudi Arabia and Yemen, 2000–01. Emerg. Infect. Dis. *8*, 1415–1420.

Simons, J.F., and Pettersson, R.F. (1991). Host-derived 5′ ends and overlapping complementary 3′ ends of the two mRNAs transcribed from the ambisense S segment of Uukuniemi virus. J. Virol. *65*, 4741–4748.

Simons, J.F., Hellman, U., and Pettersson, R.F. (1990). Uukuniemi virus S RNA segment: ambisense coding strategy, packaging of complementary strands into virions, and homology to members of the genus Phlebovirus. J. Virol. *64*, 247–255.

Simons, J.F., Persson, R., and Pettersson, R.F. (1992). Association of the nonstructural protein NSs of Uukuniemi virus with the 40S ribosomal subunit. J. Virol. *66*, 4233–4241.

Sissoko, D., Giry, C., Gabrie, P., Tarantola, A., Pettinelli, F., Collet, L., D'Ortenzio, E., Renault, P., and Pierre, V. (2009). Rift Valley fever, Mayotte, 2007–2008. Emerg. Infect. Dis. *15*, 568–570.

Smith, D.R., Steele, K.E., Shamblin, J., Honko, A., Johnson, J., Reed, C., Kennedy, M., Chapman, J.L., and Hensley, L.E. (2010). The pathogenesis of Rift Valley fever virus in the mouse model. Virology *407*, 256–267.

Smith, J.F., and Pifat, D.Y. (1982). Morphogenesis of sandfly viruses (*Bunyaviridae* family). Virology *121*, 61–81.

Smithburn, K. (1949). Rift valley fever: the neurotropic adaptation of the virus and the experimental use of his modified virus a s a vaccine. Br. J. Exp. Pathol. *30*, 1–16.

Smithburn, K.C., Mahaffy, A.F., Haddow, A.J., Kitchen, S.F., and Smith, H.H. (1949). Rift Valley fever; accidental infections among laboratory workers. J. Immunol. *62*, 213–227.

Soldan, S.S., and Gonzalez-Scarano, F. (2005). Emerging infectious diseases: the *Bunyaviridae*. J. NeuroVirol. *11*, 412–423.

Soldan, S.S., Plassmeyer, M.L., Matukonis, M.K., and Gonzalez-Scarano, F. (2005). La Crosse virus nonstructural protein NSs counteracts the effects of short interfering RNA. J. Virol. *79*, 234–244.

Spik, K., Shurtleff, A., McElroy, A.K., Guttieri, M.C., Hooper, J.W., and SchmalJohn, C. (2006). Immunogenicity of combination DNA vaccines for Rift Valley fever virus, tick-borne encephalitis virus, Hantaan virus, and Crimean Congo hemorrhagic fever virus. Vaccine *24*, 4657–4666.

Stertz, S., Dittmann, J., Blanco, J.C., Pletneva, L.M., Haller, O., and Kochs, G. (2007). The antiviral potential of interferon-induced cotton rat Mx proteins against orthomyxovirus (influenza), rhabdovirus, and bunyavirus. J. Interferon. Cytokine Res. *27*, 847–855.

Struthers, J.K., and Swanepoel, R. (1982). Identification of a major non-structural protein in the nuclei of Rift Valley fever virus-infected cells. J. Gen. Virol. *60*, 381–384.

Subramanian, G.M., Moore, P.A., Gowen, B.B., Olsen, A.L., Barnard, D.L., Paragas, J., Hogan, R.J., and Sidwell, R.W. (2008). Potent *in vitro* activity of the albumin fusion type 1 interferons (albumin-interferon-alpha and albumin-interferon-beta) against RNA viral agents of bioterrorism and the severe acute respiratory syndrome (SARS) virus. Chemotherapy *54*, 176–180.

Suzich, J.A., Kakach, L.T., and Collett, M.S. (1990). Expression strategy of a phlebovirus: biogenesis of proteins from the Rift Valley fever virus M segment. J. Virol. *64*, 1549–1555.
Swanepoel, R., and Blackburn, N.K. (1977). Demonstration of nuclear immunofluorescence in Rift Valley fever infected cells. J. Gen. Virol. *34*, 557–561.
Swanepoel, R., and Coetzer, J.A.W. (1994). Rift Valley Fever (Oxford University Press, Cape Town).
Swanepoel, R., and Coetzer, J.A.W. (2004). Rift Valley fever. In Infectious Diseases of Livestock, Coetzer, J., and Tustin, eds. R., (Oxford University Press, Cape Town), pp. 1037–1070.
Takeda, A., Sugiyama, K., Nagano, H., Mori, M., Kaido, M., Mise, K., Tsuda, S., and Okuno, T. (2002). Identification of a novel RNA silencing suppressor, NSs protein of *Tomato spotted wilt virus*. FEBS Lett *532*, 75–79.
Terasaki, K., Murakami, S., Lokugamage, K.G., and Makino, S. (2011). Mechanism of tripartite RNA genome packaging in Rift Valley fever virus. Proc. Natl Acad. Sci. USA *108*, 804–809.
Tesh, R.B. (1988a). Phlebotomus fevers (CRC Press. Boca Raton, FL).
Tesh, R.B. (1988b). The genus *Phlebovirus* and its vectors. Annu. Rev. Entomol. *33*, 169–181.
Theiler, M., and Smith, H.H. (1937). The use of yellow fever virus modified by *in vitro* cultivation for human immunization. J. Exp. Med. *65*, 787–800.
Tomori, O. (1979). Clinical, virological and serological response of the West African dwarf sheep to experimental infection with different strains of Rift Valley fever virus. Res. Vet. Sci. *26*, 152–159.
Travassos da Rosa, A.P., Tesh, R.B., Pinheiro, F.P., Travassos da Rosa, J.F., and Peterson, N.E. (1983). Characterization of eight new phlebotomus fever serogroup arboviruses (*Bunyaviridae*: Phlebovirus) from the Amazon region of Brazil. Am. J. Trop. Med. Hyg. *32*, 1164–1171.
Valassina, M., Cuppone, A.M., Bianchi, S., Santini, L., and Cusi, M.G. (1998). Evidence of Toscana virus variants circulating in Tuscany, Italy, during the summers of 1995 to 1997. J. Clin. Microbiol. *36*, 2103–2104.
Valassina, M., Meacci, F., Valensin, P.E., and Cusi, M.G. (2000). Detection of neurotropic viruses circulating in Tuscany: the incisive role of Toscana virus. J. Med. Virol. *60*, 86–90.
Valassina, M., Cusi, M.G., and Valensin, P.E. (2003). A Mediterranean arbovirus: the Toscana virus. J. Neurovirol. *9*, 577–583.
do Valle, T.Z., Billecocq, A., Guillemot, L., Alberts, R., Gommet, C., Geffers, R., Calabrese, K., Schughart, K., Bouloy, M., Montagutelli, X., and Panthier, J.J. (2010). A new mouse model reveals a critical role for host innate immunity in resistance to Rift Valley fever. J. Immunol. *185*, 6146–6156.
Venturi, G., Ciccozzi, M., Montieri, S., Bartoloni, A., Francisci, D., Nicoletti, L., Fortuna, C., Marongiu, L., Rezza, G., and Ciufolini, M.G. (2007). Genetic variability of the M genome segment of clinical and environmental Toscana virus strains. J. Gen. Virol. *88*, 1288–1294.
Verani, P., and Nicoletti, L. (1995). Phlebovirus Infections (Chapman & Hall, London).
Verani, P., Nicoletti, L., and Ciufolini, M.G. (1984a). Antigenic and biological characterization of Toscana virus, a new Phlebotomus fever group virus isolated in Italy. Acta Virol. *28*, 39–47.
Verani, P., Nicoletti, L., and Marchi, A. (1984b). Establishment and maintenance of persistent infection by the Phlebovirus Toscana in Vero cells. J. Gen. Virol. *65 (Pt 2)*, 367–375.
Vialat, P., Muller, R., Vu, T.H., Prehaud, C., and Bouloy, M. (1997). Mapping of the mutations present in the genome of the Rift Valley fever virus attenuated MP12 strain and their putative role in attenuation. Virus Res. *52*, 43–50.
Vialat, P., Billecocq, A., Kohl, A., and Bouloy, M. (2000). The S segment of Rift valley fever phlebovirus (*Bunyaviridae*) carries determinants for attenuation and virulence in mice. J. Virol. *74*, 1538–1543.
Wallace, D.B., Ellis, C.E., Espach, A., Smith, S.J., Greyling, R.R., and Viljoen, G.J. (2006). Protective immune responses induced by different recombinant vaccine regimes to Rift Valley fever. Vaccine *24*, 7181–7189.
Wasmoen, T.L., Kakach, L.T., and Collett, M.S. (1988). Rift Valley fever virus M segment: cellular localization of M segment-encoded proteins. Virology *166*, 275–280.
Weber, F., Dunn, E.F., Bridgen, A., and Elliott, R.M. (2001). The Bunyamwera virus nonstructural protein NSs inhibits viral RNA synthesis in a minireplicon system. Virology *281*, 67–74.
Weber, F., Kochs, G., and Haller, O. (2004). Inverse interference: how viruses fight the interferon system. Viral Immunol. *17*, 498–515.
Weber, F., Kochs, G., Haller, O., and Staeheli, P. (2003). Viral evasion of the interferon system: old viruses, new tricks. J. Interferon. Cytokine Res. *23*, 209–213.
Won, S., Ikegami, T., Peters, C.J., and Makino, S. (2006). NSm and 78-kilodalton proteins of Rift Valley fever virus are nonessential for viral replication in cell culture. J. Virol. *80*, 8274–8278.

Won, S., Ikegami, T., Peters, C.J., and Makino, S. (2007). NSm protein of Rift Valley fever virus suppresses virus-induced apoptosis. J. Virol. *81*, 13335–13345.

Xu, F., Liu, D., Nunes, M.R., AP, D.A.R., Tesh, R.B., and Xiao, S.Y. (2007). Antigenic and genetic relationships among Rift Valley fever virus and other selected members of the genus Phlebovirus (*Bunyaviridae*). Am. J. Trop. Med. Hyg. *76*, 1194–1200.

Xu, F., Liang, X., Tesh, R.B., and Xiao, S.Y. (2008). Characterization of cell-death pathways in Punta Toro virus-induced hepatocyte injury. J. Gen. Virol. *89*, 2175–2181.

Yang, H., Bell, T.A., Churchill, G.A., and Pardo-Manuel de Villena, F. (2007). On the subspecific origin of the laboratory mouse. Nat. Genet. *39*, 1100–1107.

Yoneyama, M., and Fujita, T. (2007). RIG-I family RNA helicases: cytoplasmic sensor for antiviral innate immunity. Cytokine Growth Factor Rev. *18*, 545–551.

Yu, X.J., Liang, M.F., Zhang, S.Y., Liu, Y., Li, J.D., Sun, Y.L., Zhang, L., Zhang, Q.F., Popov, V.L., Li, C., *et al.* (2011). Fever with thrombocytopenia associated with a novel bunyavirus in China. N. Engl. J. Med. *364*, 1523–1532.

Zamoto-Niikura, A., Terasaki, K., Ikegami, T., Peters, C.J., and Makino, S. (2009). Rift Valley Fever virus L protein forms a biologically active oligomer. J. Virol. *83*, 12779–12789.

Zhang, Y.Z., Zhou, D.J., Xiong, Y., Chen, X.P., He, Y.W., Sun, Q., Yu, B., Li, J., Dai, Y.A., Tian, J.H., *et al.* (2011). Hemorrhagic fever caused by a novel tick-borne Bunyavirus in Huaiyangshan, China. Zhonghua Liu Xing Bing Xue Za Zhi *32*, 209–220.

Zhioua, E., Moureau, G., Chelbi, I., Ninove, L., Bichaud, L., Derbali, M., Champs, M., Cherni, S., Salez, N., Cook, S., *et al.* Punique virus, a novel phlebovirus, related to sandfly fever Naples virus, isolated from sandflies collected in Tunisia. J. Gen. Virol. *91*, 1275–1283.

Nairovirus Molecular Biology and Interaction with Host Cells

5

Natalia Frias-Staheli, Rafael A. Medina and Anne Bridgen

Abstract

A decade ago nairoviruses were the least understood members of the *Bunyaviridae* family. Today this situation is totally different, with huge advances having been made in the field. These have been in three major areas: (1) sequencing complete genomes, as well as conserved regions of additional viruses, thus allowing phylogenetic analysis, (2) understanding the processes of glycoprotein processing and (3) dissecting three ways in which these viruses avoid the innate immune response, viz. removal of 5′ triphosphate groups from viral RNAs to avoid RIG-I activation, the action of the OTU domain in the viral polymerase and delaying activation of the interferon pathway. In addition, two new viruses have been discovered, Kupe virus and Finch Creek virus. However, fundamental gaps in our knowledge still remain. The difficulty of manipulating the viruses through reverse genetic experiments has hampered our understanding of the processes of transcription and replication, and individual gene function. Much also remains to be learnt in terms of pathogenesis and in particular the genomic regions required to induce severe disease, as well as in the area of vector interactions, drug therapy and vaccine design. If current progress continues, the next decade will provide many more exciting insights.

Introduction to the genus

Nairoviruses are small, enveloped viruses of approximately 100 nm diameter that form one genus within the *Bunyaviridae* family. They are transmitted by ticks or, in some cases, by mosquitoes or culicoid flies, and have a widespread distribution related to that of their host (Marriott and Nuttall, 1996a). Thirty-six viruses have been identified belonging to this genus that have been classified into one of seven different sero-groups on the basis of haemagglutination inhibition and neutralization tests and, more recently, on molecular data (Casals and Tignor, 1980; Elliott *et al.*, 2000; Nichol, 2001) (Table 5.1). Two nairoviruses are serious pathogens: Crimean–Congo haemorrhagic fever virus (CCHFV) causes haemorrhagic disease, with up to 50% mortality in man, while Nairobi sheep disease virus (NSDV, also known as Ganjam virus in India) causes severe gastroenteritis in sheep and goats with up to 90% mortality (Marczinke and Nichol, 2002; Honig *et al.*, 2004a; Whitehouse, 2004). Several other viruses have also been associated with disease in humans: Dugbe virus (DUGV) may cause thrombocytopenia (Burt *et al.*, 1996), while Erve virus has been associated with neurological conditions including thunderclap headaches that may be due to subarachnoid

Table 5.1 Summary of the known nairoviruses allocated to their serogroups. The table also shows the tick vector from which the virus was isolated and important diseases caused by the viruses. Virus lists and serogroups modified from Elliott *et al.* (2000) and Honig *et al.* (2004b)

Serogroup	Disease	Viruses	Isolation (vector, location)	Reference
Crimean–Congo haemorrhagic fever	Haemorrhagic fever in man	Crimean–Congo haemorrhagic fever (CCHFV)	*Hyalomma* and other ticks, worldwide	Clerx *et al*. (1981)
		Hazara (HAZV)	*I. redikorzevi*, Pakistan	Begum *et al*. (1970)
		Khazan (KHAV)		
Dera Ghazi Khan		Abu Hammad virus (AHV)		
		Abu Mina virus (ABMV)	*A. persicas*, Egypt	Clerx *et al*. (1981)
		Dera Ghazi Khan virus (DGKV)	*H. dromedanii*, Pakistan	Clerx *et al*. (1981)
		Kao Shuan virus (KSV)		
		Pathum Thani virus (PTHV)		
		Pretoria virus (PREV)		
Hughes	Several of these viruses cause pruritis, fever and headache in man	Farallon virus (FARV)	*O. capensis*,	Gould *et al*. (1983)
		Fraser Point virus (FPV)	*O. capensis*, USA	Clerx *et al*. (1981)
		Great Saltee virus (GRSV)	*O. maritimas*	Gould *et al*. (1983)
		Hughes virus (HUGV)	*O. capensis*	Gould *et al*. (1983)
		Puffin Island virus (PIV)	*O. capensis*, O. maritimus	Gould *et al*. (1983)
		Punta Salinas virus (PSV)	Multiple isolations worldwide	Chastel *et al*. (1983)
		Raza virus (RAZAV)	*O. capensis*	Gould *et al*. (1983)
		Sapphire II virus (SAPV)		
		Soldado virus (SOLV)		
		Zirqa virus (ZIRV)		
Nairobi sheep disease	Gastroenteritis in sheep and goats; febrile illness in man	Nairobi sheep disease (NSDV)	*R. appendiculatus, (A. variegatum)* East Africa; India: *H. intermedia, R. hemaphysaloides, H. wellingtoni*	Daubney and Hudson (1934)

	Thrombocytopenia in man	Ganjam (GANV)	*A. variegatum* and other ticks	Joshi *et al*. (2005)
		Dugbe (DUGV)	*A. gemma, R. pulchellus*, Kenya	Crabtree *et al*. (2009)
		Kupe		
Qalyub		Bakel virus (BAKV)		
		Bandia virus (BDAV)	*Mastomys* spp. (Senegal	Clerx *et al*. (1981)
		Omo virus (OMOV)	*Mastomys* spp. (Ethiopia	Clerx *et al*. (1981)
		Qalyub virus (QYBV)	*O. erraticus*, Egypt	Clerx *et al*. (1981)
Sakhalin		Avalon virus (AVAV)	*I. uriae*, Newfoundland	Clerx *et al*. (1981)
		(Paramushir virus)	Ixodid ticks, USSR	Main *et al*. (1976)
		Clo Mor virus (CMV)	*I. uriae*, Scotland	Main *et al*. (1976)
		Finch creek virus	*I. uriae*, Macquarie Island	Major *et al*. (2009)
		Kachemak Bay virus (KBV)	*I. uriae*, France	Doherty *et al*. (1975)
		Sakhalin virus (SAKV)	*I. uriae*, Macquarie Island	
		Taggert virus (TAGV)	*I. uriae*, USA	
		Tillamook virus (TILLV)		
Thiafora group	Neurological disorder in man	Erve virus (ERVEV)	Unknown, France	
		Thiafora virus (TFAV)		Zeller *et al*. (1989)

haemorrhaging (Treib *et al.*, 1998). Several of the Hughes group viruses such as the Punta Salinas, Soldado virus, or Zirqa viruses can cause pruritus, fever and headaches (Gould *et al.*, 1983).

Nairoviruses are RNA viruses consisting of three segments of single stranded, negative sense RNA, designated large (L), medium (M) and small (S). These encode four structural proteins: the L segment codes for the RNA dependent RNA polymerase (L) protein, the M segment codes for two structural membrane glycoproteins Gn and Gc and the S segment encodes the nucleocapsid (N) protein (Schmaljohn and Hooper, 2001) (Fig. 5.1). There are also additional non-structural proteins encoded by the M segment, and different protein domains within the L segment. The RNA itself is not infectious; the minimal infectious unit is the RNA encapsidated by the N protein together with the L polymerase protein. The genome size is almost 50% larger than that of most members of the *Bunyaviridae* family, totalling approximately 19 kbp, with both the M and L segments being considerably larger than those of other members of the family (Marriott and Nuttall, 1996a). There are conserved sequences at the termini of each segment that are complementary, which may indicate the formation of pan-handle structures or reflect highly conserved polymerase binding sites at the 3′ ends of the genome and antigenome (Marriott and Nuttall, 1996a; Schmaljohn and Hooper, 2001).

Viral replication occurs in the cytoplasm of infected cells, and viral budding though the Golgi. Fig.5.2 shows an abbreviated replication cycle of the virus and the regions of the cell in which these occur. Viral mRNAs are capped and DUGV and NSDV have been shown to 'cap snatch' host cell derived caps to stabilize the viral messages (Jin, and Elliott, 1993; Holzer, Bridgen and Baron, unpublished). Like other members of the *Bunyaviridae* family, viral messages are not polyadenylated (Blakqori *et al.*, 2009).

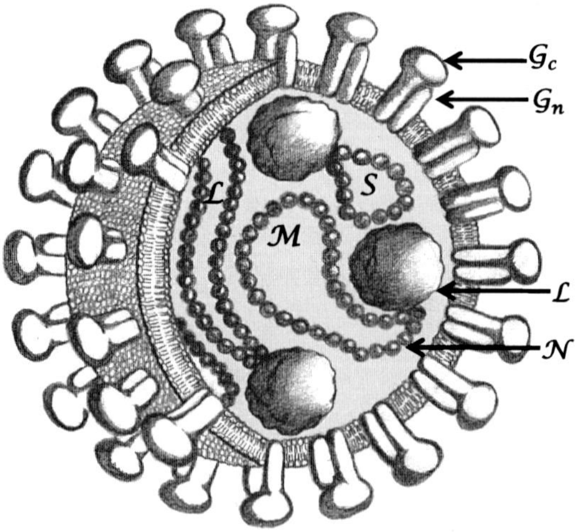

Figure 5.1 Schematic representation of a typical nairovirus. The three single-stranded RNA genomic segments (S, M and L) are shown tightly encapsidated by the nucleocapsid (N) protein. Associated with N is the L protein, and the lipid envelope contains the glycoproteins (Gn and Gc). The nature of the possible interactions between the membrane glycoproteins Gn and Gc has not been fully elucidated and no attempt has been made to illustrate them.

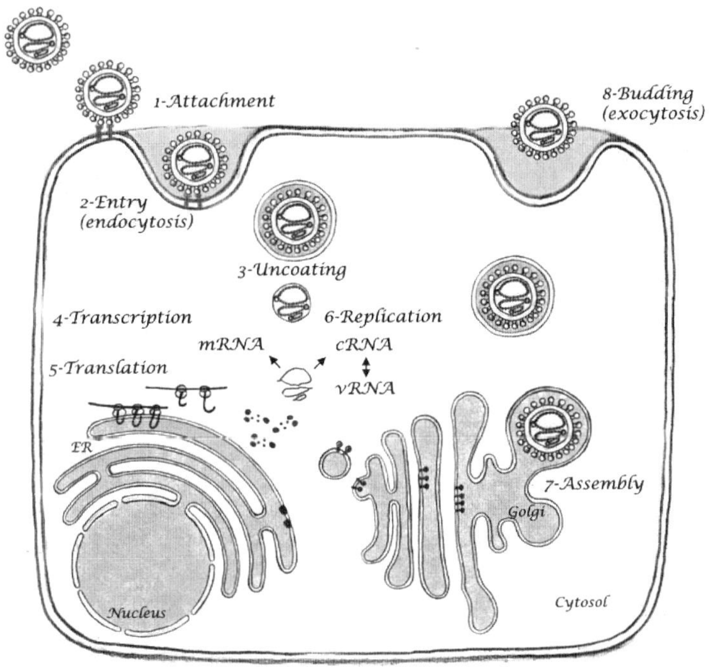

Figure 5.2 Nairovirus life cycle within the cell. The replication of nairoviruses occurs in the cytoplasm of infected cells and can be summarized in eight steps: (1) attachment to the cell, (2) entry by receptor mediated endocytosis, (3) uncoating in a pH-dependent manner, (4) transcription of the viral RNA, (5) translation of viral mRNAs, (6) membrane-associated RNA replication, (7) assembly in the Golgi apparatus, and (8) release of infectious virions. Host cell receptors have not been identified for the members of this genus.

Ten years ago only DUGV had been sequenced completely (Marriott *et al.*, 1992; Marriott and Nuttall, 1996b; Bridgen *et al.*, 2002). Now more than 15 isolates of CCHFV have been sequenced completely, as well as Kupe, Hazara, NSD and Ganjam viruses (Marriott and Nuttall, 1992; Marczinke and Nichol, 2002; Deyde *et al.*, 2006; Duh *et al.*, 2008; Crabtree *et al.*, 2009; Yadav *et al.*, 2011). In addition, conserved regions of the L segment have been amplified and sequenced for representative members of each serogroup (Honig *et al.*, 2004b), thus considerably enlarging our sequence information and allowing phylogenetic analysis.

Much of the recent interest in nairoviruses has stemmed from the greatly increased incidence of CCHFV infection in Europe in the last decade, as well as the potential use of CCHFV as a bioterrorist weapon and increased funding into such agents. There have been over 1100 cases of CCHFV infection in Turkey since 2002 (Estrada-Pena *et al.*, 2007; Ergonul, 2008). The increase in cases has been attributed to global warming, which has affected the tick vector distribution (Vorou, 2009), although other authors have warned against interpreting such data too simplistically (Elliott, 2009). Despite the difficulty of working with a virus that can only be studied in category 4 containment, much of our increase in understanding of nairoviruses comes from experiments with CCHFV. These findings, treatment options and perspectives have been summarized in a number of excellent reviews (Whitehouse,

2004; Flick and Whitehouse, 2005; Ergonul, 2006, 2008; Morikawa *et al.*, 2007; Weber and Mirazimi, 2008). In contrast, there has been little research on many other members of the genus, with the exception of a few studies on DUGV (Bridgen *et al.*, 2002, 2004; Boyd *et al.*, 2006), Kupe virus (Crabtree *et al.*, 2009) and Ganjam virus (Marczinke and Nichol, 2002; Yadav *et al.*, 2011). We predict that the study of additional viruses will be a major theme of the next decade.

In this chapter we are not attempting to describe everything known about nairoviruses, but instead wish to highlight some interesting findings from the last decade of research. We will describe two areas of research in which our understanding has markedly developed (interferon evasion and M segment processing), and two areas which show some development but have a long way to go (reverse genetics and pathogenesis). Finally, we will put forward our ideas as to how research on nairoviruses may develop over the next 10 years.

Nairoviruses and innate immunity

Cellular responses to viral infection

Organisms have evolved sophisticated mechanisms to prevent and control viral infections. Type I IFNs (or IFN-α/β) are antiviral cytokines that are produced by eukaryotic cells and represent one of the first and most effective lines of defence against viral infections. Viral replication products, such as double stranded RNA or other pathogen-associated molecular patterns (PAMPs) are detected by the cell through its membrane-associated receptors (toll-like receptors (TLRs)) or cytoplasmic sensors (Yoneyama *et al.*, 2004, 2008; Garcia-Sastre and Biron, 2006; Yoneyama and Fujita, 2007a–c). These activated cellular sensors induce complex signalling pathways that lead to the activation of transcription factors, which results in the transcriptional activation of the IFN-β promoter and in the synthesis and secretion of IFN-α/β. Secreted IFN then acts in an autocrine or paracrine manner to transcriptionally upregulate the expression of hundreds of interferon-stimulated genes (ISGs) (Der *et al.*, 1998; de Veer *et al.*, 2001; Prescott *et al.*, 2005). The individual or coordinated action of these ISGs results in the activation of a complex host defence programme that establishes an antiviral state within the infected or neighbouring cell. Among these ISGs, those encoding the dsRNA-activated protein kinase (PKR), the 2′–5′ oligoadenylate synthetase (OAS), viperin, the RNA-specific adenosine deaminase 1 (ADAR) and the MxA protein, have the best characterized antiviral activities (Chin and Cresswell, 2001; Samuel, 2001; Garcia-Sastre and Biron, 2006; Wang *et al.*, 2007; Weber and Mirazimi, 2008; George *et al.*, 2009). However, mutant cells deficient in one or more of these ISGs still retain the ability to resolve viral infections after IFN treatment, supporting the notion that other uncharacterized ISGs also have important antiviral activities (Zhou *et al.*, 1997, 1999; Ryman *et al.*, 2005). Besides their role as antiviral cytokines, IFNs possess a wide range of other biological activities including inhibition of cell proliferation, induction of apoptosis and cellular resistance to viral infection. IFN production triggered by the first contact with the invading pathogen slows down or even stops virus multiplication, activates natural killer (NK) cells and dendritic cells (DC) and induces the activation of the adaptive immune system (Le Bon and Tough, 2002; Koyama *et al.*, 2008).

Interferon has inhibitory effects on several bunyaviruses (Tamura *et al.*, 1987; Peters *et al.*, 1989; Temonen *et al.*, 1995; Andersson *et al.*, 2006), including nairoviruses (Andersson *et al.*, 2006; Boyd *et al.*, 2006). DUGV infection of IFN-α/β receptor null IFN-α/β R–/– mice resulted in a faster mortality rate as compared with infected wild type mice, highlighting the importance of having an active IFN system to protect from DUGV infection (Boyd *et al.*, 2006). In a separate study, it was shown that treatment of cultured human endothelial and hepatoma cells with IFN-α 24 h prior to infection reduced CCHFV yields by approximately two to three logs by an as yet unidentified mechanism (Andersson *et al.*, 2006). A better understanding, however, has been achieved with respect to the mechanisms of some ISGs affecting bunyavirus replication.

The groups of Otto Haller and Georg Kochs have studied extensively the effects of the ISG MxA, a small GTPase involved in intracellular trafficking, on La Crosse orthobunyavirus (LACV) replication (Kochs *et al.*, 2002; Reichelt *et al.*, 2004; Haller *et al.*, 2007). They demonstrated that the MxA protein is able to bind to the nucleocapsid protein of LACV thereby sequestering it to perinuclear inclusions in the cytoplasm of the cell. This redistribution of the viral N protein by MxA leads to depletion of N protein from the viral replication sites in the Golgi apparatus where this protein is no longer available for generating new viral particles. The MxA protein also inhibits other representative members of the *Bunyaviridae* family. When constitutively expressed in stably transfected Vero cells, MxA prevented the accumulation of viral transcripts and proteins of Rift Valley fever virus (Frese *et al.*, 1996) and hantaviruses (Frese *et al.*, 1996; Kanerva *et al.*, 1996; Khaiboullina *et al.*, 2005). Members of the *Nairovirus* genus are also susceptible to the antiviral action of the MxA GTPase. Andersson and colleagues showed that Vero cells stably expressing human MxA protein were highly impaired in supporting CCHFV replication. This inhibition appeared to be mediated by a similar mechanism to that described for LACV, as the MxA blocked CCHFV spread at the level of genome replication. This was further supported by the observation that, during CCHFV infection, N was also sequestered to a perinuclear location where it colocalized with MxA. Furthermore, the interaction between MxA and CCHFV N was demonstrated by co-immunoprecipitation assays (Andersson *et al.*, 2004). Likewise, DUGV replication was also inhibited in infected cells expressing MxA protein. Immunofluorescence and western blots assays showed a decrease in the accumulation of DUGV N protein, and Northern blot analyses confirmed a reduction in genomic DUGV RNA production (Bridgen *et al.*, 2004). Altogether, these results confirm that MxA restricts the growth of members of all animal bunyaviruses, targeting the nucleocapsid protein and, most likely, with a similar mechanism of action.

Nairoviral evasion of IFN responses

As a countermeasure, most viral families have developed mechanisms to evade the IFN-α/β-mediated antiviral state of the host. This immune evasion by viruses can occur at the level of IFN induction, IFN signalling or by preventing the activity of ISGs (Fig. 5.3).

In the *Bunyaviridae* family, IFN antagonism has been demonstrated for the non-structural (NS) proteins encoded by viruses in the *Phlebovirus* and *Orthobunyavirus* genera. Although different in sequence and coding strategy, the NS_S of both groups of viruses appear to be responsible for blocking IFN induction at the transcriptional level thereby impairing the production of IFN by the infected cells (Bouloy *et al.*, 2001; Bridgen *et al.*, 2001). It has also been proposed that the Tula strain of hantaviruses also encode a NS_S protein in their genome with the ability to partially inhibit IFN production (Jaaskelainen *et al.*, 2007, 2008). In addition other forms of IFN antagonism by

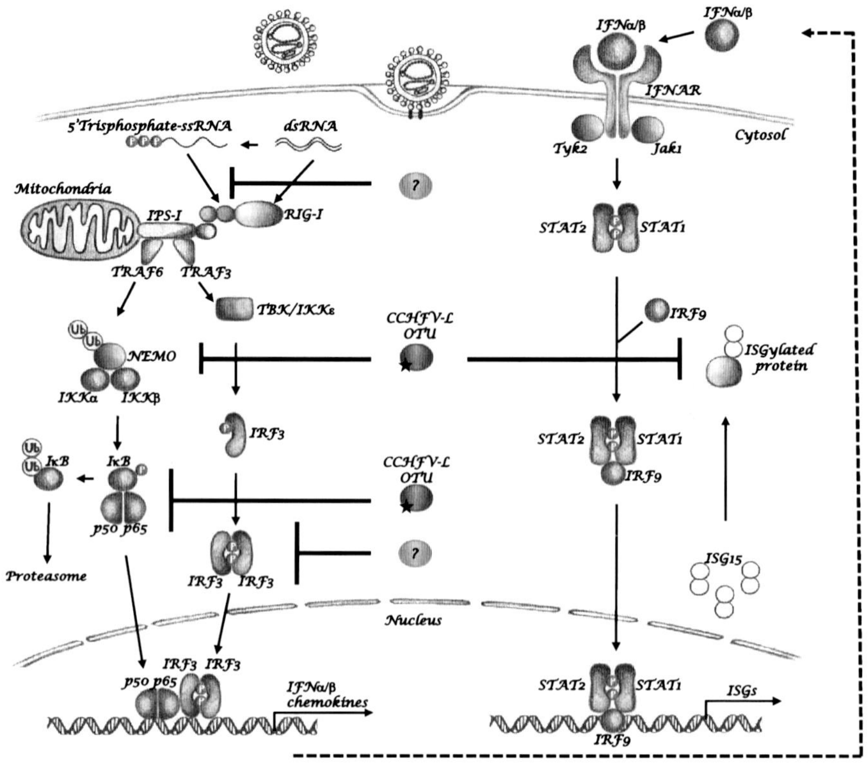

Figure 5.3 Summary of nairovirus interferon evasion strategies. CCHFV utilizes several strategies to regulate innate immunity. During viral infection, the 5′ triphosphate group of the ssRNA is removed by an as yet unidentified viral factor to prevent RIG-I activation and IRF3 activation and nuclear translocation is delayed, interfering with early type I IFN response. The OTU domain present in the L protein utilizes its ubiquitin- and ISG15-deconjugating activities to prevent activation of the NF-κB activation and disrupt the cellular antiviral response.

the glycoproteins (Alff *et al.*, 2006; Spiropoulou *et al.*, 2007) of some hantaviruses have been reported.

Nairoviruses do not encode a NS_S protein and a bona fide interferon antagonist has not yet been reported for this genus of the *Bunyaviridae* family. However they have developed alternative strategies to circumvent the IFN pathway and its effects. It is known that IFNα treatment before experimental infection with CCHFV results in reduction of viral replication, an effect that is likely attributable to a certain extent to the antiviral effect of MxA. Silencing of MxA proteins does not fully recover virus titres in IFN-pre-treated CCHFV-infected cells and, in the DUGV IFN-α/β R–/– model, MxA was not sufficient for protection, suggesting a role of other ISGs in the inhibition of nairovirus replication cycle (Andersson *et al.*, 2006; Boyd *et al.*, 2006). But why does CCHFV infection result in such a devastating and fatal disease in the human host when MxA and other antiviral ISGs are induced by IFN early in infection? A likely explanation is that CCHFV prevents IFN induction or antagonizes the downstream effects of this cytokine. In fact, the latest data demonstrate that both mechanisms seem to occur.

Delay in IFN production by CCHFV

A significant delay in type I IFN production has been demonstrated in CCHFV-infected cells by using an interferon bioassay (Park *et al.*, 2003) along with RT-PCR for detection of IFN-α and ISG56 transcripts. In this study, biologically active IFN-α could only be detected 48 h after infection (Andersson *et al.*, 2008). This delay in IFN production correlates with a delay in IRF3 translocation to the nucleus. Since early type I IFN production depends on IRF3 dimerization, nuclear translocation and binding to the IFN promoter, the delay in IRF3 activation allows CCHFV to replicate in an IFN-free host environment during the first hours of infection. Furthermore, purified genomic RNA from CCHFV is unable to induce IFN-α because of its inability to activate the cytoplasmic PAMP sensor RIG-I (Habjan *et al.*, 2008). The main ligand for RIG-I activation is the triphosphate group on the 5′-end of uncapped ssRNAs, such as those generated by many viral RNA polymerases (Hornung *et al.*, 2006; Pichlmair *et al.*, 2006; Cui *et al.*, 2008). In the case of CCHFV its genome contains a 5′ monophosphate which most likely originates from cleavage of the 5′ triphosphate group during CCHFV genome replication, thus preventing recognition of viral RNA and binding by RIG-I. This is a mechanism of immune evasion that has also been proposed for Hantaan virus (HTNV) and *Borna disease virus* of the *Bornaviridae* family (Garcin *et al.*, 1995; Schneider *et al.*, 2007; Habjan *et al.*, 2008). Despite the failure to activate RIG-I, CCHFV-infected cells still manage to produce IFN at later time points. This may be due to the presence of other molecular patterns that trigger IFN production, such as the viral nucleocapsid and dsRNA, which have been shown with other viruses to induce cytokine production (tenOever *et al.*, 2002, 2004; Weber *et al.*, 2006). Moreover, it has been described that non-replicative enveloped virus particles by themselves can activate ISGs and an antiviral response in a TLR- and RIG-I-independent manner (Paladino *et al.*, 2006; Prescott *et al.*, 2007), suggesting that the host cell can respond to several independent signatures of virus infection.

Targeting of interferon stimulated genes

ISG15 is an interferon-induced ubiquitin-like molecule with still uncharacterized molecular functions. Like ubiquitin, ISG15 is synthesized as a precursor that has to be cleaved to expose a conjugation sequence (LRLRGG) which attaches to target proteins via an isopeptide bond between the terminal carboxyl group of ISG15 and the ε-amino group of the lysine of target proteins (Ritchie and Zhang, 2004; Welchman *et al.*, 2005). ISG15 conjugation (ISGylation) is mediated by the coordinated action of an IFN-induced activating enzyme (E1), conjugating enzymes (E2) and protein ligases (E3). The reverse process, de-ISGylation, is catalysed by UBP43, a cellular ISG15-specific protease (Malakhov *et al.*, 2002). Proteomic approaches have identified hundreds of proteins that are modified by ISGylation (Giannakopoulos *et al.*, 2005; Zhao *et al.*, 2005) but the functional effects of this modification remain a mystery. Some of these proteins, such as PKR, Mx and RIG-I, are also IFN-α/β-inducible and have important antiviral functions (Meurs *et al.*, 1992; Haller *et al.*, 1995; Yoneyama *et al.*, 2004). The existence of an IFN-inducible pathway that mirrors ubiquitin conjugation suggested that protein ISGylation may be an important component of the IFN-induced innate immune response. In support of this, Lenschow and colleagues showed that heterologous expression of ISG15 in a double subgenomic Sindbis virus protected interferon-alpha receptor null (IFN-α $R^{-/-}$) mice against Sindbis virus-induced lethality (Lenschow *et al.*, 2005). Furthermore, $ISG15^{-/-}$ mice displayed increased susceptibility to infection with herpes simplex virus, influenza A virus and influenza B virus (Lenschow *et al.*, 2007), although they had no defect in the antiviral response against vesicular stomatitis virus or lymphocytic choriomeningitis virus (Osiak *et al.*, 2005). Additional studies suggested that ISGylation causes inhibition of HIV (Okumura *et al.*, 2006) and Ebola virus (Okumura *et al.*, 2008a,b) budding. Furthermore, the NS1 protein of influenza B virus was shown to bind specifically to ISG15 and block cellular protein ISGylation (Yuan and Krug, 2001; Yuan *et al.*, 2002).

The L proteins of nairoviruses contain an ovarian tumour (OTU) domain in their amino-terminal region (Honig *et al.*, 2004a). The OTU domain represents a superfamily of ubiquitin-deconjugating cysteine proteases found in prokaryotes, eukaryotes, and viruses (Makarova *et al.*, 2000). An alignment of the OTU domains present in nairoviruses and some mammalian proteins revealed limited overall identity. However a strong conservation of D^{37}, G^{38}, N^{39}, C^{40}, W^{71}, H^{151} and an aromatic amino acid at position 152 (numbering based on the CCHFV-L, strain IbAr 10200 sequence) was observed suggesting that there may be functional relationships between viral and host OTU-domain containing proteins (Frias-Staheli *et al.*, 2007). This led to the hypothesis that the OTU domains from nairoviruses may target the ubiquitin pathway and, given its similarity to ubiquitin and its antiviral role, the ISG15 conjugation pathway.

The groups of Adolfo García-Sastre and Skip Virgin addressed this hypothesis by performing *in vitro* and *in vivo* deubiquitination and de-ISGylation experiments and demonstrated that transient overexpression of the CCHFV-L protein resulted in a global reduction of ubiquitin- and ISG15-conjugated proteins in the cell. This dual effect was also observed with other viral OTU domain-containing proteins tested, such as the DUGV L and the nsp2 proteins of the arteriviruses equine arteritis virus (EAV) and porcine respiratory and reproductive syndrome virus (PRRSV). Overexpression of mammalian OTU proteins, on the contrary, only showed a reduction in overall levels of ubiquitin conjugates and had no effect on ISG15 conjugates, suggesting that the viral OTU proteins represent a distinct class of proteases that are functionally different from the mammalian OTU proteins. In the same study, it was demonstrated that the CCHFV-L OTU domain could hydrolyse *in vitro* poly and di-ubiquitin chains, ISG15 conjugates and Nedd8 and ISG15 precursors but not linear di-ubiquitin and SUMO-2 or SUMO-3 chains, suggesting that CCHFV-L OTU has the ability to recognize ubiquitin, but only acts on certain ubiquitin-like molecules. Mutation of the predicted catalytic cysteine residue to alanine abolished CCHFV-L protease activity, corroborating the notion that the OTU domain has *bona fide* deconjugating activity. Nedd8 is the ubiquitin-like molecule most closely related to ubiquitin with 60% identity to mammalian ubiquitin (Schmidt and Dikic, 2006). It regulates the rate or extent of ubiquitination by modifying the ubiquitin-ligase activity of the E3 ligase cullin (Petroski and Deshaies, 2005). SUMO is nearly identical to ubiquitin in overall structure but is more divergent in amino acid sequence and is mostly involved in transcriptional regulation and in nuclear import. The mammalian SUMO family includes three members, SUMO-1, SUMO-2 and SUMO-3. While target proteins are predominantly conjugated to monomeric SUMO-1, SUMO-2 and SUMO-3 have the potential to multimerize and form chains (Vertegaal, 2007). Ubiquitin, ISG15, Nedd8 and SUMO are synthesized as longer precursors that are processed by deubiquitinating enzymes (DUBs) into the mature form by exposing a C-terminal LRLRGG motif (ubiquitin and ISG15), LALRGG motif (Nedd8) or a QQQTGG motif (SUMO). The observation that the OTU of CCHFV-L is able to hydrolyse ubiquitin, ISG15 and

Ned88 but not SUMO chains raised the interesting possibility that sequences similar to the LRLRGG sequence may play an important role in substrate recognition and specific cleavage by the CCHFV-L OTU. Interestingly, the nsp2 of arteriviruses participates in processing of the viral polyprotein by cleaving the nsp2/3 junction at a FRLIGG (EAV) or GRLLGG (PRSSV) sequence (Snijder *et al.*, 1995, 1996; Allende *et al.*, 1999; Ziebuhr *et al.*, 2000) that is similar to the LRLRGG motif. Undergoing experiments by Marjolein Kikkert and colleagues suggest that EAV-nsp2 also has *in vitro* deconjugating activity towards K^{48}- and K^{63}-linked polyubiquitin, with a slight preference for K^{63}-linked chains. This allows us to speculate about nairovirus OTU proteases having dual functions: targeting host substrates to modulate the anti-viral response and performing essential viral polyprotein processing, although this last function still has to be demonstrated. If proved, it would be analogous to the hepatitis C virus NS3–4A protease, which is involved in both viral polyprotein processing and cleaving the cellular antiviral proteins TRIF and IPS-1 (Li *et al.*, 2005; Lin *et al.*, 2006). Another example would be the SARS coronavirus papain-like protease (SARS-CoV PLpro) which is also able to deconjungate ubiquitin and ISG15 *in vitro* in addition to its polyprotein processing activity (Lindner *et al.*, 2005, 2007).

It is important to note that the *in vitro* studies conducted by Frias-Staheli and colleagues were performed with the catalytic core domain of CCHFV-L and not the entire protein. A common aspect to cellular and viral deubiquitinating enzymes is the regulation of their activities by domains outside the protease catalytic core. For example the cellular deubiquitinating enzyme USP7 utilizes an upstream domain to bind to and achieve substrate specificity towards p53 (Hu *et al.*, 2002); and USP14 relies on proteasome binding via its amino-terminal ubiquitin-like domain to activate the autoinhibited catalytic domain (Hu *et al.*, 2005). SARS-CoV PLpro is enclosed in a larger protein, nsp3, and its polyprotein processing activity is thought to be affected by additional sequences in the nsp3 protein (Harcourt *et al.*, 2004). Finally, the deubiquitinating activity of the herpesvirus UL36 appears to be inhibited in the context of the full-length large tegument protein (Kattenhorn *et al.*, 2005).

How relevant is the presence of an OTU domain during CCHFV infection? The group of Stuart Nichol showed in a minireplicon system that the OTU domain is dispensable for the L protein polymerase function (Bergeron *et al.*, 2010). However, the lack of a reverse genetics system for nairoviruses currently hampers the possibility of rescuing a mutant virus with a catalytically inactive OTU domain. Phenotypic comparison of such a mutant virus with a wild type virus would give insights into the contribution of this domain in virus replication and pathogenesis. As an alternative strategy to study the effects of this protein in viral pathogenesis, transgenic mice expressing the CCHFV-L OTU domain were generated and the sensitivity of these transgenic mice to infection with Sindbis virus was evaluated. Infectivity of Sindbis virus correlated with expression of the OTU transgene, strongly suggesting that CCHFV-L OTU enhances susceptibility to viral disease *in vivo*. The increased pathogenicity of Sindbis virus observed in OTU-expressing mice suggested that this domain might counteract the antiviral activities of ISG15 *in vivo*. This was addressed by generating chimeric Sindbis viruses expressing ISG15 in the presence or absence of enzymatically active CCHFV-L OTU domains. It was previously shown that heterologous expression of ISG15 from a Sindbis virus protected IFN-α $R^{-/-}$ mice from Sindbis-induced lethality. This protection was dependent on the ability of ISG15 to conjugate to target proteins (Lenschow *et al.*, 2005). In this model system, co-expression of the OTU domain with ISG15 abolished ISG15's anti-viral function, while expression of an OTU catalytic mutant had no effect on ISG15-mediated protection from lethality. Altogether these experiments demonstrated that the OTU proteases play an important role in the immune evasion properties of nairoviruses.

Viral modulation of NF-κB activity

Protein ubiquitination is fundamentally important in the regulation of both the innate and the adaptive cellular immune system (Liu *et al.*, 2005). Ubiquitin regulates several aspects of antiviral immunity such as MHC class I and II antigen presentation (Loureiro *et al.*, 2006; Shin *et al.*, 2006), TLR/IL1 signalling (Chen, 2005), induction of type I IFN by the cellular helicase RIG-I (Gack *et al.*, 2007) and the nuclear factor-κB (NF-κB) signalling pathway. In unstimulated cells, NF-κB is sequestered in the cytoplasm through binding to the inhibitor of κB (IκB). Upon tumour necrosis factor α (TNF-α) binding to its receptor, or other stimuli, IκB is phosphorylated and then poly ubiquitinated with a K^{48}-linked poly ubiquitin chain and targeted for proteasomal degradation. This results in translocation of IκB-free NF-κB dimers to the nucleus to initiate the expression of NF-κB -dependent target genes. NF-κB-induced genes include type I IFNs and other cytokines, hence this ubiquitin-controlled pathway plays a vital role in both innate and adaptive immunity, as well as in inflammation (Tergaonkar, 2006). In addition to ubiquitination of IκB, multiple components and regulators of the NF-κB-dependent signalling pathway are also subject to regulation by K^{63}- and K^{48}-linked ubiquitination. The OTU domain-containing cellular protein A20 is an important down-regulator of the TNF-α signalling pathway via its deubiquitination of TRAF6 (Boone *et al.*, 2004) and its dual function of poly ubiquitin K^{63} deconjugation followed by poly ubiquitin K^{48} modification of the signalling molecule RIP (Wertz *et al.*, 2004). These activities are mediated by its amino-terminal OTU domain and its carboxy-terminal zinc finger domain, respectively.

Viral OTU-domain containing proteins also modulate the IκB-dependent signalling pathway. Expression of the OTU domain of CCHFV-L decreased in a dose-dependent manner the activation of an NF-κB-responsive promoter and inhibited endogenous p65 nuclear translocation after TNF-α treatment, demonstrating the ability of viral OTU domains to affect at least one immune pathway that is regulated by ubiquitination (Frias-Staheli *et al.*, 2007). It is still unclear at what specific level the viral OTU domains inhibit NF-κB activation, since many steps involved in this activation are dependent on ubiquitination. Mono-ubiquitination of IKK-β and poly ubiquitination of NEMO (IKKγ/NF-κB essential modulator), TRAFs and RIP with K^{63}-linked ubiquitin chains, in addition to poly ubiquitination of IκB by K^{48}-linked ubiquitin and free K^{63} poly ubiquitin chains, are required for NF-κB activation. In a recent study, CCHFV-L OTU was found to remove poly ubiquitin chains from NEMO (Xia *et al.*, 2009) suggesting a potential mechanism for the NF-κB inhibition by this viral protein. Surprisingly, the catalytically inactive CCHFV-L OTU protein also inhibits TNF-α –driven NF-κB activation, but with about ten times lower efficiency, suggesting that the OTU domain may have additional functions (Frias-Staheli *et al.*, 2007). Another explanation could be that the catalytically dead mutant, although not able to hydrolyse the substrate, is still able to bind it, thus impairing its function. It would be interesting to analyse further the precise steps in the NF-κB pathway that are being affected by CCHFV OTU and also what other ubiquitin-dependent processes and pathways are being affected by this domain.

Concluding remarks and perspectives

Despite their importance as human and veterinary pathogens, relatively little is known about the biology of the nairoviruses. For CCHFV, this is attributable in part to the requirement for BSL-4 facilities, which limits the investigation of this virus to a few research groups worldwide. In addition, the absence of an animal model that mimics the disease symptoms observed in natural hosts, along with the lack of reverse genetics systems, restricts studies on the pathogenesis and molecular biology of this group of viruses. CCHFV, and probably other nairoviruses, smartly employs a range of different strategies to evade the host's early type I IFN response. Firstly, RIG-I-mediated virus recognition is circumvented by removal of the 5′ triphosphate group from the viral genome and IRF3 activation is delayed by an as yet unknown mechanism. Furthermore, CCHFV via its OTU domain is also able to inhibit NF-κB activation. This coordinated cascade of events results in a relatively delayed type I IFN production that allows a window of IFN-free environment for virus replication and spread. The host's IFN response eventually results in the transcriptional activation of ISGs such as ISG15, which is itself efficiently targeted by CCHFV and related nairoviruses. While significant progress has been made in the last couple of years in understanding nairovirus immune evasion strategies, there are still numerous open questions in this field. IFN induction pathways have still not been characterized and no viral IFN antagonist has been identified for this group of viruses. Such data are urgently needed as a better understanding of virus-innate immune interplay may elucidate critical determinants of pathogenesis and may give insight into new therapeutic measures.

M segment analysis

Introduction

The nairoviral M segment contains the coding capacity for an unusually large polyprotein of up to approximately 186 kDa (approx. 1684 amino acids in length), approximately 30–50% larger than the M segments of members of other genera in the *Bunyaviridae* family. The polyprotein undergoes extensive and complex proteolytic processing into the two major structural glycoproteins Gn and Gc and additional non-structural proteins. At least one non-structural protein has now been described for two of its members, DUGV and CCHFV (Marriott *et al.*, 1992; Altamura *et al.*, 2007), and the presence of protein precursors and additional proteins that are secreted during viral infection have also been demonstrated. The viral glycoproteins are likely to influence cell tropism and the ability of the viruses to infect their vertebrate and tick hosts, since they are thought to be the determinants of receptor recognition sites on susceptible cells. This then leads to binding, internalization, and to the beginning of replication, which occurs entirely in the cytoplasm.

Diversity and phylogenetic relationship of the M segment of nairoviruses

Sequencing analysis of the M segment has shown a great deal of genetic diversity even among members of the same serogroup. The known coding sequences of this segment have a 7–19.3% and 6.2–14.8% difference at the nucleotide and amino acid levels, respectively (Papa *et al.*, 2002; Sanchez *et al.*, 2002). Data arising from sequences of CCHFV isolates indicate that the N-terminal regions of Gn (previously named G2) are highly variable, with fewer differences observed in the reminder of the Gn and Gc (previously named G1) (Ahmed *et al.*, 2005). Some of these changes appear to confer antigenic diversity between the different M segment groups (discussed below), even though a number of neutralizing epitopes in both Gn and Gc are shared among different strains, as shown by studies using specific monoclonal antibodies directed against each of the glycoproteins (Ahmed *et al.*, 2005).

Although a number of sequences are now available from viruses pertaining to the different nairovirus serogroups, phylogenetic relationships of the M segments have been most thoroughly investigated for the CCHFV serogroup of the *Nairovirus* genus. Phylogenetic analyses have demonstrated the presence of four M segment groups, named M1, M2, M3 and M4 (Hewson *et al.*, 2004b; Ahmed *et al.*, 2005). Although there is some degree of geographical clustering of the M segments of CCHFV strains, some viruses isolated from geographically distant locations appear to be closely related at the nucleotide level (Hewson *et al.*, 2004b; Ahmed *et al.*, 2005), a phenomenon not observed in analyses of the S and L segments. This discrepancy appears to be a result of reassortment events rather than a distinct rate of evolution occurring in this segment (Hewson *et al.*, 2004b). This implies co-infection of hosts, perhaps ticks and/or livestock, in which reassortment of the M segment might occur and from which reassortant viruses are subsequently dispersed (Hewson *et al.*, 2004b; Ahmed *et al.*, 2005). As further full-length sequences become available, we will discover whether this is a phenomenon that is exclusive to the CCHFV serogroup, or if it is more widely applicable to other nairoviruses.

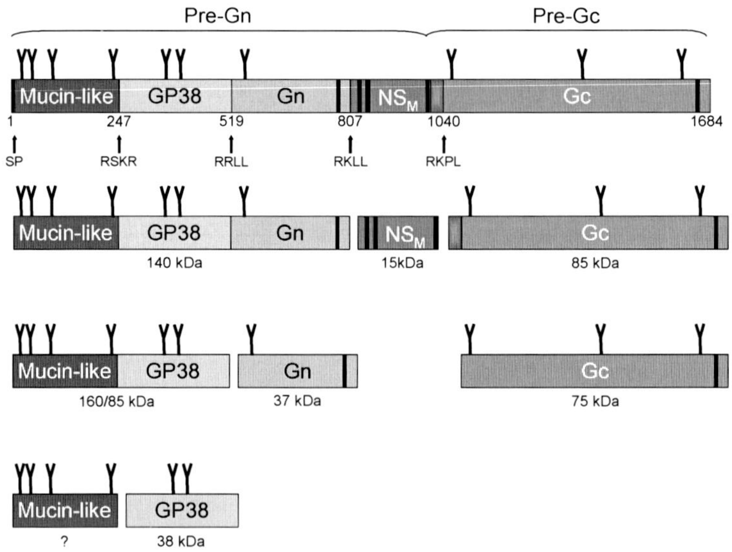

Figure 5.4 Representation of the M segment-encoded polyprotein. Amino acid positions are according to CCHFV strain IbAr 10200. Black bars represent trans-membrane (hydrophobic) regions. Arrows indicate the known and predicted cleavage site positions and the sequence motifs are shown (top panel). Y-shapes represent N-glycosylation sites present in the polyprotein. Cleavage events are represented sequentially and the protein products and their sizes are indicated (middle and bottom panels).

General features of the CCHFV M segment polyprotein

Extensive analyses of the CCHFV M segment protein have helped elucidate several key features of its sequence (Fig. 5.4). The full-length polyprotein contains an amino terminus signal peptide, which is cotranslationally cleaved by signalase. There are 10–12 potential N-glycosylation sites predicted in the entire polyprotein, six of which are conserved in all CCHFV virus strains (Sanchez *et al.*, 2002). Interestingly, a mucin-like variable region is found to be present at the N-terminus of CCHFV M. Analysis of this variable region showed high differences at the nucleotide and an amino acid levels (14.3–37.2% and 23.8–56.4% different, respectively) (Sanchez *et al.*, 2002). The unprocessed full-length polyprotein shows at least six highly hydrophobic regions, corresponding to transmembrane helices that span the membrane six times (Papa *et al.*, 2002; Sanchez *et al.*, 2002; Altamura *et al.*, 2007). In addition, four conserved cleavage sites have been predicted and described. Specific events of the biosynthesis of the viral glycoproteins are covered in the following sections.

Proteolytic processing of the CCHFV glycoproteins

Much of our current understanding of nairovirus glycoprotein processing comes from the study of CCHFV. The M segment of CCHFV encodes a polyprotein that is proteolytically processed resulting in a 140-kDa precursor protein of Gn (PreGn, formerly known as PreG2) and an 85-kDa precursor protein of Gc (PreGc, formerly known as PreG1). These in turn are further processed and give rise to the mature Gn (37 kDa) and Gc (75 kDa) glycoproteins, which constitute the major structural components of the virion (Sanchez *et al.*, 2002). The cotranslational cleavage of the precursor polyproteins by signalase and N-glycosylation, which take place in the endoplasmic reticulum (ER), are the first steps of the extensive proteolytic processes that take place during biosynthesis of these glycoproteins (Fig. 5.5). Further processing of PreGn and PreGc takes place in the ER and Golgi (Sanchez *et al.*, 2006). There is a similar glycoprotein precursor–product relationship to that observed with Clo Mor, Qalyub and Dugbe viruses, which seems to be unique to viruses of the *Nairovirus* genus.

Processing of the Gn and Gc glycoproteins

As observed with other members of the *Bunyaviridae* family, the mature Gc is primarily localized to the ER and requires Gn to enable its transport. The subtilase SKI-1 and possibly additional related cellular proteases aid the major precursor cleavage events needed for processing of the CCHFV glycoproteins during viral infection (Sanchez *et al.*, 2002; Vincent *et al.*, 2003), a phenomenon that has also been described for processing of GP-C of the arenavirus *Lassa virus* (Lenz *et al.*, 2001). The mature Gc is initially processed in the ER from PreGc at its N-terminus by signalase cleavage and subsequently by a SKI-1-like-protease at the RKPL motif (approx amino acid 1040)(Vincent *et al.*, 2003). The C-terminus of PreGn is predicted to span the membrane four times based on the presence of hydrophobic amino acid stretches. PreGn, which contains a mucin-like domain, a 38 kDa protein named GP38, and Gn is processed at motif RRLL (approximately amino acid 519) by SKI-1 in the ER/cis Golgi to generate the mature Gn and a mucin-GP38 protein complex (Vincent *et al.*, 2003).

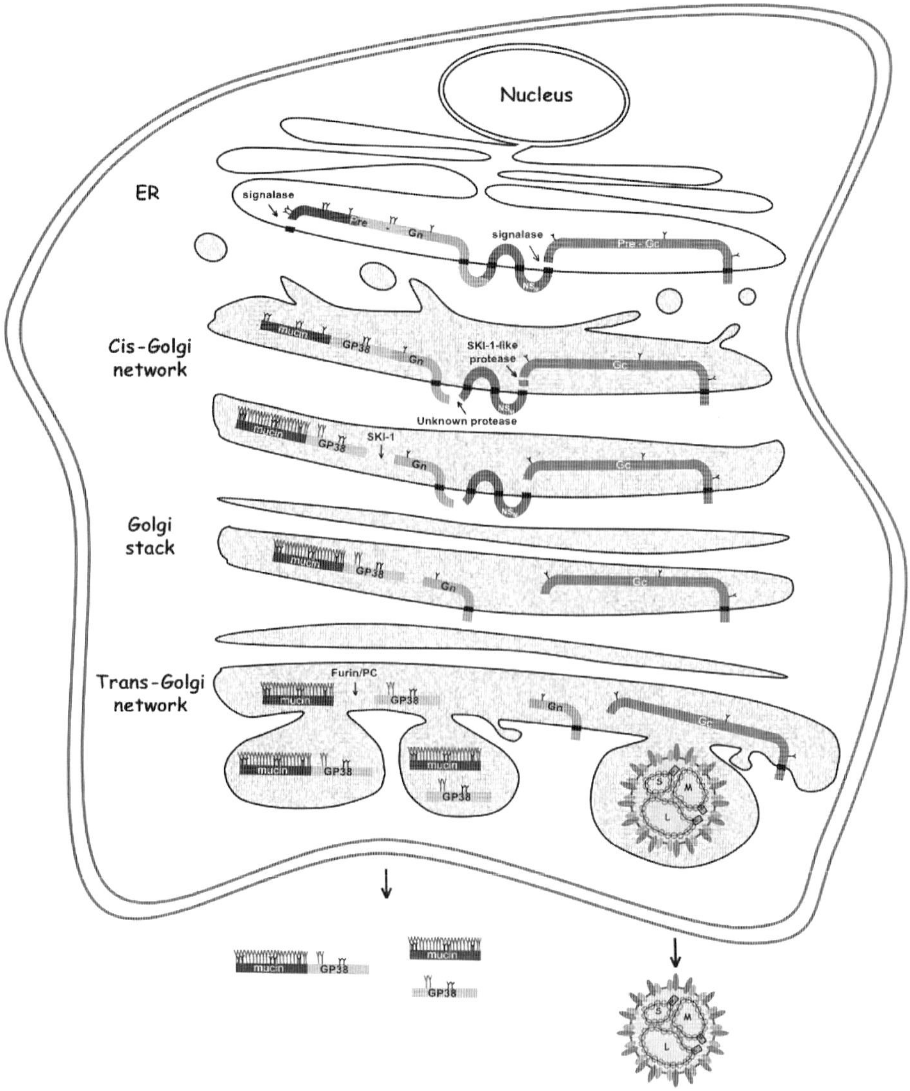

Figure 5.5 Processing of the M segment glycoproteins of CCHFV in the cell. Firstly signalase cleavage and N-linked glycosylation of the polyprotein precursor occur cotranslationally in the ER. Cleavage and release of the NS_M protein at its N-terminus by an unknown protease takes place in the cis-Golgi. Processing of the mature Gc protein by cleavage by a SKI-1-like protease follows this. Mature Gn is generated by SKI-1 in the ER/cis-Golgi, also giving rise to the mucin–GP38 complex. Extensive O-glycosylation of the mucin-like domain occurs during its transport to the secretory pathway. In the trans-Golgi network furin or furin-like PCs cleave the mucin–GP38 complex into single proteins (e.g. mucin-like and GP38 proteins). The processed and unprocessed mucin-like and GP38, as well as GP85 and GP160 are secreted in substantial amounts to the medium. Fully mature Gn and Gc interact directly and participate in viral assembly, which eventually results in virions being released into the medium. With the authors' permission, the figure has been modified and updated from the original version in Sanchez *et al.* (2006).

Secreted glycoproteins derived from PreGn and the mucin-like domain

Overall, signalase cleavage and processing of the mature Gn results in the generation of three additional proteins derived from the N-terminal 519 amino acids. During virus replication proteins of 85, 38 and 160kDa termed GP85, GP38 and GP160, respectively, are generated. These appear to be soluble proteins since they are not detected in virions, and can be immunoprecipitated from supernatants of infected cells (Sanchez *et al.*, 2006). After SKI-1 mediated cleavage of the Gn from PreGn, the mucin-GP38 protein complex is released. Sequence analysis of several CCHFV strains indicated that the RSKR motif, located between the mucin-like domain and the rest of the PreGn polyprotein (approx. amino acid position 244–247), is conserved and would be an appropriate cleavage site for furin or other furin-like protein convertases (PC) found in the trans-Golgi network of multiple tissues and cell types (Sanchez *et al.*, 2006). Cleavage by furin or furin-like PC results in the generation of a mucin-like protein and GP38 (Sanchez *et al.*, 2006). Additional immunoprecipitation analyses demonstrated that GP85 and GP160, both of which contain the mucin-like domain, and as predicted they become heavily O-linked glycosylated and moderately N-glycosylated during their transport through the Golgi secretory pathway. Overall, the currently known processing of PreGn generates four secreted proteins that can be detected in supernatant, GP38, GP85, GP160 and the mucin-like domain (Vincent *et al.*, 2003; Sanchez *et al.*, 2006). Interestingly, a similar mucin-like domain of the sGP of Ebola virus has been shown to be a major determinant of pathogenesis (Yang *et al.*, 2000; Simmons *et al.*, 2002). To date the functional significance of the CCHFV mucin-like domain remains unknown, although it likely plays a role in pathogenesis. Similarly, the functions and roles of each of these secreted glycoprotein products have not yet been investigated.

N-linked glycosylation of the CCHFV M segment polyprotein

Polypeptide sequences of CCHFV M contain at least ten potential asparagine-linked (N-linked) glycosylation sites (Papa *et al.*, 2002; Sanchez *et al.*, 2002; Bertolotti-Ciarlet *et al.*, 2005). Thus far, only residue 557N of Gn has been shown to be required for correct localization. Mutation of this glycosylation site did not affect appropriate proteolytic processing, but resulted in retention and mislocalization of all the protein products of the M ORF to the ER (including GP160, GP85, GP38, Gn and Gc) (Erickson *et al.*, 2007). Glycosylation sites in the Gc protein do not seem to be a requisite for appropriate processing and localization. Thus, correct folding, localization and transport through the ER and Golgi appear to be dependent on N-glycosylation of Gn (Erickson *et al.*, 2007). Further studies are needed to elucidate whether N-linked glycosylation sites present in the secreted mucin-like or GP38 proteins have a functional role required for adequate processing of the surface glycoproteins.

The NS_M protein of CCHFV

Recent studies have confirmed the presence of a non-structural (NS) M protein derived from processing of the PreGn polyprotein. Discrepancies in the predicted mass of Gn after processing of the PreGn polyprotein by SKI-1 and SKI-1-like proteases (58kDa), and the actual migration pattern of the mature protein (37kDa), led to the identification of a ~15kDa protein corresponding to the NS_M protein of CCHFV, a previously uncharacterized protein in the *Nairovirus* genus (Altamura *et al.*, 2007). A cleavage event at the C-terminal region of PreGn, somewhere close to amino acids 805–807, by an as yet unknown protease results in the generation NS_M, see Fig. 5.4. This event occurs cotranslationally or very soon after synthesis and prior to N-terminus cleavage of Gn by SKI-1, and the NS_M protein was found to traffic to the Golgi (Altamura *et al.*, 2007). Although NS_M proteins have been described for other genera of the *Bunyaviridae* family, very little is known about their function, thus further studies will be needed to elucidate the role of the CCHFV NS_M.

The glycoproteins of other nairoviruses

Early studies showed that Hazara virus (belonging to the CCHFV serogroup) contains three structural glycoproteins of 84, 45 and 30 kDa (Foulke *et al.*, 1981). The M segment of Clo Mor virus, a member of the Sakhalin serogroup, encodes three virion-associated glycoproteins with sizes of 90, 80 and 45 kDa (Watret *et al.*, 1985). In contrast, Qalyub virus (Qalyub serogroup) seemed to contain only two structural proteins of 75 and 40 kDa (Clerx *et al.*, 1981), a picture more typical of other members of the *Bunyaviridae* family. Similarly DUGV, a member of the Nairobi sheep disease serogroup, possesses only two structural glycoproteins of 73 and 35 kDa (Cash, 1985; el-Ghorr *et al.*, 1990; Marriott *et al.*, 1992). Studies with Clo Mor, Qalyub and Dugbe viruses also revealed that their M segments generate at least one precursor protein. In DUGV infected cells, proteins of 110 and 85 kDa have been shown to share epitopes with the Gc glycoprotein; however, only the 85 kDa protein was shown to be a precursor of Gc. Additionally, a non-structural protein of 70 kDa shares amino acid sequence with Gn, suggesting that it might be its precursor protein (Marriott *et al.*, 1992). Additional research on the molecular mechanisms involved in processing of the M polyprotein of nairoviruses, including elucidating the role of the NS_M protein and the secreted glycoproteins, is needed. Such studies would not only allow further understanding of the key steps in viral biology, but would also probably lead to the development of novel therapeutic interventions against these highly pathogenic viruses.

Reverse genetics

Reverse genetics is the manipulation of the viral genome in order to understand the effects on the viral phenotype, or the role of specific genetic sequences. Such experiments can be conducted with so called minigenomes, in which the viral replicative sequences flank a suitable reporter gene that can be replicated *in vitro* using transfected cDNAs expressing the replication proteins, or by a helper virus (either a wild type or modified virus). Alternatively, infectious virus can be rescued in entirety from cDNA clones. Since the first rescue of synthetic Bunyamwera virus (BUNV) from cDNA clones in 1996 (Bridgen and Elliott, 1996), several other bunyaviruses including the orthobunyaviruses LACV and Akabane virus and the phlebovirus RVFV have been rescued from cDNA clones (Blakqori and Weber, 2005; Ikegami *et al.*, 2006; Ogawa *et al.*, 2007).

Nairoviruses are proving more problematic to study in this way. To start with, the full genome sequences are only just becoming available for many of the nairoviruses. However, the main difficulty seems to arise from the RNA dependent RNA polymerase (L protein) which is almost twice as large as any of the other polymerases of the *Bunyaviridae* family; bacterial clones are often unstable and transfection of large quantities of the cDNA into eukaryotic cells also appears to be toxic. Thus far, the development of a full rescue system from infectious clones still proves elusive.

There have, however, been two reports of minigenome replication for CCHFV, one using the human RNA polymerase I transcription (pol I) system (Flick *et al.*, 2003) and a more recent report using the T7 system (Bergeron *et al.*, 2010). These two systems have various advantages and disadvantages. More bunyavirus reverse genetics experiments have been performed with constructs expressed from the bacteriophage T7 RNA polymerase promoter, which is expressed in the cytoplasm of cells where these viruses normally replicate. However, the T7 polymerase has to be provided in trans and most (non-toxic) expression systems result in a relatively low expression level. Additionally, the T7 polymerase favours transcription initiation from a promoter terminating in three guanine residues, resulting in additional non-templated G residues at the 5′ terminus of the bunyavirus sequence. The correct 3′ terminus can be obtained by use of the hepatitis virus δ ribozyme sequence to cleave the nascent RNA at the precise 3′ terminus. The pol I system generates nuclear transcripts, but these are successfully translocated to the cytoplasm unless there are splicing complications. It has a higher expression level than most T7 systems and can also be used to generate defined transcripts without the need for insertion of the hepatitis virus δ ribozyme sequences to cleave the nascent RNA. It is the system generally used for influenza A virus rescue (Neumann and Kawaoka, 2001), and has also been used to rescue the orthobunyavirus *Akabane virus* (Ogawa *et al.*, 2007) and the phlebovirus RVFV (Billecocq *et al.*, 2008). A recent study using RVFV demonstrated rescue using both techniques with very little difference in yield (Billecocq *et al.*, 2008).

Flick *et al.* (2003) used the pol I minigenome system to replicate minigenomes containing the green fluorescent protein (GFP) and chloramphenicol acetyl-transferase (CAT) reporter genes with CCHFV helper virus and investigate the capacity of different cell lines to support replication. This study was not however continued in further work. More recently Bergeron *et al.* (2009) have shown that CCHFV minigenomes can be replicated by replication proteins expressed from transfected viral cDNAs and that this replication was inhibited by the drug ribavirin. Undoubtedly, the development of a reliable and robust reverse genetics system for CCHFV and other nairoviruses will constitute an important advance in the field and an indispensable tool to further our understanding of the molecular biology and pathogenesis of these viruses.

Pathogenesis and antiviral therapy

Early studies of nairovirus pathogenesis involved the study of organs and tissues infected with virus. These have been conducted in mouse models for CCHFV, DUGV, and Hazara viruses, in sheep for NSDV and, in addition, human pathology samples have been studied for CCHFV. Further information on these experiments can be found in the following sources: Hazara virus (Smirnova *et al.*, 1977), DUGV (Buckley *et al.*, 1990; Coates and Sweet, 1990), CCHFV in animals (Hoogstraal, 1979; Wilson *et al.*, 1991; Tignor and Hanham, 1993; Gonzalez *et al.*, 1995) and man (Davies *et al.*, 1977; Gonzalez *et al.*, 1998). As in other areas of research, however, the last decade has seen a switch to the study of individual cells and to the molecular events taking place within these cells.

Nairoviruses have a fairly wide cellular tropism, infecting and replicating in most mammalian cells examined. They also replicate well in the *Xenopus laevis* XTC-2 cell lines (Watret *et al.*, 1985) but not in *Aedes albopictus* C6/36 cells (Crabtree *et al.*, 2009). The replication levels and ability to cause cytopathic effects in different cell lines varies with the virus, but BHK, Vero and SW-13 cells support high levels of replication of several nairoviruses.

All nairoviruses appear to be strongly neurotropic *in vivo*. Both newborn and 3 week-old outbred Swiss white (TO) mice inoculated intraperitoneally with DUGV strains including IbAr 1792 and KT 281/75 died from viral infection, and virus was isolated from the brains of these mice. Neurovirulence differed between strains, with strains IbH 11480 and IbAr 1792 being more neurovirulent; other strains such as ArD 44313 only killed newborn but not adult mice when inoculated i.p. (Buckley *et al.*, 1990). Coates and Sweet (1990), using DUGV strain KT 281/75, demonstrated lethal disease by all routes of inoculation in neonatal mice but only by intracerebral inoculation in adult mice (Coates and Sweet, 1990). Boyd *et al.* (2006) found extensive evidence of DUGV infection in the brains of DUGV-infected mice. Immunostaining was punctate, a distribution consistent with virus replication sites. DUGV infection was observed in neurones of the mitral cell layer of the olfactory bulb, as well as in neuronal Purkinje cells and in adjacent cells, probably Bergman glial cells, and cells of the molecular and granule cell layers in the cerebellum. In the cortex and thalamus, large cells consistent with a neuronal morphology were infected. This specificity of DUGV infection for neuronal cells was striking, particularly in the light of the normally wide cell tropism of bunyaviruses, and may indicate that there is a specific viral receptor in these cells.

Nairoviral effects on haemostasis

At least three viruses within the genus appear to be capable of inducing haemorrhaging: NSDV causes haemorrhagic gastroenteritis in sheep and goats with up to 90% mortality (Marczinke and Nichol, 2002); fortunately it only induces a febrile illness in man (Rao *et al.*, 1981). CCHFV causes haemorrhagic fever (HF) in humans with up to 50% mortality in Africa, southern and eastern Europe and parts of Asia (Whitehouse, 2004). Erve virus has been associated with 'thunderclap headaches' involving subarachnoid haemorrhaging (Treib *et al.*, 1998). Although DUGV is technically not a haemorrhagic fever virus, it shares considerable genetic content with CCHFV and NSDV (Marczinke and Nichol, 2002), has similar cell tropism to CCHFV and results in thrombocytopaenia in man (Burt *et al.*, 1996) so many facets of the pathogenesis are likely to be similar. In order to tackle the interesting question of what molecular mechanisms are used by nairoviruses to induce their haemorrhagic effects, we need to look for precedents in viruses from other families that are better studied. Four virus families are commonly associated with HF in humans and other animals: arenaviruses, filoviruses, bunyaviruses and flaviviruses. This HF can vary in severity from relatively mild disease to frequently fatal conditions of shock similar to bacterial induced septic shock and multi organ failure (Aleksandrowicz *et al.*, 2008). In addition, two recent emerging viruses, the SARS coronavirus and the orthomyxovirus avian influenza virus, can also induce haemorrhagic effects (Muramoto *et al.*, 2006), which highlights the need for further research into this area.

There has been a significant increase in understanding of the pathogenesis of many HF viruses over the past few years, particularly for Dengue and the filoviruses *Ebola virus* and *Marburg virus* (Geisbert *et al.*, 2003a,b; Geisbert and Jahrling, 2004; Bray and Geisbert, 2005; Hoenen *et al.*, 2006). However, pathogenic mechanisms described for filoviruses will not be universally applicable due, for example, to differences in the glycoprotein mucin domains found to be important for pathogenicity in filoviruses (Yang *et al.*, 2000; Sanchez *et al.*, 2006). Peters *et al.* (1989) indicate that there are useful similarities between different HF viruses but that there are also significant differences in pathogenesis and clinical evolution between HF viruses from different families (Peters *et al.*, 1989). Within the *Bunyaviridae* family, HF is caused by several hantaviruses such as Hantaan, Andes or Dobrava virus, the phlebovirus RVFV and nairoviruses as discussed above, but these are probably the least well characterized viral HFs. In general, however, there are several common features of viral induced HF. In particular, macrophage involvement has been suggested for all HF viruses studied including *Ebola virus*, RVFV and CCHFV (Geisbert and Jahrling, 2004). Several early events in viral infection impact on host haemostasis: infection of macrophages and dendritic cells with subsequent release of cytokines; infection of endothelial and macrophage cells, induction of tissue factor and thus initiation of the host clotting cascade, and an increase in vascular permeability brought about either directly through viral infection or indirectly through cytokine action.

Nairoviruses are arboviruses, transmitted generally by ticks but also sometimes by mosquitoes or culicoid flies, (Marriott and Nuttall, 1996a), so their first port of entry in the (mammalian) host during natural infection is the blood stream. Thus cells of the bloodstream such as macrophages and dendritic cells are the likely first targets for infection. We have recently shown that DUGV can replicate *in vitro* within the murine J774 macrophage cell line (Bridgen, unpublished results). Evidence that CCHFV infects mononuclear phagocytes *in vivo* was provided by Burt *et al.* (1997) who demonstrated by *in situ* hybridization that the main target cells in CCHFV infected patients were mononuclear phagocytes, endothelial cells and hepatocytes. High levels of the cytokines IL-6, TNF-α and IL-10 in CCHFV patient sera, all of which are secreted by macrophages, are also highly suggestive of macrophage involvement. The levels of these cytokines were also higher in fatal CCHF cases (Ergonul *et al.*, 2006).

Interestingly, Connolly Anderson *et al.* (2009) found that monocyte derived dendritic cells (moDCs) were very permissive for CCHFV infection *in vitro*, much more so than B cells, T cells, NK cells or monocytes (Connolly Andersen *et al.*, 2009). Dendritic cells are key cells in defence as they are involved in the switch from innate to adaptive immunity; infection of dendritic cells could therefore hinder an adaptive immune response. Furthermore, these infected moDCs produced significantly more IL-6 and somewhat more IL-10 and TNF-α than uninfected cells. IL-8 and IL-19 levels remained unaffected. Supernatants from these infected moDCs also activated endothelial cells as indicated by intercellular adhesion molecule 1 (ICAM-1) mRNA levels (Connolly Andersen *et al.*, 2009), suggesting that this is a contributor to endothelial damage. The findings for IL-6 and TNF-α are interesting as these are the two cytokines most commonly associated with CCHFV infection (Ergonul *et al.*, 2006; Papa *et al.*, 2006). Of these, TNF-α levels appear to be associated with disease severity whereas IL-6 levels are more variable.

Endothelial cell involvement is also typical of HF viruses (Geisbert and Jahrling, 2004), although opinions differ on whether direct infection or indirect cytokine action is responsible for subsequent vessel damage, and there is variation in the extent of endothelial involvement from virus to virus. Evidence for nairovirus endothelial involvement was indicated by several earlier studies (Joubert *et al.*, 1985; Burt *et al.*, 1997; Schnittler and Feldmann, 2003). CCHFV replicates *in vitro* in both HUVEC endothelial and HuH-7 hepatocyte cells (Andersson *et al.*, 2006). DUGV also replicates in HUVEC endothelial cells (unpublished results).

What effects does the infection of macrophages, dendritic cells and endothelial cells have? Infection of macrophages activates them to secrete cytokines such as TNF-α and IL-6.

This in turn has several effects: there is tissue damage which may lead to vascular leakage, other cells of the immune system are activated and Tissue Factor (TF) gene expression is induced in both macrophage and endothelial cells. TF expression is mediated by both IL-6 and TNF-α, but more effectively by TNF-α. Upstream elements in the TF gene promoter bind the transcription factors AP1, NF-κB, Sp1 and Egr-1. Of these, NF-κB and AP-1 induce TF expression in macrophages and endothelial cells, and can do so synergistically in human endothelial cells (Oeth *et al.*, 1997; Liu *et al.*, 2004). IL-6 only activates the NF-κB pathway, while TNF-α can activate both the NF-κB and AP-1 (c-Jun) pathways. Blood flowing over cells expressing TF initiates coagulation via interactions with factor VIIa then either factor IX or factor X (Hoffbrand *et al.*, 2005). Unfortunately, HF viruses tend to overstimulate this pathway since increased inflammation increases coagulation, which in turn increases inflammation (Esmon, 2005). This may lead to disseminated intravascular coagulation (DIC) as well as depletion of platelets and clotting factors, which then results in haemorrhaging. So far, there has been no investigation into the induction of TF expression by nairoviruses.

Finally, direct infection of the liver also has effects on haemostasis. Nairoviruses can infect hepatocytes (Burt *et al.*, 1997). Liver damage could result in loss of clotting factors, particularly factors II, VII and X (Hoffbrand *et al.*, 2005) and reduced thrombopoietin levels, reducing platelet synthesis and perturbing blood clotting even further. Liver enzymes and other acute phase proteins are raised during CCHFV infections (Joubert *et al.*, 1985).

Drug and supportive therapy to treat nairoviral infection

The only nairovirus for which drug or treatment therapy has been investigated is CCHFV. Cases of CCHFV may be treated by supportive therapy, use of specific IgG and/or corticosteroids (Ergonul, 2008), as well as by ribavarin or interferon therapy. Ribavirin, (1-β-d-ribofuranosyl-1,2,4-triazole-3-carboxamide), is a ribonucleoside analogue used in the treatment of hepatitis C and respiratory syncital virus. It is the only drug that has been used extensively against any nairovirus to date. It interferes with RNA synthesis by inhibiting the conversion of inosine 5′-monophosphate to xanthothene 5′- monophosphate resulting in a reduced supply of guanosine 5′-monophosphate and thus GTP, affecting RNA synthesis (Leyssen *et al.*, 2008). However, the story is not this simple, since its efficacy is very different for different virus families. Ribavirin inhibits CCHFV *in vitro* (Watts *et al.*, 1989); recently it was shown to inhibit minigenome replication (Bergeron *et al.*, 2010) and to be protective at low dose in a small animal model (Bente *et al.*, 2010). It has been used in multiple small *in vivo* observational studies using either oral or intravenous drug where its use appears to be beneficial (Ergonul, 2008; Leyssen *et al.*, 2008). However, a recent quasi-experimental study from Turkey suggested that the drug is of limited benefit in patients (Elaldi *et al.*, 2009).

The more we understand about the processes of nairoviral entry, replication, and egress, the more we can develop inhibitors that are very specific to nairoviruses or the cell processes used by these viruses in entering and replicating within cells. The first studies along these lines came from Nichol's group at the CDC studying M protein processing. Researchers in this group suggested that the SKI-1/S1P enzyme represents a good candidate to be targeted by antivirals, as it catalyses a fundamental step in the processing of the surface glycoproteins of CCHFV and other viruses (Vincent *et al.*, 2003; Pullikotil *et al.*, 2004; Sanchez *et al.*, 2006; Bergeron *et al.*, 2007). A protein-based inhibitor of the cellular SKI-1 enzyme has already been reported and its use results in blockage of the proteolytic cleavage of the viral glycoprotein and inhibition of CCHFV spread (Pullikotil *et al.*, 2004). The caveat of using antivirals based on cellular proteins is that they have to target the function of this protein used by the virus without affecting its cellular functions. Thus, although promising, antivirals based on the SKI-1 enzyme are still in their very early stages of development.

The characterization of the nairovirus OTU domain and the development of *in vitro* assays for enzymatic activities as described above offer a new viral target to screen for potential inhibitors specific for CCHFV-L and other OTU-containing viral proteins. High-throughput screening of chemical compound libraries has proved to be a valuable tool in the identification of small molecules inhibitors of other viral proteases such as HCV-NS3-NS4A (Sudo *et al.*, 2005) and SARS-3CLpro (Blanchard *et al.*, 2004) and are currently being performed for CCHFV-L. Difficulties may arise in trying to develop OTU inhibitors that are specific for the viral protease and do not affect cellular DUBs. However, since mammalian DUBs seem to have different specificities, it might be possible to design specific inhibitors of the viral OTU domains. For this reason, structural studies of CCHFV-L OTU are ongoing and analysis of the structural data may potentially reveal differential mechanisms of recognition or catalysis between cellular and viral OTU proteins. This information could direct a rational approach to design structure-based drugs that specifically inhibit the viral OTU proteases, which may result in the specific inhibition of CCHFV replication.

Simon *et al.* (2009b) have investigated CCHFV entry, finding that the virus uses clathrin-dependent endocytosis for entry. siRNA against clathrin had a marked effect on CCHFV N protein levels and also S RNA levels, indicating the role of clathrin coated pits in virus entry. They made use of this information to inhibit clathrin pit formation using 0.3M sucrose or 10 μg/ml chlorpromazine, which both reduce the number of coated pit-associated receptors at the cell surface; both of these drugs reduced CCHFV N protein expression. Drugs that prevent endosome acidification (50 nM bafilomycin A1 and 25 mM NH_4Cl) were also found to inhibit CCHFV. In addition, they also found that cholesterol depletion in the cell plasma membrane significantly reduced CCHFV infection. Two other studies by the same group (Simon *et al.*, 2006, 2009a), describe investigations into the effects of exogenously added nitric oxide and microtubule inhibitors, finding both to be effective in reducing CCHFV yield or egress, respectively.

Other drugs which may be developed in the future could include those targeting the viral endonuclease responsible for 'cap-snatching' as has been tried for *Influenza A virus* (Tomassini *et al.*, 1996). Another influenza A drug, T-705, which inhibits the RNA-dependent RNA polymerase and is less toxic and more active than ribavirin, inhibits the bunyaviruses La Crosse virus, RVFV and Punto Toro virus, though no nairoviruses have been tested (Furuta *et al.*, 2009). Finally, it is likely that RNA based therapies such as RNAi or antisense RNAs, with viral or cellular gene targets, will be developed for use *in vivo* as delivery methods improve (Spurgers *et al.*, 2008).

For all these studies we urgently need a good *in vivo* model to study the course of the disease and to test therapeutic interventions and vaccine candidates. Although CCHFV can cause disease and death in rodents, it does not induce haemorrhaging. Recently, a stat-1 knockout mouse model has been developed for CCHFV (Bente *et al.*, 2010). Drs Baron and Bridgen are using an alternative approach using sheep to study HF in NSDV-infected animals at the Institute of Animal Health at Pirbright, Surrey, UK, funded by the BBSRC.

Perspectives

The next decade of nairovirus research promises to be an exciting one, with current investigations being completed more fully and with new areas of research opening out. These are our predictions for areas of development.

- Reverse genetics: there are now two reports of mini-genome replication. The most recent utilized a cloned L segment cDNA indicating that it is possible to maintain a stable clone. It should thus be just a matter of time before whole virus is rescued from cDNA clones, thus opening up the possibility of investigating individual gene function.
- Following on from the above, it should also be possible to dissect the processes of replication and transcription, as well as other putative roles of the L segment such as the role of the OTU domain *in vivo* and mRNA cap snatching, which by analogy with orthomyxoviruses is likely to be an L protein function (Rao *et al.*, 2003; Guilligay *et al.*, 2008; Dias *et al.*, 2009). The processes by which the 5′ triphosphate is removed to avoid RIG-I activation may also receive some attention.
- Pathogenesis, in particular an understanding of pathogenic determinants, for example the role of the furin domain in pathogenesis: this will also include development of suitable animal models of disease, as well as genomic and proteomic means of studying host responses activated by the viruses.
- Drug therapies targeting both viral proteins and cellular processes used by the virus. The latter may employ genome wide RNAi screens or similar. The development of suitable animal models also allows potential targets to be additionally screened *in vivo*.
- Imaging of whole virions: this has recently been achieved at high resolution for two other members of the *Bunyaviridae* family, the phleboviruses RVFV and Uukuniemi virus (Freiberg *et al.*, 2008; Overby *et al.*, 2008; Huiskonen *et al.*, 2009) and will likely be extended to a representative virus within the *Nairovirus* genus for comparative purposes.
- Increase in the breadth of viruses studied: many may be capable of causing disease that we are currently unaware of, and it is likely that more than just NSDV and CCHFV affect the haemostatic response.
- Vector–host interactions: the increase in the understanding of arthropod gene function is making analysis of virus–arthropod host interactions more possible to study.

In their book chapter on nairoviruses, Marriott and Nuttall pointed to the importance of determining further nairoviral sequences and also the pathways of M processing, two areas which have largely been fulfilled (Marriott and Nuttall, 1996a). We hope that we may be as successful in our predictions. Whether we are or not, it is clear that our understanding of nairovirus molecular biology in another decade will far exceed what we now know, and we look forward to more exciting and sometimes unexpected findings such as has been the case for the M segment processing and IFN evasion strategies that have been uncovered during this past decade.

Acknowledgements

The authors wish to thank Claudio Alarcón for his artistic work on Figs. 5.1–5.3, Dr Michael Baron for his comments on the manuscript and Dr Stuart Nichol for permission to modify and update a figure from Sanchez *et al.* (2006).

References

Ahmed, A.A., McFalls, J.M., Hoffmann, C., Filone, C.M., Stewart, S.M., Paragas, J., Khodjaev, S., Shermukhamedova, D., Schmaljohn, C.S., Doms, R.W., *et al.* (2005). Presence of broadly reactive and group-specific neutralizing epitopes on newly described isolates of Crimean–Congo hemorrhagic fever virus. J. Gen. Virol. *86*, 3327–3336.

Aleksandrowicz, P., Wolf, K., Falzarano, D., Feldmann, H., Seebach, J., and Schnittler, H. (2008). Viral haemorrhagic fever and vascular alterations. Hamostaseologie *28*, 77–84.

Alff, P.J., Gavrilovskaya, I.N., Gorbunova, E., Endriss, K., Chong, Y., Geimonen, E., Sen, N., Reich, N.C., and Mackow, E.R. (2006). The pathogenic NY-1 hantavirus G1 cytoplasmic tail inhibits RIG-I- and TBK-1-directed interferon responses. J. Virol. *80*, 9676–9686.

Allende, R., Lewis, T.L., Lu, Z., Rock, D.L., Kutish, G.F., Ali, A., Doster, A.R., and Osorio, F.A. (1999). North American and European porcine reproductive and respiratory syndrome viruses differ in non-structural protein coding regions. J. Gen. Virol. *80*, 307–315.

Altamura, L.A., Bertolotti-Ciarlet, A., Teigler, J., Paragas, J., Schmaljohn, C.S., and Doms, R.W. (2007). Identification of a novel C-terminal cleavage of Crimean–Congo hemorrhagic fever virus PreGN that leads to generation of an NSM protein. J. Virol. *81*, 6632–6642.

Andersson, I., Bladh, L., Mousavi-Jazi, M., Magnusson, K.E., Lundkvist, A., Haller, O., and Mirazimi, A. (2004). Human MxA protein inhibits the replication of Crimean–Congo hemorrhagic fever virus. J. Virol. *78*, 4323–9.

Andersson, I., Lundkvist, A., Haller, O., and Mirazimi, A. (2006). Type I interferon inhibits Crimean–Congo hemorrhagic fever virus in human target cells. J. Med. Virol. *78*, 216–222.

Andersson, I., Karlberg, H., Mousavi-Jazi, M., Martinez-Sobrido, L., Weber, F., and Mirazimi, A. (2008). Crimean–Congo hemorrhagic fever virus delays activation of the innate immune response. J. Med. Virol. *80*, 1397–404.

Begum, F., Wisseman, C.L.,Jr, and Casals, J. (1970). Tick-borne viruses of West Pakistan. II. Hazara virus, a new agent isolated from Ixodes redikorzevi ticks from the Kaghan Valley, W. Pakistan. Am. J. Epidemiol. *92*, 192–194.

Bente, D.A., Alimonti, J.B., Shieh, W.J., Camus, G., Ströher U, Zaki, S. and Jones, S.M. (2010). Pathogenesis and immune response of Crimean-Congo hemorrhagic fever virus in a STAT-1 knockout mouse model. J. Virol. *84*, 11089–11100.

Bergeron, E., Vincent, M.J., and Nichol, S.T. (2007). Crimean–Congo hemorrhagic fever virus glycoprotein processing by the endoprotease SKI-1/S1P is critical for virus infectivity. J. Virol. *81*, 13271–13276.

Bergeron, E., Albariño, C.G., Khristova, M.L. and Nichol, S.T. (2010). Crimean-Congo hemorrhagic fever virus-encoded ovarian tumor protease activity is dispensable for virus RNA polymerase function. J Virol. *84*, 216–226.

Bertolotti-Ciarlet, A., Smith, J., Strecker, K., Paragas, J., Altamura, L.A., McFalls, J.M., Frias-Staheli, N., Garcia-Sastre, A., Schmaljohn, C.S., and Doms, R.W. (2005). Cellular localization and antigenic characterization of Crimean–Congo hemorrhagic fever virus glycoproteins. J. Virol. *79*, 6152–6161.

Billecocq, A., Gauliard, N., Le May, N., Elliott, R.M., Flick, R., and Bouloy, M. (2008). RNA polymerase I-mediated expression of viral RNA for the rescue of infectious virulent and avirulent Rift Valley fever viruses. Virology *378*, 377–384.

Blakqori, G., and Weber, F. (2005). Efficient cDNA-based rescue of La Crosse bunyaviruses expressing or lacking the nonstructural protein NSs. J. Virol. *79*, 10420–10428.

Blakqori, G., van Knippenberg, I., and Elliott, R.M. (2009). Bunyamwera orthobunyavirus S-segment untranslated regions mediate poly(A) tail-independent translation. J. Virol. *83*, 3637–3646.

Blanchard, J.E., Elowe, N.H., Huitema, C., Fortin, P.D., Cechetto, J.D., Eltis, L.D., and Brown, E.D. (2004). High-throughput screening identifies inhibitors of the SARS coronavirus main proteinase. Chem Biol *11*, 1445–1453.

Boone, D.L., Turer, E.E., Lee, E.G. Ahmad, R.C., Wheeler, M.T., Tsui, C., Hurley, P., Chien, M., Chai, S., Hitotsumatsu, O., *et al.* (2004). The ubiquitin-modifying enzyme A20 is required for termination of Toll-like receptor responses. Nat. Immunol. *5*, 1052–1060.

Bouloy, M., Janzen, C., Vialat, P., Khun, H., Pavlovic, J., Huerre, M., and Haller, O. (2001). Genetic evidence for an interferon-antagonistic function of rift valley fever virus nonstructural protein NSs. J. Virol. *75*, 1371–1377.

Boyd, A., Fazakerley, J.K., and Bridgen, A. (2006). Pathogenesis of Dugbe virus infection in wild-type and interferon-deficient mice. J. Gen. Virol. *87*, 2005–2009.

Bray, M., and Geisbert, T.W. (2005). Ebola virus: the role of macrophages and dendritic cells in the pathogenesis of Ebola hemorrhagic fever. Int. J. Biochem. Cell Biol. *37,* 1560–1566.
Bridgen, A., and Elliott, R.M. (1996). Rescue of a segmented negative-strand RNA virus entirely from cloned complementary DNAs. Proc. Natl. Acad. Sci. U.S.A. *93,* 15400–15404.
Bridgen, A., Weber, F., Fazakerley, J.K., and Elliott, R.M. (2001). Bunyamwera bunyavirus nonstructural protein NSs is a nonessential gene product that contributes to viral pathogenesis. Proc. Natl. Acad. Sci. U.S.A. *98,* 664–9.
Bridgen, A., Dalrymple, D.A., and Elliott, R.M. (2002). Dugbe nairovirus S segment: correction of published sequence and comparison of five isolates. Virology *294,* 364–371.
Bridgen, A., Dalrymple, D.A., Weber, F., and Elliott, R.M. (2004). Inhibition of Dugbe nairovirus replication by human MxA protein. Virus Res. *99,* 47–50.
Buckley, A., Higgs, S., and and Gould, E.A. (1990). Dugbe virus susceptibility to neutralization by monoclonal antibodies as a marker of virulence in mice. Arch. Virol (*Suppl 1*), 197–205.
Burt, F.J., Spencer, D.C., Leman, P.A., Patterson, B., and Swanepoel, R. (1996). Investigation of tick-borne viruses as pathogens of humans in South Africa and evidence of Dugbe virus infection in a patient with prolonged thrombocytopenia. Epidemiol. Infect. *116,* 353–361.
Burt, F.J., Swanepoel, R., Shieh, W.J., Smith, J.F., Leman, P.A., Greer, P.W., Coffield, L.M., Rollin, P.E., Ksiazek, T.G., Peters, C.J., *et al.* (1997). Immunohistochemical and *in situ* localization of Crimean–Congo hemorrhagic fever (CCHF) virus in human tissues and implications for CCHF pathogenesis. Arch. Pathol. Lab. Med. *121,* 839–846.
Casals, J., and Tignor, G.H. (1980). The Nairovirus genus: serological relationships. Intervirology *14,* 144–147.
Cash, P. (1985). Polypeptide synthesis of Dugbe virus, a member of the Nairovirus genus of the *Bunyaviridae*. J. Gen. Virol. *66 (Pt 1),* 141–148.
Chastel, C., Le Goff, F., and Le Lay, G. (1983). Antigenic variants of the Soldado virus (Nairovirus, *Bunyaviridae*) isolated in different parts of the world. Acta Virol. *27,* 51–58.
Chen, Z.J. (2005). Ubiquitin signalling in the NF-kappaB pathway. Nat. Cell Biol. *7,* 758–765.
Chin, K.C., and Cresswell, P. (2001). Viperin (cig5), an IFN-inducible antiviral protein directly induced by human cytomegalovirus. Proc. Natl. Acad. Sci. U.S.A. *98,* 15125–15130.
Clerx, J.P., Casals, J., and Bishop, D.H. (1981). Structural characteristics of nairoviruses (genus Nairovirus, *Bunyaviridae*). J. Gen. Virol. *55,* 165–178.
Coates, D.M., and Sweet, C. (1990). Studies on the pathogenicity of a nairovirus, Dugbe virus, in normal and immunosuppressed mice. J. Gen. Virol. *71,* 325–332.
Connolly Andersen, A.M., Douagi, I., Kraus, A.A., and Mirazimi, A. (2009). Crimean Congo hemorrhagic fever virus infects human monocyte-derived dendritic cells. Virology *390,* 157–162.
Crabtree, M.B., Sang, R., and Miller, B.R. (2009). Kupe virus, a new virus in the family bunyaviridae, genus nairovirus, kenya. Emerg. Infect. Dis. *15,* 147–154.
Cui, S., Eisenacher, K., Kirchhofer, A., Brzozka, K., Lammens, A., Lammens, K., Fujita, T., Conzelmann, K.K., Krug, A., and Hopfner, K.P. (2008). The C-terminal regulatory domain is the RNA 5′-triphosphate sensor of RIG-I. Mol. Cell *29,* 169–179.
Daubney, R., and Hudson, J.R. (1934). Nairobi sheep disease. Parassitologia *23,* 507–524.
Davies, F.G., Otieno, S., and Jessett, D.M. (1977). The antibody response in sheep vaccinated with experimental Nairobi sheep disease vaccines. Trop. Anim. Health Prod. *9,* 181–183.
Der, S.D., Zhou, A., Williams, B.R., and Silverman, R.H. (1998). Identification of genes differentially regulated by interferon alpha, beta, or gamma using oligonucleotide arrays. Proc. Natl. Acad. Sci. U.S.A. *95,* 15623–15628.
Deyde, V.M., Khristova, M.L., Rollin, P.E., Ksiazek, T.G., and Nichol, S.T. (2006). Crimean–Congo hemorrhagic fever virus genomics and global diversity. J. Virol. *80,* 8834–8842.
Dias, A., Bouvier, D., Crepin, T., McCarthy, A.A., Hart, D.J., Baudin, F., Cusack, S., and Ruigrok, R.W. (2009). The cap-snatching endonuclease of influenza virus polymerase resides in the PA subunit. Nature *458,* 914–918.
Doherty, R.L., Carley, J.G., Murray, M.D., Main, A.J.,Jr, Kay, B.H., and Domrow, R. (1975). Isolation of arboviruses (Kemerovo group, Sakhalin group) from Ixodes uriae collected at Macquarie Island, Southern ocean. Am. J. Trop. Med. Hyg. *24,* 521–526.
Duh, D., Nichol, S.T., Khristova, M.L., Saksida, A., Hafner-Bratkovic, I., Petrovec, M., Dedushaj, I., Ahmeti, S., and Avsic-Zupanc, T. (2008). The complete genome sequence of a Crimean–Congo hemorrhagic fever virus isolated from an endemic region in Kosovo. Virol. J. *5,* 7.

Elaldi, N., Bodur, H., Ascioglu, S., Celikbas, A., Ozkurt, Z., Vahaboglu, H., Leblebicioglu, H., Yilmaz, N., Engin, A., Sencan, M., *et al.* (2009). Efficacy of oral ribavirin treatment in Crimean–Congo haemorrhagic fever: a quasi-experimental study from Turkey. J. Infect. *58,* 238–244.

Elliott, R.M. (2009). Bunyaviruses and climate change. Clin. Microbiol. Infect. *15,* 510–517.

Elliott, R.M., Bouloy, M., Calisher, C.H., Goldbach, R., Moyer, J.T., Nichol, S.T., Pettersson, R., Plyusnin, A., and Schmaljohn, C.S. (2000). *Bunyaviridae.* In Virus Taxonomy. Seventh Report of the International Committee on Taxonomy of Viruses. (San Diego: Academic Press), pp. 599–621.

Ergonul, O. (2006). Crimean–Congo haemorrhagic fever. Lancet Infect. Dis. *6,* 203–214.

Ergonul, O. (2008). Treatment of Crimean–Congo hemorrhagic fever. Antiviral Res. *78,* 125–131.

Ergonul, O., Tuncbilek, S., Baykam, N., Celikbas, A., and Dokuzoguz, B. (2006). Evaluation of serum levels of interleukin (IL)–6, IL-10, and tumor necrosis factor-alpha in patients with Crimean–Congo hemorrhagic fever. J. Infect. Dis. *193,* 941–944.

Erickson, B.R., Deyde, V., Sanchez, A.J., Vincent, M.J., and Nichol, S.T. (2007). N-linked glycosylation of Gn (but not Gc) is important for Crimean Congo hemorrhagic fever virus glycoprotein localization and transport. Virology *361,* 348–355.

Esmon, C.T. (2005). The interactions between inflammation and coagulation. Br. J. Haematol. *131,* 417–430.

Estrada-Pena, A., Zatansever, Z., Gargili, A., Aktas, M., Uzun, R., Ergonul, O., and Jongejan, F. (2007). Modeling the spatial distribution of Crimean–Congo hemorrhagic fever outbreaks in Turkey. Vector Borne Zoonotic Dis. *7,* 667–678.

Flick, R., and Whitehouse, C.A. (2005). Crimean–Congo hemorrhagic fever virus. Curr. Mol. Med. *5,* 753–760.

Flick, R., Flick, K., Feldmann, H., and Elgh, F. (2003). Reverse genetics for Crimean–Congo hemorrhagic fever virus. J. Virol. *77,* 5997–6006.

Foulke, R.S., Rosato, R.R., and French, G.R. (1981). Structural polypeptides of Hazara virus. J. Gen. Virol. *53,* 169–172.

Freiberg, A.N., Sherman, M.B., Morais, M.C., Holbrook, M.R., and Watowich, S.J. (2008). Three-dimensional organization of Rift Valley fever virus revealed by cryoelectron tomography. J. Virol. *82,* 10341–10348.

Frese, M., Kochs, G., Feldmann, H., Hertkorn, C., and Haller, O. (1996). Inhibition of bunyaviruses, phleboviruses, and hantaviruses by human MxA protein. J. Virol. *70,* 915–923.

Frias-Staheli, N., Giannakopoulos, N.V., Kikkert, M., Taylor, S.L., Bridgen, A., Paragas, J., Richt, J.A., Rowland, R.R., Schmaljohn, C.S., Lenschow, D.J., *et al.* (2007). Ovarian tumor domain-containing viral proteases evade ubiquitin- and ISG15-dependent innate immune responses. Cell Host Microbe *2,* 404–416.

Furuta, Y., Takahashi, K., Shiraki, K., Sakamoto, K., Smee, D.F., Barnard, D.L., Gowen, B.B., Julander, J.G., and Morrey, J.D. (2009). T-705 (favipiravir) and related compounds: Novel broad-spectrum inhibitors of RNA viral infections. Antiviral Res. *82,* 95–102.

Gack, M.U., Shin, Y.C., Joo, C.H., Urano, T., Liang, C., Sun, L., Takeuchi, O., Akira, S., Chen, Z., Inoue, S., *et al.* (2007). TRIM25 RING-finger E3 ubiquitin ligase is essential for RIG-I-mediated antiviral activity. Nature *446,* 916–920.

Garcia-Sastre, A., and Biron, C.A. (2006). Type 1 interferons and the virus–host relationship: a lesson in detente. Science *312,* 879–882.

Garcin, D., Lezzi, M., Dobbs, M., Elliott, R.M., Schmaljohn, C., Kang, C.Y., and Kolakofsky, D. (1995). The 5′ ends of Hantaan virus (*Bunyaviridae*) RNAs suggest a prime-and-realign mechanism for the initiation of RNA synthesis. J. Virol. *69,* 5754–5762.

Geisbert, T.W., and Jahrling, P.B. (2004). Exotic emerging viral diseases: progress and challenges. Nat. Med. *10,* S110–121.

Geisbert, T.W., Hensley, L.E., Jahrling, P.B., Larsen, T., Geisbert, J.B., Paragas, J., Young, H.A., Fredeking, T.M., Rote, W.E., and Vlasuk, G.P. (2003a). Treatment of Ebola virus infection with a recombinant inhibitor of factor VIIa/tissue factor: a study in rhesus monkeys. Lancet *362,* 1953–1958.

Geisbert, T.W., Young, H.A., Jahrling, P.B., Davis, K.J., Kagan, E., and Hensley, L.E. (2003b). Mechanisms underlying coagulation abnormalities in ebola hemorrhagic fever: overexpression of tissue factor in primate monocytes/macrophages is a key event. J. Infect. Dis. *188,* 1618–1629.

George, C.X., Li, Z., Okonski, K.M., Toth, A.M., Wang, Y., and Samuel, C.E. (2009). Tipping the balance: antagonism of PKR kinase and ADAR1 deaminase functions by virus gene products. J. Interferon Cytokine Res. *29,* 477–87.

Giannakopoulos, N.V., Luo, J.K., Papov, V., Zou, W., Lenschow, D.J., Jacobs, B.S., Borden, E.C., Li, J., Virgin, H.W., and Zhang, D.E. (2005). Proteomic identification of proteins conjugated to ISG15 in mouse and human cells. Biochem. Biophys. Res. Commun. *336,* 496–506.

el-Ghorr, A.A., Marriott, A.C., Ward, V.K., Booth, T.F., Higgs, S., and Nuttall, P.A. (1990). Characterisation of Dugbe virus (*Nairovirus, Bunyaviridae*) by biochemical and immunological procedures using monoclonal antibodies. Arch. Virol. (Suppl .1) 169–179.

Gonzalez, J.P., Wilson, M.L., Cornet, J.P., and Camicas, J.L. (1995). Host-passage-induced phenotypic changes in Crimean–Congo haemorrhagic fever virus. Res. Virol. *146,* 131–140.

Gonzalez, J.P., Camicas, J.L., Cornet, J.P., and Wilson, M.L. (1998). Biological and clinical responses of west African sheep to Crimean–Congo haemorrhagic fever virus experimental infection. Res. Virol. *149,* 445–455.

Gould, E.A., Chanas, A.C., Buckley, A., and Varma, M.G. (1983). Immunofluorescence studies on the antigenic interrelationships of the Hughes virus serogroup (genus Nairovirus) and identification of a new strain. J. Gen. Virol. *64 (Pt 3),* 739–742.

Guilligay, D., Tarendeau, F., Resa-Infante, P., Coloma, R., Crepin, T., Sehr, P., Lewis, J., Ruigrok, R.W., Ortin, J., Hart, D.J., *et al.* (2008). The structural basis for cap binding by influenza virus polymerase subunit PB2. Nat. Struct. Mol. Biol. *15,* 500–506.

Habjan, M., Andersson, I., Klingstrom, J., Schumann, M., Martin, A., Zimmermann, P., Wagner, V., Pichlmair, A., Schneider, U., Muhlberger, E., *et al.* (2008). Processing of genome 5′ termini as a strategy of negative-strand RNA viruses to avoid RIG-I-dependent interferon induction. PLoS One *3,* e2032.

Haller, O., Frese, M., Rost, D., Nuttall, P.A., and Kochs, G. (1995). Tick-borne thogoto virus infection in mice is inhibited by the orthomyxovirus resistance gene product Mx1. J. Virol. *69,* 2596–2601.

Haller, O., Staeheli, P., and Kochs, G. (2007). Interferon-induced Mx proteins in antiviral host defense. Biochimie *89,* 812–818.

Harcourt, B.H., Jukneliene, D., Kanjanahaluethai, A., Bechill, J., Severson, K.M., Smith, C.M., Rota, P.A., and Baker, S.C. (2004). Identification of severe acute respiratory syndrome coronavirus replicase products and characterization of papain-like protease activity. J. Virol. *78,* 13600–13612.

Hewson, R., Chamberlain, J., Mioulet, V., Lloyd, G., Jamil, B., Hasan, R., Gmyl, A., Gmyl, L., Smirnova, S.E., Lukashev, A., *et al.* (2004a). Crimean–Congo haemorrhagic fever virus: sequence analysis of the small RNA segments from a collection of viruses world wide. Virus Res. *102,* 185–189.

Hewson, R., Gmyl, A., Gmyl, L., Smirnova, S.E., Karganova, G., Jamil, B., Hasan, R., Chamberlain, J., and Clegg, C. (2004b). Evidence of segment reassortment in Crimean–Congo haemorrhagic fever virus. J. Gen. Virol. *85,* 3059–3070.

Hoenen, T., Groseth, A., Falzarano, D., and Feldmann, H. (2006). Ebola virus: unravelling pathogenesis to combat a deadly disease. Trends Mol. Med. *12,* 206–215.

Hoffbrand, V.A., Catovsky, D., and Tuddenham, E.G.T. (2005). Postgraduate Haematology, 5th edition. (Blackwell Publishing, Oxford, UK).

Honig, J.E., Osborne, J.C., and Nichol, S.T. (2004a). Crimean–Congo hemorrhagic fever virus genome L RNA segment and encoded protein. Virology *321,* 29–35.

Honig, J.E., Osborne, J.C., and Nichol, S.T. (2004b). The high genetic variation of viruses of the genus Nairovirus reflects the diversity of their predominant tick hosts. Virology *318,* 10–16.

Hoogstraal, H. (1979). The epidemiology of tick-borne Crimean–Congo hemorrhagic fever in Asia, Europe, and Africa. J. Med. Entomol. *15,* 307–417.

Hornung, V., Ellegast, J., Kim, S., Brzozka, K., Jung, A., Kato, H., Poeck, H., Akira, S., Conzelmann, K.K., Schlee, M., *et al.* (2006). 5′-Triphosphate RNA is the ligand for RIG-I. Science *314,* 994–997.

Hu, M., Li, P., Li, M., Li, W., Yao, T., Wu, J.W., Gu, W., Cohen, R.E., and Shi, Y. (2002). Crystal structure of a UBP-family deubiquitinating enzyme in isolation and in complex with ubiquitin aldehyde. Cell *111,* 1041–1054.

Hu, M., Li, P., Song, L., Jeffrey, P.D., Chenova, T.A., Wilkinson, K.D., Cohen, R.E., and Shi, Y. (2005). Structure and mechanisms of the proteasome-associated deubiquitinating enzyme USP14. EMBO J. *24,* 3747–56.

Huiskonen, J.T., Overby, A.K., Weber, F., and Grunewald, K. (2009). Electron cryo-microscopy and single-particle averaging of Rift Valley fever virus: evidence for GN-GC glycoprotein heterodimers. J. Virol. *83,* 3762–3769.

Ikegami, T., Won, S., Peters, C.J., and Makino, S. (2006). Rescue of infectious rift valley fever virus entirely from cDNA, analysis of virus lacking the NSs gene, and expression of a foreign gene. J. Virol. *80,* 2933–2940.

Jaaskelainen, K.M., Kaukinen, P., Minskaya, E.S., Plyusnina, A., Vapalahti, O., Elliott, R.M., Weber, F., Vaheri, A., and Plyusnin, A. (2007). Tula and Puumala hantavirus NSs ORFs are functional and the products inhibit activation of the interferon-beta promoter. J. Med. Virol. *79,* 1527–36.

Jin, H., and Elliott, R.M. (1993). Non-viral sequences at the 5′ ends of Dugbe nairovirus S mRNAs. J. Gen. Virol. *74 (Pt 10),* 2293–2297.

Joshi, M.V., Geevarghese, G., Joshi, G.D., Ghodke, Y.S., Mourya, D.T., and Mishra, A.C. (2005). Isolation of Ganjam virus from ticks collected off domestic animals around Pune, Maharashtra, India. J. Med. Entomol. *42,* 204–206.

Joubert, J.R., King, J.B., Rossouw, D.J., and Cooper, R. (1985). A nosocomial outbreak of Crimean–Congo haemorrhagic fever at Tygerberg Hospital. Part III. Clinical pathology and pathogenesis. S. Afr. Med. J. *68,* 722–728.

Kanerva, M., Melen, K., Vaheri, A., and Julkunen, I. (1996). Inhibition of puumala and tula hantaviruses in Vero cells by MxA protein. Virology *224,* 55–62.

Kattenhorn, L.M., Korbel, G.A., Kessler, B.M., Spooner, E., and Ploegh, H.L. (2005). A deubiquitinating enzyme encoded by HSV-1 belongs to a family of cysteine proteases that is conserved across the family Herpesviridae. Mol. Cell *19,* 547–557.

Khaiboullina, S.F., Rizvanov, A.A., Deyde, V.M., and St Jeor, S.C. (2005). Andes virus stimulates interferon-inducible MxA protein expression in endothelial cells. J. Med. Virol. *75,* 267–275.

Kochs, G., Janzen, C., Hohenberg, H., and Haller, O. (2002). Antivirally active MxA protein sequesters La Crosse virus nucleocapsid protein into perinuclear complexes. Proc. Natl. Acad. Sci. U.S.A. *99,* 3153–3158.

Koyama, S., Ishii, K.J., Coban, C., and Akira, S. (2008). Innate immune response to viral infection. Cytokine *43,* 336–341.

Le Bon, A., and Tough, D.F. (2002). Links between innate and adaptive immunity via type I interferon. Curr Opin Immunol *14,* 432–436.

Lenschow, D.J., Giannakopoulos, N.V., Gunn, L.J., Johnston, C., O'Guin, A.K., Schmidt, R.E., Levine, B., and Virgin, H.W.t. (2005). Identification of interferon-stimulated gene 15 as an antiviral molecule during Sindbis virus infection *in vivo*. J. Virol. *79,* 13974–13983.

Lenschow, D.J., Lai, C., Frias-Staheli, N., Giannakopoulos, N.V., Lutz, A., Wolff, T., Osiak, A., Levine, B., Schmidt, R.E., Garcia-Sastre, A., *et al.* (2007). IFN-stimulated gene 15 functions as a critical antiviral molecule against influenza, herpes, and Sindbis viruses. Proc. Natl. Acad. Sci. U.S.A. *104,* 1371–1376.

Lenz, O., ter Meulen, J., Klenk, H.D., Seidah, N.G., and Garten, W. (2001). The Lassa virus glycoprotein precursor GP-C is proteolytically processed by subtilase SKI-1/S1P. Proc. Natl. Acad. Sci. U.S.A. *98,* 12701–12705.

Leyssen, P., De Clercq, E., and Neyts, J. (2008). Molecular strategies to inhibit the replication of RNA viruses. Antiviral Res. *78,* 9–25.

Li, K., Foy, E., Ferreon, J.C., Nakamura, M., Ferreon, A.C., Ikeda, M., Ray, S.C., M., G.,Jr, and Lemon, S.M. (2005). Immune evasion by hepatitis C virus NS3/4A protease-mediated cleavage of the Toll-like receptor 3 adaptor protein TRIF. Proc. Natl. Acad. Sci. U.S.A. *102,* 2992–2997.

Lin, R., Lacoste, J., Nakhaei, P., Sun, Q., Yang, L., Paz, S., Wilkinson, P., Julkunen, I., Vitour, D., Meurs, E., *et al.* (2006). Dissociation of a MAVS/IPS-1/VISA/Cardif-IKKepsilon molecular complex from the mitochondrial outer membrane by hepatitis C virus NS3-4A proteolytic cleavage. J. Virol. *80,* 6072–6083.

Lindner, H.A., Fotouhi-Ardakani, N., Lytvyn, V., Lachance, P., Sulea, T., and Menard, R. (2005). The papain-like protease from the severe acute respiratory syndrome coronavirus is a deubiquitinating enzyme. J. Virol. *79,* 15199–15208.

Lindner, H.A., Lytvyn, V., Qi, H., Lachance, P., Ziomek, E., and Menard, R. (2007). Selectivity in ISG15 and ubiquitin recognition by the SARS coronavirus papain-like protease. Arch Biochem Biophys *466,* 8–14.

Liu, Y., Pelekanakis, K., and Woolkalis, M.J. (2004). Thrombin and tumor necrosis factor alpha synergistically stimulate tissue factor expression in human endothelial cells: regulation through c-Fos and c-Jun. J. Biol. Chem. *279,* 36142–36147.

Liu, Y.C., Penninger, J., and Karin, M. (2005). Immunity by ubiquitylation: a reversible process of modification. Nat. Rev. Immunol. *5,* 941–952.

Loureiro, J., Lilley, B.N., Spooner, E., Noriega, V., Tortorella, D., and Ploegh, H.L. (2006). Signal peptide peptidase is required for dislocation from the endoplasmic reticulum. Nature *441,* 894–897.

Lvov, D.K., Timopheeva, A.A., Smirnov, V.A., Gromashevsky, V.L., Sidorova, G.A., Nikiforov, L.P., Sazonov, A.A., Andreev, A.P., Skvortzova, T.M., Beresina, L.K., *et al.* (1975). Ecology of tick-borne viruses in colonies of birds in the USSR. Med. Biol. *53,* 325–330.
Main, A.J., Downs, W.G., Shope, R.E., and Wallis, R.C. (1976). Avalon and Clo Mor: two new Sakhalin group viruses from the North Atlantic. J. Med. Entomol. *13,* 309–315.
Major, L., Linn, M.L., Slade, R.W., Schroder, W.A., Hyatt, A.D., Gardner, J., Cowley, J., and Suhrbier, A. (2009). Ticks associated with macquarie island penguins carry arboviruses from four genera. PLoS One *4,* e4375.
Makarova, K.S., Aravind, L., and Koonin, E.V. (2000). A novel superfamily of predicted cysteine proteases from eukaryotes, viruses and Chlamydia pneumoniae. Trends Biochem Sci *25,* 50–52.
Malakhov, M.P., Malakhova, O.A., Kim, K.I., Ritchie, K.J., and Zhang, D.E. (2002). UBP43 (USP18) specifically removes ISG15 from conjugated proteins. J. Biol. Chem. *277,* 9976–9981.
Marczinke, B.I., and Nichol, S.T. (2002). Nairobi sheep disease virus, an important tick-borne pathogen of sheep and goats in Africa, is also present in Asia. Virology *303,* 146–151.
Marriott, A.C., and Nuttall, P.A. (1992). Comparison of the S RNA segments and nucleoprotein sequences of Crimean–Congo hemorrhagic fever, Hazara, and Dugbe viruses. Virology *189,* 795–799.
Marriott, A.C., and Nuttall, P.A. (1996a). Molecular biology of nairoviruses. In The *Bunyaviridae,* Elliott, R.M. ed. (Plenum Press, New York), pp. 91–104.
Marriott, A.C., and Nuttall, P.A. (1996b). Large RNA segment of Dugbe nairovirus encodes the putative RNA polymerase. J. Gen. Virol. *77 (Pt 8),* 1775–1780.
Marriott, A.C., el-Ghorr, A.A., and Nuttall, P.A. (1992). Dugbe Nairovirus M RNA: nucleotide sequence and coding strategy. Virology *190,* 606–615.
Meurs, E.F., Watanabe, Y., Kadereit, S., Barber, G.N., Katze, M.G., Chong, K., Williams, B.R., and Hovanessian, A.G. (1992). Constitutive expression of human double-stranded RNA-activated p68 kinase in murine cells mediates phosphorylation of eukaryotic initiation factor 2 and partial resistance to encephalomyocarditis virus growth. J. Virol. *66,* 5805–5814.
Morikawa, S., Saijo, M., and Kurane, I. (2007). Recent progress in molecular biology of Crimean–Congo hemorrhagic fever. Comp. Immunol. Microbiol. Infect. Dis. *30,* 375–389.
Muramoto, Y., Ozaki, H., Takada, A., Park, C.H., Sunden, Y., Umemura, T., Kawaoka, Y., Matsuda, H., and Kida, H. (2006). Highly pathogenic H5N1 influenza virus causes coagulopathy in chickens. Microbiol. Immunol. *50,* 73–81.
Neumann, G., and Kawaoka, Y. (2001). Reverse genetics of influenza virus. Virology *287,* 243–250.
Nichol, S.T. (2001). Bunyaviruses. In Fields Virology, Knipe, D.M., and Howley, P.M., eds. (Lippincott Williams and Wilkins, Philadelphia), pp. 1603–1633.
Oeth, P., Parry, G.C., and Mackman, N. (1997). Regulation of the tissue factor gene in human monocytic cells. Role of AP-1, NF-kappa B/Rel, and Sp1 proteins in uninduced and lipopolysaccharide-induced expression. Arterioscler. Thromb. Vasc. Biol. *17,* 365–374.
Ogawa, Y., Sugiura, K., Kato, K., Tohya, Y., and Akashi, H. (2007). Rescue of Akabane virus (family *Bunyaviridae*) entirely from cloned cDNAs by using RNA polymerase I. J. Gen. Virol. *88,* 3385–3390.
Okumura, A., Lu, G., Pitha-Rowe, I., and Pitha, P.M. (2006). Innate antiviral response targets HIV-1 release by the induction of ubiquitin-like protein ISG15. Proc. Natl. Acad. Sci. U.S.A. *103,* 1440–1445.
Okumura, A., Alce, T., Lubyova, B., Ezelle, H., Strebel, K., and Pitha, P.M. (2008a). HIV-1 accessory proteins VPR and Vif modulate antiviral response by targeting IRF-3 for degradation. Virology *373,* 85–97.
Okumura, A., Pitha, P.M., and Harty, R.N. (2008b). ISG15 inhibits Ebola VP40 VLP budding in an L-domain-dependent manner by blocking Nedd4 ligase activity. Proc. Natl. Acad. Sci. U.S.A. *105,* 3974–3979.
Osiak, A., Utermohlen, O., Niendorf, S., Horak, I., and Knobeloch, K.P. (2005). ISG15, an interferon-stimulated ubiquitin-like protein, is not essential for STAT1 signaling and responses against vesicular stomatitis and lymphocytic choriomeningitis virus. Mol. Cell Biol. *25,* 6338–6345.
Overby, A.K., Pettersson, R.F., Grunewald, K., and Huiskonen, J.T. (2008). Insights into bunyavirus architecture from electron cryotomography of Uukuniemi virus. Proc. Natl. Acad. Sci. U.S.A. *105,* 2375–2379.
Paladino, P., Cummings, D.T., Noyce, R.S., and Mossman, K.L. (2006). The IFN-independent response to virus particle entry provides a first line of antiviral defense that is independent of TLRs and retinoic acid-inducible gene I. J. Immunol. *177,* 8008–8016.

Papa, A., Ma, B., Kouidou, S., Tang, Q., Hang, C., and Antoniadis, A. (2002). Genetic characterization of the M RNA segment of Crimean–Congo hemorrhagic fever virus strains, China. Emerg. Infect. Dis. *8,* 50–53.

Papa, A., Bino, S., Velo, E., Harxhi, A., Kota, M., and Antoniadis, A. (2006). Cytokine levels in Crimean–Congo hemorrhagic fever. J. Clin. Virol. *36,* 272–276.

Park, M.S., Shaw, M.L., Munoz-Jordan, J., Cros, J.F., Nakaya, T., Bouvier, N., Palese, P., Garcia-Sastre, A., and Basler, C.F. (2003). Newcastle disease virus (NDV)-based assay demonstrates interferon-antagonist activity for the NDV V protein and the Nipah virus V, W, and C proteins. J. Virol. *77,* 1501–1511.

Peters, C.J., Liu, C.T., Anderson, G.W.,Jr, Morrill, J.C., and Jahrling, P.B. (1989). Pathogenesis of viral hemorrhagic fevers: Rift Valley fever and Lassa fever contrasted. Rev. Infect. Dis. *11 (Suppl 4),* S743–749.

Petroski, M.D., and Deshaies, R.J. (2005). Function and regulation of cullin-RING ubiquitin ligases. Nat. Rev. Mol. Cell Biol. *6,* 9–20.

Pichlmair, A., Schulz, O., Tan, C.P., Naslund, T.I., Liljestrom, P., Weber, F., and Reis e Sousa, C. (2006). RIG-I-mediated antiviral responses to single-stranded RNA bearing 5′-phosphates. Science *314,* 997–1001.

Prescott, J., Ye, C., Sen, G., and Hjelle, B. (2005). Induction of innate immune response genes by Sin Nombre hantavirus does not require viral replication. J. Virol. *79,* 15007–15015.

Prescott, J.B., Hall, P.R., Bondu-Hawkins, V.S., Ye, C., and Hjelle, B. (2007). Early innate immune responses to Sin Nombre hantavirus occur independently of IFN regulatory factor 3, characterized pattern recognition receptors, and viral entry. J. Immunol. *179,* 1796–1802.

Pullikotil, P., Vincent, M., Nichol, S.T., and Seidah, N.G. (2004). Development of protein-based inhibitors of the proprotein of convertase SKI-1/S1P: processing of SREBP-2, ATF6, and a viral glycoprotein. J. Biol. Chem. *279,* 17338–17347.

Rao, C.V., Dandawate, C.N., Rodrigues, J.J., Rao, G.L., Mandke, V.B., Ghalsasi, G.R., and Pinto, B.D. (1981). Laboratory infections with Ganjam virus. Indian J. Med. Res. *74,* 319–324.

Rao, P., Yuan, W., and Krug, R.M. (2003). Crucial role of CA cleavage sites in the cap-snatching mechanism for initiating viral mRNA synthesis. EMBO J. *22,* 1188–1198.

Reichelt, M., Stertz, S., Krijnse-Locker, J., Haller, O., and Kochs, G. (2004). Missorting of LaCrosse virus nucleocapsid protein by the interferon-induced MxA GTPase involves smooth ER membranes. Traffic *5,* 772–784.

Ritchie, K.J., and Zhang, D.E. (2004). ISG15: the immunological kin of ubiquitin. Semin. Cell Dev. Biol. *15,* 237–246.

Ryman, K.D., Meier, K.C., Nangle, E.M., Ragsdale, S.L., Korneeva, N.L., Rhoads, R.E., MacDonald, M.R., and Klimstra, W.B. (2005). Sindbis virus translation is inhibited by a PKR/RNase L-independent effector induced by alpha/beta interferon priming of dendritic cells. J. Virol. *79,* 1487–1499.

Samuel, C.E. (2001). Antiviral actions of interferons. Clin. Microbiol. Rev. *14,* 778–809.

Sanchez, A.J., Vincent, M.J., and Nichol, S.T. (2002). Characterization of the glycoproteins of Crimean–Congo hemorrhagic fever virus. J. Virol. *76,* 7263–7275.

Sanchez, A.J., Vincent, M.J., Erickson, B.R., and Nichol, S.T. (2006). Crimean–Congo hemorrhagic fever virus glycoprotein precursor is cleaved by Furin-like and SKI-1 proteases to generate a novel 38-kilodalton glycoprotein. J. Virol. *80,* 514–525.

Schmaljohn, C.S., and Hooper, J.W. (2001). *Bunyaviridae*: The viruses and their replication. In Fields Virology, Knipe, D.M., and Howley, P.M., eds. (Lippincott Williams and Wilkins, Philadelphia), pp. 1581–1602.

Schmidt, M.H., and Dikic, I. (2006). Ubiquitin and NEDD8: brothers in arms. Sci STKE *362,* pe 50.

Schneider, U., Martin, A., Schwemmle, M., and Staeheli, P. (2007). Genome trimming by Borna disease viruses: viral replication control or escape from cellular surveillance? Cell Mol Life Sci *64,* 1038–42.

Schnittler, H.J., and Feldmann, H. (2003). Viral hemorrhagic fever--a vascular disease? Thromb. Haemost. *89,* 967–972.

Shin, J.S., Ebersold, M., Pypaert, M., Delamarre, L., Hartley, A., and Mellman, I. (2006). Surface expression of MHC class II in dendritic cells is controlled by regulated ubiquitination. Nature *444,* 115–118.

Simmons, G., Wool-Lewis, R.J., Baribaud, F., Netter, R.C., and Bates, P. (2002). Ebola virus glycoproteins induce global surface protein down-modulation and loss of cell adherence. J. Virol. *76,* 2518–2528.

Simon, M., Falk, K.I., Lundkvist, A., and Mirazimi, A. (2006). Exogenous nitric oxide inhibits Crimean–Congo hemorrhagic fever virus. Virus Res. *120,* 184–190.

Simon, M., Johansson, C., Lundkvist, A., and Mirazimi, A. (2009a). Microtubule-dependent and microtubule-independent steps in Crimean–Congo hemorrhagic fever virus replication cycle. Virology *385,* 313–322.

Simon, M., Johansson, C., and Mirazimi, A. (2009b). Crimean–Congo hemorrhagic fever virus entry and replication is clathrin-, pH- and cholesterol-dependent. J. Gen. Virol. *90,* 210–215.

Smirnova, S.E., Shestopalova, N.M., Reingold, V.N., Zubri, G.L., and Chumakov, M.P. (1977). Experimental Hazara Virus infection in mice. Acta Virol. *21,* 128–132.

Snijder, E., Wassenaar, A.L., Den Boon, J.A., and Spaan, W.J. (1995). Proteolytic processing of the arterivirus replicase. Adv. Exp. Med. Biol. *380,* 443–451.

Snijder, E.J., Wassenaar, A.L., van Dinten, L.C., Spaan, W.J., and Gorbalenya, A.E. (1996). The arterivirus nsp4 protease is the prototype of a novel group of chymotrypsin-like enzymes, the 3C-like serine proteases. J. Biol. Chem. *271,* 4864–4871.

Spiropoulou, C.F., Albarino, C.G., Ksiazek, T.G., and Rollin, P.E. (2007). Andes and Prospect Hill hantaviruses differ in early induction of interferon although both can downregulate interferon signaling. J. Virol. *81,* 2769–2776.

Spurgers, K.B., Sharkey, C.M., Warfield, K.L., and Bavari, S. (2008). Oligonucleotide antiviral therapeutics: antisense and RNA interference for highly pathogenic RNA viruses. Antiviral Res. *78,* 26–36.

Sudo, K., Yamaji, K., Kawamura, K., Nishijima, T., Kojima, N., Aibe, K., Shimotohno, K., and Shimizu, Y. (2005). High-throughput screening of low molecular weight NS3-NS4A protease inhibitors using a fluorescence resonance energy transfer substrate. Antivir. Chem. Chemother. *16,* 385–392.

Tamura, M., Asada, H., Kondo, K., Takahashi, M., and Yamanishi, K. (1987). Effects of human and murine interferons against hemorrhagic fever with renal syndrome (HFRS) virus (Hantaan virus). Antiviral Res. *8,* 171–178.

Temonen, M., Lankinen, H., Vapalahti, O., Ronni, T., Julkunen, I., and Vaheri, A. (1995). Effect of interferon-alpha and cell differentiation on Puumala virus infection in human monocyte/macrophages. Virology *206,* 8–15.

tenOever, B.R., Servant, M.J., Grandvaux, N., Lin, R., and Hiscott, J. (2002). Recognition of the measles virus nucleocapsid as a mechanism of IRF-3 activation. J. Virol. *76,* 3659–3669.

tenOever, B.R., Sharma, S., Zou, W., Sun, Q., Grandvaux, N., Julkunen, I., Hemmi, H., Yamamoto, M., Akira, S., Yeh, W.C., *et al.* (2004). Activation of TBK1 and IKKvarepsilon kinases by vesicular stomatitis virus infection and the role of viral ribonucleoprotein in the development of interferon antiviral immunity. J. Virol. *78,* 10636–10649.

Tergaonkar, V. (2006). NFkappaB pathway: a good signaling paradigm and therapeutic target. Int. J. Biochem. Cell Biol. *38,* 1647–1653.

Tignor, G.H., and Hanham, C.A. (1993). Ribavirin efficacy in an *in vivo* model of Crimean–Congo hemorrhagic fever virus (CCHF) infection. Antiviral Res. *22,* 309–325.

Tomassini, J.E., Davies, M.E., Hastings, J.C., Lingham, R., Mojena, M., Raghoobar, S.L., Singh, S.B., Tkacz, J.S., and Goetz, M.A. (1996). A novel antiviral agent which inhibits the endonuclease of influenza viruses. Antimicrob. Agents Chemother. *40,* 1189–1193.

Treib, J., Dobler, G., Haass, A., von Blohn, W., Strittmatter, M., Pindur, G., Froesner, G., and Schimrigk, K. (1998). Thunderclap headache caused by Erve virus? Neurology *50,* 509–511.

de Veer, M.J., Holko, M., Frevel, M., Walker, E., Der, S., Paranjape, J.M., Silverman, R.H., and Williams, B.R. (2001). Functional classification of interferon-stimulated genes identified using microarrays. J. Leukoc. Biol. *69,* 912–920.

Vertegaal, A.C. (2007). Small ubiquitin-related modifiers in chains. Biochem. Soc. Trans. *35,* 1422–1423.

Vincent, M.J., Sanchez, A.J., Erickson, B.R., Basak, A., Chretien, M., Seidah, N.G., and Nichol, S.T. (2003). Crimean–Congo hemorrhagic fever virus glycoprotein proteolytic processing by subtilase SKI-1. J. Virol. *77,* 8640–8649.

Vorou, R.M. (2009). Crimean–Congo hemorrhagic fever in southeastern Europe. Int. J. Infect. Dis. *13,* 659–662.

Wang, X., Hinson, E.R., and Cresswell, P. (2007). The interferon-inducible protein viperin inhibits influenza virus release by perturbing lipid rafts. Cell Host Microbe *2,* 96–105.

Watret, G.E., Pringle, C.R., and Elliott, R.M. (1985). Synthesis of bunyavirus-specific proteins in a continuous cell line (XTC-2) derived from Xenopus laevis. J. Gen. Virol. *66 (Pt 3),* 473–482.

Watts, D.M., Ussery, M.A., Nash, D., and Peters, C.J. (1989). Inhibition of Crimean–Congo hemorrhagic fever viral infectivity yields *in vitro* by ribavirin. Am. J. Trop. Med. Hyg. *41,* 581–585.

Weber, F., and Mirazimi, A. (2008). Interferon and cytokine responses to Crimean Congo hemorrhagic fever virus; an emerging and neglected viral zonoosis. Cytokine Growth Factor Rev. *19,* 395–404.

Weber, F., Wagner, V., Rasmussen, S.B., Hartmann, R., and Paludan, S.R. (2006). Double-stranded RNA is produced by positive-strand RNA viruses and DNA viruses but not in detectable amounts by negative-strand RNA viruses. J. Virol. *80,* 5059–5064.

Welchman, R.L., Gordon, C., and Mayer, R.J. (2005). Ubiquitin and ubiquitin-like proteins as multifunctional signals. Nat. Rev. Mol. Cell Biol. *6,* 599–609.

Wertz, I.E. O'Rourke, K.M., Zhou, H., Eby, M., Aravind, L., Seshagiri, S., Wu, P., Wiesmann, C., Baker, R., Boone, D.L., *et al.* (2004). De-ubiquitination and ubiquitin ligase domains of A20 downregulate NF-kappaB signalling. Nature *430,* 694–699.

Whitehouse, C.A. (2004). Crimean–Congo hemorrhagic fever. Antiviral Res. *64,* 145–160.

Wilson, M.L., Gonzalez, J.P., Cornet, J.P., and Camicas, J.L. (1991). Transmission of Crimean–Congo haemorrhagic fever virus from experimentally infected sheep to Hyalomma truncatum ticks. Res. Virol. *142,* 395–404.

Xia, Z.P., Sun, L., Chen, X., Pineda, G., Jiang, X., Adhikari, A., Zeng, W., and Chen, Z.J. (2009). Direct activation of protein kinases by unanchored polyubiquitin chains. Nature *461,* 114–119.

Yadav, P.D., Vincent, M.J., Khristova, M., Kale. C., Nichol, S.T., Mishra, A.C., Mourya, D.T. (2011). Genomic analysis reveals Nairobi sheep disease virus to be highly diverse and present in both Africa, and in India in the form of the Ganjam virus variant. Infect Genet Evol. *11.* doi:10.1016/j.meegid.2011.04.001.

Yang, Z.Y., Duckers, H.J., Sullivan, N.J., Sanchez, A., Nabel, E.G., and Nabel, G.J. (2000). Identification of the Ebola virus glycoprotein as the main viral determinant of vascular cell cytotoxicity and injury. Nat. Med. *6,* 886–889.

Yoneyama, M., and Fujita, T. (2007a). Cytoplasmic double-stranded DNA sensor. Nat. Immunol. *8,* 907–908.

Yoneyama, M., and Fujita, T. (2007b). Function of RIG-I-like receptors in antiviral innate immunity. J. Biol. Chem. *282,* 15315–15318.

Yoneyama, M., and Fujita, T. (2007c). RIG-I family RNA helicases: cytoplasmic sensor for antiviral innate immunity. Cytokine Growth Factor Rev. *18,* 545–551.

Yoneyama, M., Kikuchi, M., Natsukawa, T., Shinobu, N., Imaizumi, T., Miyagishi, M., Taira, K., Akira, S., and Fujita, T. (2004). The RNA helicase RIG-I has an essential function in double-stranded RNA-induced innate antiviral responses. Nat. Immunol. *5,* 730–737.

Yoneyama, M., Onomoto, K., and Fujita, T. (2008). Cytoplasmic recognition of RNA. Adv. Drug Deliv. Rev. *60,* 841–846.

Yuan, W., Aramini, J.M., Montelione, G.T., and Krug, R.M. (2002). Structural basis for ubiquitin-like ISG 15 protein binding to the NS1 protein of influenza B virus: a protein–protein interaction function that is not shared by the corresponding N-terminal domain of the NS1 protein of influenza A virus. Virology *304,* 291–301.

Yuan, W., and Krug, R.M. (2001). Influenza B virus NS1 protein inhibits conjugation of the interferon (IFN)-induced ubiquitin-like ISG15 protein. EMBO J. *20,* 362–371.

Zeller, H.G., Karabatsos, N., Calisher, C.H., Digoutte, J.P., Cropp, C.B., Murphy, F.A., and Shope, R.E. (1989). Electron microscopic and antigenic studies of uncharacterized viruses. II. Evidence suggesting the placement of viruses in the family *Bunyaviridae.* Arch. Virol. *108,* 211–227.

Zhao, C., Denison, C., Huibregtse, J.M., Gygi, S., and Krug, R.M. (2005). Human ISG15 conjugation targets both IFN-induced and constitutively expressed proteins functioning in diverse cellular pathways. Proc. Natl. Acad. Sci. U.S.A. *102,* 10200–10205.

Zhou, A., Paranjape, J., Brown, T.L., Nie, H., Naik, S., Dong, B., Chang, A., Trapp, B., Fairchild, R., Colmenares, C., *et al.* (1997). Interferon action and apoptosis are defective in mice devoid of 2′,5′-oligoadenylate-dependent RNase L. EMBO J. *16,* 6355–6363.

Zhou, A., Paranjape, J.M., Der, S.D., Williams, B.R., and Silverman, R.H. (1999). Interferon action in triply deficient mice reveals the existence of alternative antiviral pathways. Virology *258,* 435–440.

Ziebuhr, J., Snijder, E.J., and Gorbalenya, A.E. (2000). Virus-encoded proteinases and proteolytic processing in the Nidovirales. J. Gen. Virol. *81,* 853–879.

The Molecular Biology of Tospoviruses and Resistance Strategies

6

Richard Kormelink

Abstract

Within the family of *Bunyaviridae* tospoviruses represent the genus of plant-pathogenic viruses. Viruses of this genus are only transmitted by thrips in a propagative manner. Their host range includes economically important agricultural and horticultural crops but also ornamentals. The type species of the tospoviruses is *Tomato spotted wilt virus* (TSWV), from which the siglum to spo has been derived. TSWV is not only the best studied but also the most important member of this genus, having a host range that expands more than 800 different plant species within more than 80 different families, monocots as well as dicots (Goldbach and Peters, 1994; Peters, 1998). Tospoviruses have a worldwide distribution and not only occur in (sub)tropical regions, but also in temperate climate regions where they are more restricted to greenhouse cultivations. Among plant viruses, tospoviruses rank among the top ten of economically most important viruses wordwide with annual crop losses up to more than one billion US$. (Goldbach and Peters, 1994). In this chapter the current status will be presented on the molecular biology of Tospoviruses and disease management strategies using natural, RNAi- and protein-mediated transgenic resistance strategies.

Taxonomy, vector transmission and geographical distribution

While for a long time TSWV was thought to be the sole member of this genus, the availability of more powerful diagnostics and extensive surveys revealed the existence of an increasing number of additional species. In 1996 tospovirologists agreed to set N protein sequence divergence (10%) as one of the most important demarcation criteria for the establishment of a new tospovirus species (Goldbach and Kuo, 1996). Next to this, host rang and vector species are important biological denominators. Currently, 21 tospovirus species are being recognized (Table 6.1). From all these, eight have been listed as established species by the ICTV (2005) and six, i.e. *Capsicum chlorosis virus* (CaCV), *Chrysanthemum stem necrosis virus* (CSNV), *Iris yellow spot virus* (IYSV), *Melon yellow spot virus* (MYSV), *Peanut chlorotic fanspot virus* (PCFV), and *Watermelon bud necrosis virus* (WBNV), are still listed as tentative (Yeh *et al.*, 1992; Chen and Chui, 1996; Resence *et al.*, 1996; Cortez *et al.*, 1998, 2001; Kato and Hanada, 2000; McMichael *et al.*, 2002). Next to these, *Alstroemeria necrotic spot virus*

In memory of Rob Goldbach, who died 7 April 2009.

Table 6.1 Tospovirus species with their thrips vectors and geographical distribution

Tospovirus species	Vector species	Geographical distribution
Alstroemeria necrotic streak virus (ANSV)[45]	*Frankliniella occidentalis*	Colombia
Capsicum chlorosis virus (CaCV)[1]	*Ceratothripoides claratris*[42]	Australia, Thailand
Calla lily chlorotic spot virus	?[51]	Taiwan
Chrysanthemum stem necrosis virus (CSNV)[2]	*F. occidentalis*[24]	Brazil
Groundnut bud necrosis virus (GBNV)[3,4]	*Thrips palmi*[25–27]; *F. schultzei*[25,28]	India[4], South East Asia
Groundnut ringspot virus (GRSV)[5]	*F. gemina*[43]; *F. occidentalis*[29]; *F. schultzei*[29]	South America[15], South Africa
Impatiens necrotic spot virus (INSV)	*F. occidentalis*[30,31]	USA[7], west and south Europe[6,16–18]
Iris yellow spot virus (IYSV)[8]	*T. tabaci* [32]	Brazil[2], Israel[19], the Netherlands[8], USA
Melon severe mosaic virus (MeSMV)[46]	?[51]	Mexico
Melon yellow spot virus (MYSV)[9]	*T. palmi*[32]	Taiwan[9], Japan[20]
Peanut chlorotic fan-spot virus (PCFV)[10]	*Scirtothrips dorsalis*[10]	Taiwan
Peanut yellow spot virus (PYSV)[11]	?[51]	India[11], Thailand[11]
Polygonum ringspot virus (PolRSV)[47]	?[51]	Italy
Tomato chlorotic spot virus (TCSV)[5]	*F. occidentalis*[29], *F. intonsa*[29], *F. schultzei*[29]	South America[12,21]
Tomato necrotic ringspot virus (TNRV)[48]	?[51]	Thailand
Tomato spotted wilt virus (TSWV)[12,13]	*F. bispinosa*[34]; *F. cephalica*[44]; *F. gemina*[43]; *F. fusca*[33]; *F. intonsa*[29]; *F. occidentalis*[35]; *F. schultzei*[29,36]; *F. setosus*[37,38]; *T. tabaci*[39]	Worldwide[22]

Tomato yellow ring virus (TYRV)[49]	*T. tabaci*	Iran
Tomato zonate spot virus (TZSV)[50]	?[51]	China
Watermelon bud necrosis virus (WBNV)	*T. palmi*[40]	India
Watermelon silver mottle virus (WSMoV)[14]	*T. palmi*[40]	Japan[23], Taiwan[14]
Zucchini lethal chlorotic virus (ZLCV)[2]	*F. zucchini*[41]	Brazil[2]

[1]McMichael *et al.* (2002); [2]Resende *et al.* (1996); [3]Reddy *et al.* (1992); [4]Satyanarayana *et al.* (1996a,b); [5]de Ávila *et al.* (1993); [6]de Ávila *et al.* (1992); [7]Law and Moyer (1990); [8]Cortês *et al.* (1998); [9]Kato and Hanada (2000); [1°]Chen and Chiu (1996); [11]Reddy *et al.* (1991); [12]Francki *et al.* (1991); [13]de Haan (1991); [14]Yeh and Chang (1995); [15]de Ávila *et al.* (1990); [16]Vaira *et al.* (1993); [17]Marchoux *et al.* (1991); [18]Louro (1996); [19] Gera *et al.* (1998a); [20]Takeuchi *et al.* (2001; [21]Granval de Millan *et al.* (1998); [22]Goldbach and Peters (1994);[23] Kameya-Iwaki *et al.* (1984); Nagata and de Ávila (2000); [25]Lakshmi *et al.* (1995); [26] Palmer *et al.* (1990); [27]Vijayalakshmi (1994); [28]Amin *et al.* (1981); [29]Wijkamp *et al.* (1995); [30]DeAngellis *et al.* (1994); [31]Wijkamp and Peters (1993); [32]Gera *et al.* (1998b); [33]Sakimura (1963); [34]Webb *et al.* (1998); [35]Gardner *et al.* (1935); [36]Samuel *et al.* (1930); [37]Fujisawa *et al.* (1988); [38]Tsuda *et al.* (1996); [39]Pittman (1927); [40]Yeh *et al.* (1992); [41]Nakahara and Monteiro (1999); [42]Premachandra *et al.* (2005);[43] Borbón *et al.* (2006); [44]Ohnishi *et al.* (2006); [45]Hassani-Mehraban *et al.* (2010) (GenBank accession no. GQ478668); [46]Ciuffo *et al.* (2009); [47]Ciuffo *et al.* (2008); [48]Hassani-Mehraban *et al.* (2011) (GenBank accession no. FJ946835); [49]Hassani-Mehraban *et al.* (2005); [50]Dong *et al.* (2008; Lin *et al.* (2005); [51]unknown.

The table presented here is a modified version from I. Wijkamp, PhD thesis (1995). Tospovirus species listed tentative or not listed yet by the ICTV are indicated in grey.

(ANSV), *Calla lily chlorotic spot virus (CCSV), Melon severe mosaic virus* (MeSMV), *Polygonum ringspot virus* (PolRSV), *Tomato necrotic ringspot virus* (TNRV), *Tomato yellow ring virus* (TYRV) and *Tomato zonate spot virus* (TZSV) have not or only recently been reported and hence are not listed yet (Hassani-Mehraban *et al.*, 2005, 2010, 2011; Lin *et al.*, 2005; Ciuffo *et al.*, 2008, 2009; Dong *et al.*, 2008). Some tospovirus species have a very broad host range and worldwide distribution, with TSWV as the best example, whereas others, such as *Iris yellow spot virus* (IYSV) and *Melon yellow spot virus* (MYSV), have a more limited host range and only seem to distribute more locally (Table 6.1).

Tospoviruses are transmitted in a propagative manner by 13 different thrips species (family *Thripidae;* Premachandra *et al.*, 2005; Whitfield *et al.*, 2005; de Borbón *et al.*, 2006; Ohnishi *et al.*, 2006) in which the virus also replicates (Ullman *et al.*, 1993; Wijkamp *et al.*, 1993). One of the most important thrips vectors is *Frankliniella occidentalis* Pergande, also known as California thrips, Western flower thrips or Alfalfa thrips. Global expansion of the latter species has assisted in the revival and worldwide spread of TSWV since 1980. The virus can only be transmitted after acquisition in the larval L1 and L2 stages. After a latency period, during which virus replication takes place, the virus is transmitted, mostly in the adult stages for the rest of the thrips life span (Wijkamp *et al.*, 1993; Wijkamp and Peters, 1993; Van de Wetering *et al.*, 1996). Adult thrips that have fed on infected plants, even after prolonged feeding periods, do not become viruliferous (Ullman *et al.*, 1989, 1992; Van de Wetering *et al.*, 1996). For a more detailed (and recent) description on the interaction of Tospoviruses and thrips see Whitfield *et al.* (2005). An overview of current tospoviruses, their vector species and geographical distribution is presented in Table 6.1.

Particle morphology and genome expression strategy

Like all members of the *Bunyaviridae*, TSWV virions are membrane bound and spherical with a diameter of 80–120 nm. The lipid envelop contains two spike proteins that consist of the viral glycoproteins Gn (58 kDa) and Gc (78 kDa). The core contains ribonucleoproteins (RNPs, or sometimes also referred to as nucleocapsids) which are composed of the viral genomic RNA tightly associated with nucleocapsid (N; 29 kDa) protein and small amounts of the viral RNA-dependent RNA polymerase (L protein; 331.5 kDa) (Mohamed *et al.*, 1973; Tas *et al.*, 1977; Van den Hurk *et al.*, 1977; Van Mohamed, 1981; Verkleij and Peters, 1983; Poelwijk *et al.*, 1993). Owing to complementary sequences at the genomic termini, the RNA segments form a panhandle structure and give the RNPs a pseudo-circular appearance (De Haan *et al.*, 1989; Kellman *et al.*, 2001). The genomic RNA itself is not infectious but requires at least the N and L protein to initiate a virus multiplication cycle.

The RNA genome of TSWV is single-stranded and tripartite, of which the RNA segments, according to their size, are denoted large (L), medium (M) and small (S) (Fig. 6.1). (Van den Hurk *et al.*, 1977; Verkleij *et al.*, 1982; De Haan *et al.*, 1989). The L RNA is 8.9 kb and of entire negative polarity, containing one large open reading frame (ORF) in the complementary strand that codes for the large RNA-dependent RNA polymerase (RdRp or L protein) of 331.5 kDa (De Haan *et al.*, 1991). The M and S RNA both have an ambisense gene arrangement, coding for two non-overlapping ORFs on opposite strands separated by a non-coding intergenic region (IR). The M RNA (4.8 kb) codes for the viral cell-to-cell movement protein (NSm; 33.6 kDa) in viral sense and in the viral complementary sense for the glycoprotein precursor (127.4 kDa) which is post-translationally cleaved into two

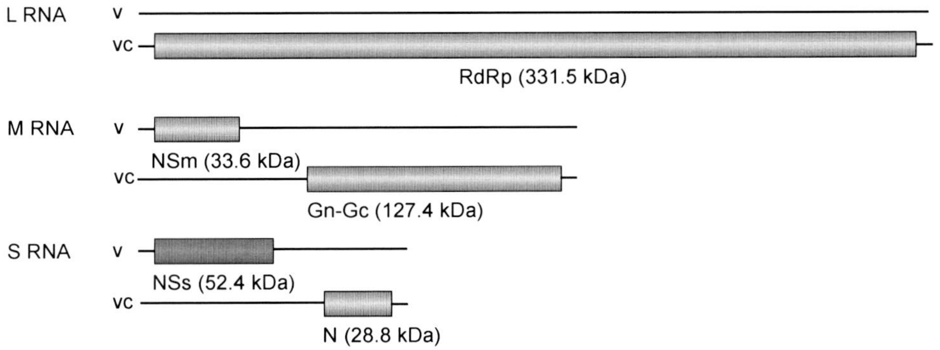

Figure 6.1 Genome organization of *Tomato spotted wilt virus*.

glycoproteins Gn (46 kDa) and Gc (75 kDa) (Kormelink *et al.*, 1992a, 1994; Storms *et al.*, 1995, 1998). Gn and Gc refer to the amino and carboxyl terminal topology within the precursor, in older literature formerly referred to as G2 (Gn) and G1 (Gc). The S RNA (2.9 kb) codes for the suppressor of RNA silencing NSs (52.4 kDa) in viral sense and the N protein (28.8 kDa) in viral complementary sense (Fig. 6.1) (De Haan *et al.*, 1990; Kormelink *et al.*, 1991; Takeda *et al.*, 2002; Bucher *et al.*, 2003).

The first eight nucleotides (nts) at the 3′ terminal end of all tospoviral RNA segments are conserved and complementary to the 5′ end (5′-AGAGCAAU and 3′-UCUGCUUA), a feature that is common for all segmented, negative-strand RNA viruses of the *Bunyaviridae, Orthomyxoviridae* and *Arenaviridae* (De Haan *et al.*, 1989). Moreover, the nucleotide sequence of the L, M and S RNA segments shows total nucleotide sequence complementarity of the 5′- and 3′-terminal ends for the first 15 nts (De Haan *et al.*, 1990, 1991; Kormelink *et al.*, 1992a). For each individual segment the terminal complementarity is extended to ~65 nts, allowing the formation of a so-called panhandle structure and thereby accounting for the pseudocircular appearance of the RNPs. The IR of several hundred nts within the M and S RNAs contains stretches highly rich in A and U residues and are predicted to fold into a stable hairpin structure (Fig. 6.2; De Haan *et al.*, 1990; Kormelink *et al.*, 1992a).

All ORFs of the M and S RNA become expressed via the synthesis of subgenomic-length mRNA molecules, which contain a cap-structure at their 5′- end followed by a heterogeneous, non-viral leader sequence (Fig. 6.2A; Kormelink *et al.*, 1992b,c; Van Poelwijk *et al.*, 1996; Okuda *et al.*, 2006). Since the virus is not capable to cap its own transcripts, initiation of genome transcription, typical for segmented negative-strand RNA viruses, occurs by cap-snatching. This process has been described first for Influenza virus (Bouloy *et al.*, 1978; Plotch *et al.*, 1979, 1981) but meanwhile demonstrated for viruses infecting hosts from both the plant and animal kingdoms and belonging to the *Bunyaviridae, Arenaviridae* and the floating genus *Tenuivirus* (Bellocq *et al.*, 1987; Raju *et al.*, 1990; Simons and Pettersson, 1991; Kormelink *et al.*, 1992c; Huiet *et al.*, 1993). During TSWV cap-snatching the viral polymerase complex cleaves cellular mRNAs and uses the capped-RNA leader sequence, 12–20 nts in size, to prime genome transcription (Duijsings *et al.*, 1999). Initial *in vivo* studies have shown that cellular mRNA molecules containing residues with base complementarity to the 3′ terminal residues of the viral RNA template positioned at around 16 nt from the capped 5′-end are most favourably being used during cap-snatching (Fig. 6.2B) (Duijsings *et al.*,

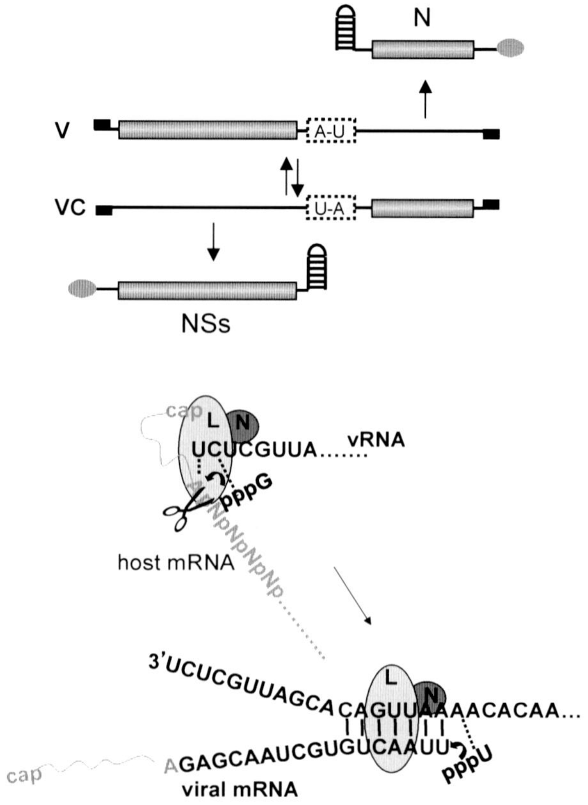

Figure 6.2 Structural features within the S RNA segment involved in replication and transcription (initiation/termination) (upper panel) and a model for transcription initiation (cap-snatching) through which *Tomato spotted wilt virus* (TSWV) renders translatable (capped) viral mRNAs (lower panel). Back boxes at the termini of the viral (v) and viral complementary (vc) genomic RNA strands represent the conserved complementary termini. The dashed A-U/U-A box within the intergenic region (IR) of the genomic RNA indicates the stretches highly rich in A and U residues. The grey circles at the 5′ ends of the NSs and N subgenomic length mRNA molecules represent the non-viral heterogeneous capped-leader sequences obtained by cap-snatching. The hairpin structure at the 3′ end of the viral mRNAs is the predicted folding of the A–U rich intergenic region sequence. Scissors indicate viral endonuclease cleavage of the cellular RNA leader sequence.

1999, 2001). *In vitro* studies on transcriptionally active virus preparations have supported these findings and, furthermore, showed that when cap donors with either single, double or multiple complementarity to the viral template are offered in pair-wise competition to TSWV *in vitro* transcription assays, the virus exhibits a strong preference for multiple-base pairing cap-donors (Van Knippenberg *et al.*, 2002, 2005a). Capped-RNA leader sequences, lacking a 3′ terminal A residue, but instead harbouring a 3′ terminal G residue are able to internally prime on the penultimate viral (C) residue. As a consequence from this internal priming, viral transcripts lack the first viral-templated A residue after the non-viral leader sequence (Duijsings *et al.*, 2001). Prime-and-realignment of capped-RNA leader sequences, during which a shortly extended capped-RNA leader sequence releases and re-aligns again by virtue of base pairing (Garcin and Kolakofsky, 1990), also occurs and leads to the presence of a repetitive AG-dinucleotide at the 5′-end of viral transcripts (Duijsings *et al.*, 1999, 2001; Van Knippenberg *et al.*, 2002, 2005a). Viral transcripts, in contrast to (anti)genomic RNA strands, are not being encapsidated by N protein (Kormelink *et al.*, 1992b).

The 3′ ends of viral mRNAs do not contain a common poly(A)-tail, and for TSWV S RNA encoded transcripts have shown the presence of a major part of the IR, including the A/U-rich hairpin structure sequence (van Knippenberg *et al.*, 2005b, Fig. 6.2A). The fact that transcription termination occurs at the end of the hairpin structure suggests that the mechanism of termination of these mRNAs is based on structure formation in the nascent transcript, in the same manner as prokaryotic transcription termination occurs (Iapalucci *et al.*, 1991; van Knippenberg *et al*, 2005b). Although the L RNA segment is of entire negative polarity and thus lacks an intergenic region, folding predictions indicated formation of a (bulged) hairpin structure 146–168 nts downstream the stop codon, supporting the idea that termination of all viral mRNAs is regulated in a similar fashion (Van Knippenberg *et al.*, 2005b).

To compensate for the absence of a poly(A)-tail in viral mRNAs, the (intergenic) hairpin structure is additionally speculated to be involved in stimulation of viral mRNA translation through 5′–3′ terminal interactions, as has been observed for 3′-end structural features of several other plant-infecting RNA viruses that lack a poly(A)-tail (Gallie and Kobayashi, 1994; Gallie, 1998; Neeleman *et al.*, 2001; Fabian and White, 2004; Leonard *et al.*, 2004; Matsuda and Dreher, 2004; Meulewaeter *et al.*, 2004).

Cytopathology, virus assembly and movement

Once viruliferous thrips feed on healthy plants, infectious virus particles become injected into the cytoplasm of the host cell. After RNP release, genome replication and transcription starts in the cytoplasm, likely in the areas that show up as moderately dense, amorphous masses and that have been reported as viroplasm (VP). Viroplasms often contain smaller complexes of high electron density either arranged in apparent chains or together in clusters (Fig. 6.3; Kitajima, 1965; Francki and Grivell, 1970; Milne, 1970; Ie, 1971; Francki and Hatta, 1981; Francki *et al.*, 1984; Kitajima *et al.*, 1992). These complexes have a diameter ranging from 30 to 120 nm and are thought to represent nucleocapsid aggregates (NCA). The composition of these complexes as functional and infectious RNPs has not been demonstrated, nor reasons for their apparent accumulation and failure to become enveloped into mature virus particles (Milne, 1970; Ie, 1971, 1982; Verkleij and Peters, 1983; Resende *et al.*, 1991). Whereas immunostaining has confirmed the presence of N protein in viroplasm, virtually no glycoproteins (Gn and Gc) are detected.

Newly formed viral RNPs either become membrane enveloped or move from cell-to-cell, facilitated by the NSm movement protein, to allow systemic spread of the infection. Envelopment of RNPs takes place at the Golgi that often shows up as curved and dilated stacks, and are referred to as paired parallel membranes (PPM; Kikkert *et al.*, 1997, 1999; Lawson *et al.*, 1996). RNPs concentrate at these stacks in concert with viral glycoproteins and likely triggered by an interaction between the cytoplasmic tails of the glycoproteins and the N protein from RNPs, leads to enwrapment of RNPs with an entire Golgi stack giving rise to doubly enveloped particles (DEVs; Milne, 1970; Francki *et al.*, 1984; Kitajima *et al.*, 1992;

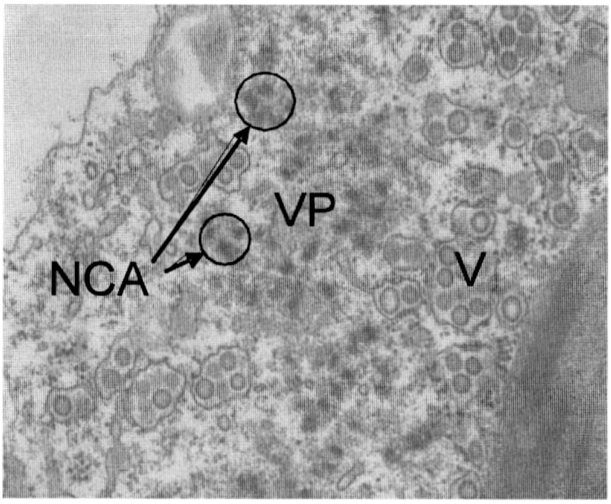

Figure 6.3 Electron micrograph showing an accumulation of *Tomato spotted wilt virus* (TSWV) particles (V) in large ER-derived vesicles. Viroplasm (VP) and nucleocapsid aggregates (NCA) are indicated.

Lawson *et al.*, 1996; Kikkert *et al.*, 1997, 1999). The latter are found in the cytoplasm but mostly only during early stages of infection (Milne, 1970; Francki *et al.*, 1985; Kitajima *et al.*, 1992; Lawson *et al.*, 1996). DEVs eventually fuse together and with ER-derived membranes and this leads to an accumulation of mature, singly enveloped particles (SEVs) clustered within large ER vesicles (Fig. 6.3) (Best and Katekar, 1964; Best and Palk, 1964; Ie, 1964; Martin, 1964; Kitajima, 1965, 1992; Van Kammen *et al.*, 1966; Francki and Grivell, 1970; Milne, 1970; Francki and Hatta, 1981; Francki *et al.*, 1984; Lawson *et al.*, 1996; Kikkert *et al.*, 1997, 1999). In plants systemically infected with TSWV, virus can be found in almost all tissues and organs (Kitajima, 1965; Francki and Grivell, 1970; Ie, 1973; Kitajima *et al.*, 1992). Mature virions accumulated and retained within these subcellular (ER-derived) compartments serve as source from which thrips acquire the virus by feeding prior to transmission.

During the very early stages of the infection, and prior to the appearance of virus particles, a (sub)population of cytoplasmic RNPs can be discriminated by the additional association of NSm (Kormelink *et al.*, 1994). The NSm protein on itself is able to form tubular structures from plasmodesmate into neighbouring healthy cells and modifies the size exclusion limit of these channels (Kormelink *et al.*, 1994; Storms *et al.*, 1995, 1998). In this way, the NSm protein is thought to facilitate the cell-to-cell movement of non-enveloped infectious RNP structures of TSWV through plasmodesmata Since plasmodesmata have an effective size exclusion limit of only 5 nm, enveloped tospovirus particles (80–120 nm in diameter) are too large to pass these. The observation that different RNP fractions are distinguished is also supported by *in vitro* transcription/replication studies. Whereas virus-derived RNPs are highly transcriptionally active, cytoplasmic RNP fractions on the other hand more support genome replication (Van Knippenberg *et al.*, 2002 and PhD thesis, 2005). Further analysis has shown that cytoplasmic RNPs of TSWV contain (higher) amounts of NSs protein than RNPs purified from enveloped virus particles (Van Knippenberg, PhD thesis, 2005), and support the idea that during the virus life cycle various compositions of RNPs exist. So far, the exact composition of RNPs to become membrane bound remains unclear. Although the maturation of TSWV particles, in specific the formation of DEVs and their subsequent fusion, is somewhat different from that of the animal-infecting bunyaviruses, they both share in common the Golgi system as being the site of particle assembly.

Other cytopathological structures in plant cells related to tospovirus infections are the cytoplasmic fibrillar and paracrystalline arrays (Francki and Grivell, 1970; Francki *et al.*, 1984; Kormelink *et al.*, 1991; Kitajima *et al.*, 1992). Not all tospovirus infections reveal their appearance, but these structures are shown to be (partly) composed of NSs protein, the viral RNAi suppressor protein (Kormelink *et al.*, 1991; Kitajima *et al.*, 1992).

Besides plants, tospoviruses also replicate in their insect vector thrips (Wijkamp *et al.*, 1993). Thrips do not seem to suffer from this infection and have been suggested to respond with an immune response, based on the observed upregulation of several genes coding for anti-microbial peptides, lectins and receptors (Medeiros *et al.*, 2004).

Upon ingestion of the virus during the larval L1 and L2 stages, a viral infection starts in the mid-gut epithelium and the associated muscle cells. From there, the virus disseminates into the thrips body and the salivary glands, the major site of virus replication (Ullman *et al.*, 1992, 1993, 1995; Wijkamp *et al.*, 1993; Tsuda *et al.*, 1996; Filho *et al.*, 2002; Nagata *et al.*, 2002). Immuno-localization studies of TSWV in viruliferous thrips *F. occidentalis* have shown the presence of both N and NSs in these tissues, the NSs protein being indicative for virus replication. The N protein mostly confined to electron-dense aggregates (viroplasm) and the NS_S protein found dispersed throughout the cytoplasm (Ullman *et al.*, 1992, 1993, 1995; Wijkamp *et al.*, 1993). Paracrystalline arrays and amorphous inclusions have additionally been observed in infected thrips cells. The paracrystalline arrays resemble the ones observed in infected plant cells and shown to contain the NS_S protein (Kormelink *et al.*, 1991; Kitajima *et al.*, 1992), whereas the amorphous inclusions labelled with antiserum directed to the Gc glycoprotein (Ullman *et al.*, 1995). So far, no stages of virus assembly in the Golgi complex have been observed but the observation of (low amounts of) virus particles in saliva secretory vesicles and their release during subsequent secretion of saliva into the ducts supports the involvement of Golgi in particle assembly. Particle assembly has not been observed in any other organ of the thrips and indicates the salivary glands as the major site of virus replication.

In thrips that have acquired TSWV while feeding on infected plants during the adult stage the virus has been observed not to disseminate beyond the mid-gut epithelium, suggesting that a mid-gut escape barrier likely exists in adult thrips (Ullman *et al.*, 1992). The viral glycoproteins are assumed to determine vector specificity (Table 6.1) but are also required for acquisition of the virus by thrips, most likely by receptor-mediated endocytosis in the mid-gut epithelium (Wijkamp, 1995; Sin *et al.*, 2005). Sofar, several attempts have been made to identify the receptor (Bandla *et al.*, 1998; Kikkert *et al.*, 1998; Medeiros *et al.*, 2000) but cloning of candidate receptor genes have failed. Whereas Bandla *et al.* (1998) showed the involvement of both Gn and Gc in virus attachment to the thrips mid-gut, the inhibition of TSWV thrips transmission by a soluble form of Gn strongly points to Gn as the candidate viral ligand for binding to a mid-gut epithelium receptor (Whitfield *et al.*, 2004, 2008). On the other hand, the presence of a cellular attachment site (the RGD-binding motif) in Gn (Kormelink *et al.*, 1992a; Law *et al.*, 1992), and a conserved amino acid sequence within the core of Gc (Cortez *et al.*, 2002) which has also been suggested to play a role in cell attachment (Fig. 6.4) still does not rule out a concerted action of Gn and Gc in the uptake of the virus by thrips.

The failure to set up a thrips cell culture has lead to attempts to study tospovirus infections in primary thrips cells, but the absence of any glycoprotein synthesis or virus accumulation

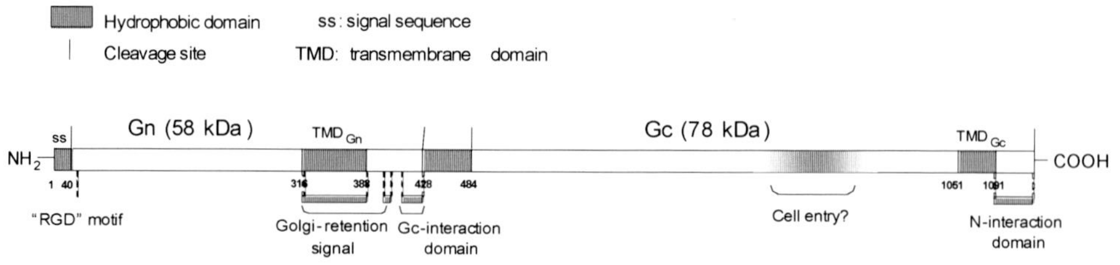

Figure 6.4 Topology of the *Tomato spotted wilt virus* (TSWV) glycoprotein precursor and identified domains indicated.

and the low percentage of virus transmission (1%) from culture fluids of infected primary cells so far has hampered further studies (Nagata *et al.*, 1997).

Despite the lack of more detailed information on the life cycle of tospoviruses in the thrips vector, the morphogenesis pathway of TSWV in thrips cells shows some differences from the pathway in plant cells. Whereas in plant cells TSWV particles are clearly retained in ER-derived vesicles, they are secreted from thrips cells in which they more resemble the animal infecting bunyaviruses.

For several animal-infecting bunyaviruses a reverse genetics system has greatly advanced the insight into the function of viral genes and sequences in the viral infection cycle (Jin and Elliott, 1993; Dunn *et al.*, 1995; Lopez *et al.*, 1995; Prehaud *et al.*, 1997; Flick and Pettersson, 2001; Flick *et al.*, 2002, 2003a,b, 2004; Blakqori *et al.*, 2003); however, no such system is yet available for tospoviruses. To further dissect the process of TSWV particle assembly in plant and insect cells, and identify the roles (and domains) of the structural proteins N, Gn and Gc involved in this, transient expression studies have been performed in plant and mammalian cell systems. In both cell types, transiently expressed N shows up as large oligomers that result from N–N protein interactions, as confirmed by fluorescence resonance energy transfer (FRET) and fluorescence life time imaging microscopy (FLIM) (Snippe *et al.*, 2005a,b; Ribeiro *et al.*, 2009a). No co-localization of N oligomers is observed with endoplasmic reticulum (ER), ER-Golgi intermediate compartment (ERGIC) or with Golgi specific marker proteins, suggesting a cytoplasmic localization (Snippe *et al.*, 2005a; Ribeiro *et al.*, 2009a). In mammalian cells N-oligomers accumulate peri-nuclear, as also reported for several animal infecting bunyaviral RNPs/N proteins (Ravkov and Compans, 2001; Andersson *et al.*, 2004a,b; Reichelt *et al.*, 2004). The perinuclear localization of TSWV N protein involves minus-end directed transport on microtubuli, as demonstrated by a change into a randomized distribution of N oligomers throughout the cytoplasm after disruption of microtubuli (Snippe *et al.*, 2005a). In plants cell, though, N oligomers show up randomly dispersed throughout the cytoplasm and instead of microtubuli only depend on actin for their oligomerization (Ribeiro, 2007). The peri-nuclear accumulation of N/RNPs has been suggested to represent aggresomes (Kochs *et al.*, 2002; Andersson *et al.*, 2004b), structures that relate to the microtubule organizing centre (MTOC) and suggested as aggregation sites for misfolded proteins (Johnston *et al.*, 1998). Whether in mammalian cells the peri-nuclear localization nearby the MTOC and in vicinity of the Golgi, is a pre-requisite to create a viral factory for particle assembly (Heath *et al.*, 2001) or merely results from a host defence response to sequester N/RNPs and avoid these from taking part in viral replication (Frese *et al.*, 1996; Kanerva *et al*, 1996; Ravkov *et al.*, 1998; Flohr *et al.*, 1999; Ravkov and Compans,

2001; Kochs *et al.*, 2002; Andersson *et al.*, 2004a,b; Bridgen *et al.*, 2004; Reichelt *et al.*, 2004), remains to be investigated.

Why the N protein behaves and localizes differently in plant and mammalian cells is not clear yet, but the organization of the Golgi complex and the microtubuli cytoskeleton is very different in both cell types and therefore requires the results also to be analysed in view of this (Kikkert *et al.*, 2001; Ward and Brandizzi, 2004;Snippe *et al.*, 2005a,b, 2007a; Ribeiro *et al.*, 2008, 2009a,b). In mammalian cells, the Golgi appears as one major organelle over the MTOC in the perinuclear region, and ER-to-Golgi transport of protein cargo involves vesicles that use dynein-driven motors to move along the microtubuli cytoskeleton. In addition, the microtubuli are organized in a polarized manner with all minus-ends of the microtubuli coming together in (mostly) one, perinuclear localized MTOC. In plants cells, on the other hand, microtubuli are not organized in a polarized manner, and the Golgi apparatus is more scattered in the cell as small cisternal stacks closely associated with the cortical ER network. Golgi stacks move over the ER in an actin-myosin dependent manner, and ER-to-Golgi trafficking does not require microtubules in contrast to mammalian cells (Ward and Brandizzi, 2004). Hence, whether the differences in localization of N oligomers in mammalian versus plant cells is due to a difference in their cellular architecture or reflects an adaptation of TSWV to replicate and form virus particles in both cell types still remains to be elucidated.

Upon co-expression of the structural Gn and Gc glycoproteins in plant and mammalian cells, either from separate constructs or from the precursor, both glycoproteins traffic and accumulate in the Golgi complex. In plant cells this occurs in association to the cortical ER-network (Ribeiro *et al.*, 2008), and in mammalian cells perinuclear (Kikkert *et al.*, 2001). Upon separate expression, Gn is still able to traffic to the Golgi, though less efficient when compared with the expression in the presence of Gc. Single expression of Gc, on the other hand, only shows ER localization (Kikkert *et al.*, 2001; Ribeiro *et al.*, 2008). In contrast to the observations in mammalian cells, transient expression of the individual Gc and Gn proteins in plant cells revealed the additional formation of pseudo-circular and pleomorphic structures that derive from ER and Golgi membranes (Fig. 6.5; Ribeiro *et al.*, 2008).

Whether these represent virus like particles (VLPs) remains a matter of speculation. The structures are relatively large in size (from 0.5 up to 4 μm), and their formation may reflect the capacity of the glycoproteins to provoke membrane curving during a natural infection,

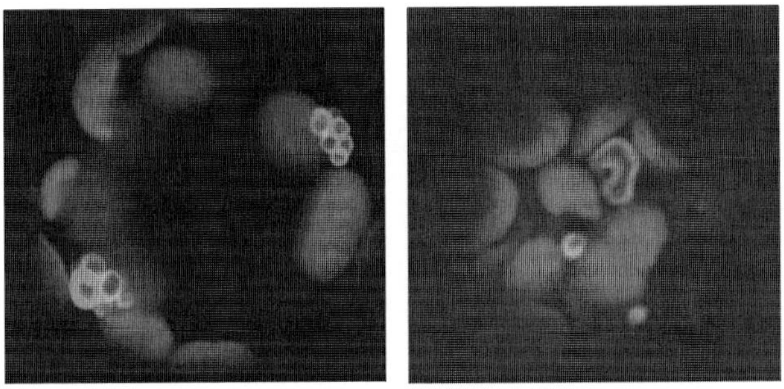

Figure 6.5 Formation of (pseudo-) circular (left panel) and pleomorphic (right panel) membrane structures upon single expression of the viral Gn and Gc glycoproteins in plant *Nicotiana benthamiana* cells.

and thereby enable Golgi enwrapment of RNPs into DEVs, a phenomenon that has not been observed in insect/mammalian cells.

The localization of Gn in the Golgi suggested the presence of a Golgi retention signal. Since no conserved Golgi retention motif exists, C-terminal truncations of Gn have been expressed in mammalian cells and mapped the Golgi retention signal to a stretch of 10 amino acids, about 20 residues downstream the transmembrane domain (TMD; Fig. 6.4). Analysis of chimeric Gc and Vesicular stomatitis virus glycoprotein (VSV-G) in which the cytoplasmic tail and/or TMD have been exchanged for those from Gn, has demonstrated the additional requirement of the TMD of Gn for Golgi targeting/retention (Fig. 6.4; Snippe *et al.*, 2007a). Truncations of Gn lacking the C-terminal 20 aa but still localizing to the Golgi are no longer able to redirect Gc to the Golgi (Snippe *et al.*, 2007a), suggesting the requirement of this domain for interaction with Gc (Fig. 6.4).

The inability of Gc to transport to the Golgi in the absence of Gn (Kikkert *et al.*, 2001; Ribeiro *et al.*, 2008) has often been explained as a result from unproper folding and requiring heterodimerization with Gn for proper folding and to allow subsequent ER exit. However, transient expression of chimeric Gc in plant cells, in which the transmembrane domain (TMD) and/or cytoplasmic tail (CT) have been swapped for those from Gn, showed that only the TMD of Gn is sufficient to allow ER exit and transport to the Golgi (Ribeiro *et al*, 2009b). Since studies in mammalian cells showed that next to the TMD, an additional stretch of 10 aa within the CT of Gn is required for Golgi localization of this protein (Snippe *et al.*, 2007a) it is likely that ER export and Golgi retention is the result of a concerted action of the TMD of Gn and one of the functionally equivalent cytoplasmic tails of Gn or Gc. Primary or secondary sequence structure analyses of these CTs, however, have not immediately revealed similarities.

The observation that ER retention of Gc is apparently not due to misfolding is furthermore supported by the absence of ER chaperone proteins Bip, calnexin and calreticulin from co-immunoprecipitation experiments on transiently expressed Gc (Ribeiro *et al.*, 2009b). The lack of a proposed Golgi retention motif within Gn, and the observation that its TMD appears a major determinant for this is in agreement with reports on type I membrane proteins that have shown the importance of TMD length in determining the final destination of type I membrane proteins within the secretory pathway (Brandizzi *et al.*, 2002a). One of the current models for this suggests that type I membrane proteins migrate within the secretory pathway by means of diffusion until it runs into a membrane with the right thickness to mask its hydrophobic transmembrane domain. In general, shorter TMDs restrict the proteins to the ER and longer TMDs allowing their migration to the Golgi or elsewhere (Brandizzi *et al.*, 2002a; Moreau *et al.*, 2007). The localization behaviour of TSWV Gn and Gc could very well be explained by this model since the TMD for Gn is much larger in size (70 aa) compared with the one of Gc (40 aa) (Ribeiro *et al.*, 2009b).

Upon co-expression of the individual glycoproteins with N, N co-localizes with Gc in both plant and mammalian cells (Snippe *et al.*, 2007b; Ribeiro *et al.*, 2009a), but only in plant cells has lead to a (visual) co-localization of Gn and N (Ribeiro *et al.*, 2009a). In all these cases the interaction has been confirmed by FRET and FLIM. In mammalian cells, the co-localization of N with Gc occurs in the peri-nuclear area, and appears similar to the perinuclear accumulation after singly expressing N (Snippe *et al.*, 2007b). In plant cells co-expression of N and Gc causes a dramatic change in distribution of Gc from an ER-reticular appearance to distinct spots within the ER that co-localized with a marker for ER-export sites (ERES). The cytoplasmic tail of Gc has been shown to be required for the interaction with N (Ribeiro, 2007; Fig. 6.4). The cytosolic N protein apparently is able to interact and subsequently concentrate Gc into ERES, and supports the observation from ER-Golgi transport inhibitor studies that ER to Golgi transport of tospoviral glycoproteins involves COPII vesicles that exit from ERES (Ribeiro, 2007). The observation that no microtubuli-dependent transport is involved in plant cells, is in agreement with the absence for a microtubuli requirement during ER-Golgi cargo transport in plant cells (Brandizzi *et al.*, 2002b; Ward and Brandizzi, 2004).

Transient expression of the cell-to-cell movement protein (NSm) in plant cells has shown that this protein, in the absence of other viral proteins, is able to form tubular structures that emerge from the cell surface (Storms *et al.*, 1995). Similar results have been obtained in *Spodoptera frugiperda* insect cells, in addition to a localization in plasma membrane and cytoplasmic vesicles (Storms *et al.*, 1995). The capacity to form tubular structures that emerge from the cell surface on both cell types indicates that no plant-specific structures or proteins are involved in their formation. Transgenic expression of NSm in *Nicotiana tabacum* plants has shown that this protein targets to the plasmodesmata were it alters the size exclusion limits (Storms *et al.*, 1998). Simultaneously, its expression provokes severe, infection-like symptoms, including deficiencies in root and shoot development (Prins *et al.*, 1997b; Rinne *et al.*, 2005) which have been attributed to the obstruction of NSm-targeted mesophyll plasmodesmata in source tissues by depositions of 1,3-β-D-glucan (GLU) or callose (Rinne *et al.*, 2005). Temperature-shift treatments of the transgenic NSm plants abolished chlorotic local lesions and rescued plant development, by restoring the transport capacity of Pd through the action of 1,3-β-D-glucanase (GLU-h) or callase, and suggests that the symptoms of NSm transgenic plants are self-inflicted and due to a basal defence response to counteract interference of the movement protein with plasmodesmate functioning (Rinne *et al.*, 2005).

Whereas the NSm protein facilitates movement of TSWV RNPs through plasmodesmata, expression of NSm from a movement-defective Tobacco mosaic virus (TMV) replicon has shown complementation of TMV cell-to-cell movement and long distance of naked viral RNA, demonstrating that TSWV NSm can functionally substitute for TMV MP (Lewandowski and Adkins, 2005). The absence of complementation with a NSm C-terminal deletion mutant (107 aa), pointed towards the importance of this domain in movement. Using the same approach, deletion and alanine-substitution mutants of TSWV NSm have been expressed and this has identified domains involved in tubule formation, movement and symptomatology (Fig. 6.6; Li *et al.*, 2009). *In vitro* protein-protein and protein–RNA interaction studies have shown that NSm interacts with the N protein and has affinity to single stranded RNA in a sequence non-specific manner (Soellick *et al.*, 2000). Using a Yeast-two-Hybrid screen, NSm showed an interaction with plant proteins showing homologies to myosin, kinesin and DnaJ-like chaperones (von Bargen *et al.*, 2001; Paape *et al.*, 2006), in line with the general idea of molecular chaperones and the cytoskeleton being involved in intracellular trafficking. One of the interactors, At-4/1, has been observed in ER-derived membrane structures in vicinity to plasmodesmata and shown to move from cell-to-cell by its own (Paape *et al.*, 2006). If and how this protein is involved in NSm-mediated cell-to-cell movement still remains unclear.

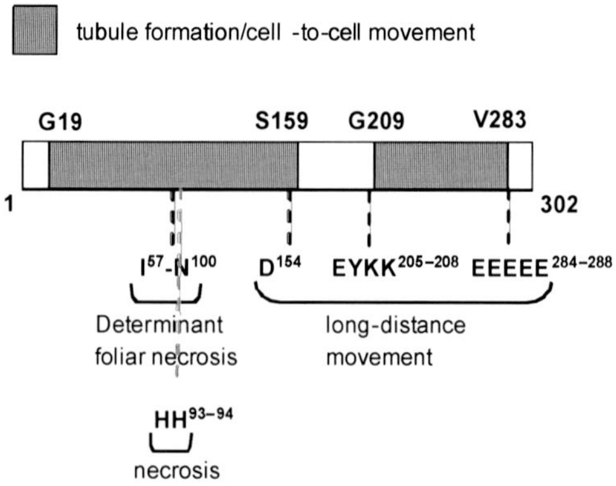

Figure 6.6 Topology of the *Tomato spotted wilt virus* (TSWV) cell-to-cell movement (NSm) protein and identified domains indicated (Li *et al.* 2009).

Antiviral RNAi, viral counterdefence and RNAi-mediated resistance

RNA silencing (Baulcombe, 1996), also referred to as post-transcriptional gene silencing (PTGS) or RNAi (from RNA interference; Fire *et al.*, 1998), is an evolutionary conserved process in which double-stranded (ds) RNA and imperfect hairpin RNAs trigger a chain of processes leading to sequence-specific breakdown of RNA. The mechanism nowadays is generally accepted as an innate immunity mechanism against viral invaders in plants and insects. During this process, dsRNA is processed into small interfering RNAs (siRNAs) or microRNAs (miRNAs) of 21–26 nucleotides in length by RNase-III-type enzymes called Dicer (DCL) (Fire *et al*, 1998; Bernstein *et al.*, 2001; Lee *et al.*, 2003). One strand of the dsRNA, the so-called guide strand, is incorporated into the RNA-induced silencing complex (RISC) based on thermodynamic stabilities at the two ends (Khvorova *et al.*, 2003; Schwarz *et al.*, 2003). The RISC complex, being activated with the guide strand and a member of the Argonaute (Ago) protein family, continuously mediates recognition and subsequent cleavage of (m)RNA target sequences with complementarity to the guide RNA, leading to silencing of the gene.

Plant viruses induce RNA silencing, often referred to as virus-induced gene silencing (VIGS), during genome replication, as can be observed by the generation of viral specific siRNA molecules during the infection process. For RNA viruses it is generally assumed that ds replicative intermediates play a role in the induction of RNAi. Recent deep sequencing of plant viral siRNAs from several plant viruses have revealed that siRNAs showed a genome-wide distribution and for several viruses with a strong dominance of RNA species originating from viral positive strands (Donaire *et al*, 2009), in agreement with other studies that showed more siRNAs from the (+) strand (Szittya *et al.*, 2002; Molnár *et al.*, 2005). In case of the latter, the siRNAs mapped to regions within the genomic RNA containing intramolecular hairpin structures.

To escape from this antiviral defence mechanism viruses have developed ways to counteract or evade from this. Most plant viruses counteract RNAi by coding for a so-called RNA silencing suppressor (RSS) protein that is able to interfere in the cascade of reactions that lead to viral RNA degradation. Some of the identified plant viral RSSs have been studied in more detail and shown to counteract RNAi by sequestering siRNAs and thereby prevent their incorporation into RISC, or they avoid cleavage of long dsRNA into siRNAs and/or systemic transport of siRNAs (Guo and Ding, 2002; Llave *et al.*, 2002; Mallory *et al.*, 2002; Lakatos *et al.*, 2004, 2006; Zhang *et al.*, 2006; Diaz-Pendon *et al.*, 2007; Goto *et al.*, 2007; Hemmes *et al.*, 2007).

For Tospoviruses, the RNA silencing suppressor activity has been identified using the *Agrobacterium*-mediated transient expression system as reported by Johanson and Carrington (2001). Whereas agroinfiltration of a functional GFP copy on *N. benthamiana* shows GFP silencing after several days (due to a conversion of aberrant GFP RNA transcripts into dsRNA by a host-RNA-dependent RNA polymerase (RdRP)), the additional infection with TSWV has shown to reverse GFP silencing and to reveal high fluorescence in the infected leaves. For other tospoviruses similar results have been obtained. During a co-infiltration of a NSs gene construct, but not with a construct coding for the N, NSm or the precursor to the glycoproteins, GFP silencing is suppressed, demonstrating that the tospoviral NSs protein contains RSS activity (Fig. 6.7; Takeda *et al.*, 2002; Bucher *et al.*, 2003). Initial studies using purified His-tagged NSs protein have shown that TSWV NSs exhibits affinity to small and large (114 nt) RNA duplex molecules (Hemmes, 2007). During more recent studies using extracts from tospovirus-infected plants, or insect cells expressing NSs from recombinant baculoviruses, this finding has been confirmed. Furthermore and surprisingly, comparative analysis of different tospoviruses additionally has shown that the NSs protein from TSWV, INSV and GRSV exhibited affinity to short (si- and mi-) and long dsRNAs while from TYRV, a distinct tospovirus, only showed affinity to short dsRNAs (Schnettler *et al.*, 2010). These results support the idea that NSs likely exerts its RSS function by sequestering these molecules to prevent their uploading into RISC or cleavage by Dicer.

For over a decade, the antiviral RNA silencing mechanism has been successfully exploited to generate (RNAi-mediated) transgenic virus resistance in plants. Also for TSWV and other tospoviruses transgenic resistance has been generated initially without the awareness of dealing with RNAi-mediated resistance (Gielen *et al.*, 1991; De Haan *et al.*, 1992; Prins and Goldbach, 1998). For tospoviruses resistance has been achieved after

Figure 6.7 Suppression of GFP silencing by TSWV NSs during a co-ATTA. The left leaf has been infiltrated with a GFP construct and shows silencing after 5 days. In the additional presence of NSs (right leaf) RNA silencing of GFP is suppressed.

transformation of (partial) N and NSm gene sequences to *N. benthamiana* (Gielen *et al.*, 1991; De Haan *et al.*, 1992; Prins *et al.*, 1995a), but initially only with low efficiency in the F2 progeny. Nuclear run-on transcription assays have shown transcriptional expression of the transgenes, whereas hardly or no cytoplasmic transgene RNA could be detected, supporting the idea of cytoplasmic RNA breakdown (Prins *et al.*, 1996). Detection of virus specific siRNA molecules has eventually demonstrated the involvement of RNAi in resistance (Bucher *et al.*, 2006).

To activate RNAi-mediated resistance against tospoviruses, a minimal length of the N transgene sequence (236–387 bp) is required but when fused to a non-target gene a shorter sequence (59–110 bp) is sufficient (Pang *et al.*, 1997; Jan *et al.*, 2000). Analyses of transgenic resistant plants containing sequences from the TSWV NSm gene have shown that the resistance does not hold on single cell (protoplast) level. Protoplasts from these lines still support replication of TSWV, suggesting that in case of tospoviruses antiviral RNA silencing targets the viral transcripts rather than the (N protein-encapsidated) viral genome (Prins *et al.*, 1997a).

A major drawback of RNA-mediated virus resistance is its high specificity, i.e. resistance generated against TSWV does not hold against the related TCSV, GRSV (~78–80% genome sequence homology) or other tospoviruses. To circumvent this problem broad tospovirus resistance has been created by designing a chimeric transgene cassette containing multiple (partial) N gene sequences derived from different tospoviruses and triggering the silencing of all tospoviral targets simultaneously (Prins *et al.*, 1995a). More recently this resistance has been further expanded to other tomato-infecting tospoviruses WSMoV and TYRV. By additionally cloning this (partial) N gene cassette as an inverted repeat, transgene transcripts formed into ds hairpin RNA molecules and have lead to rapid RNAi induction. In this way, resistance has been obtained against TSWV, GRSV, TCSV, WSMoV and TYRV with high efficiency in the progeny of transformed plants (Bucher *et.al*, 2006, Hassani-Mehraban *et al.*, 2009b).

Meanwhile, transgenic resistance raised against the tomato strain of TYRV (TYRV-t) has been broken during a mixed inoculation with the soybean strain (TYRV-s) of this virus (Hassani-Mehraban *et al.*, 2008, 2009b), which only exhibited 14.4% nucleotide sequence divergence in its N gene. The presence of N protein from both strains during the infection has suggested suppression of TYRV-t N gene breakdown by RNAi and implied a trans complementation event with the RNAi suppressor protein NSs from TYRV-s. Proof for this has been provided by delivery of TYRV-s NSs via a PVX replicon, which enabled TYRV-t to break the resistance. Surprisingly, delivery of its own RNAi suppressor in an analogous way has also supported TYRV-t rescuing and suggested that resistance breaking is most likely due to a general reduction in TYRV-t sensing RISC complexes in the transgenic lines caused by the *a priori* presence of NS_S (Hassani-Mehraban *et al.*, 2009b). The failure to rescue TYRV-t from these resistant plants during a mixed inoculations with other tospoviruses remains unclear, but timing and amounts of NSs expression might play a role.

The observation that during a mixed-infection of closely related viruses RNAi-mediated resistance can be broken, limits the use of this resistance strategy during natural field situations, especially when such viruses have overlap in geographic range. For this reason, a more durable resistance strategy may be generated using a combined approach of several resistance strategies.

Transgenic and natural resistance strategies

Although the RNAi-mediated resistance strategy has been most widely exploited, alternative protein-mediated transgenic resistance strategies have been developed successfully against tospoviruses. In one approach, single-chain variable antibody fragments (scFv) specifically recognizing the TSWV N protein have been selected. Their coding sequence transformed to *N. benthamiana* and shown to confer high levels of resistance against TSWV (Griep *et al.*, 2000; Prins *et al.*, 2005b). Transgenic plants expressing scFv against the NSm movement protein show delayed symptom development upon challenging with TSWV (Zhang *et al.*, 2008). In contrast, transformation of sequences coding for scFv directed against the TSWV glycoproteins has not lead to inhibition of virus transmission (Prins *et al.*, 2005b).

In a second approach, the coding sequence for a small dominant interfering peptide of 29 amino acids ('aptamer') selected from a yeast-two-hybrid screen with strong affinity to the N protein of different tospoviruses, has been transformed and shown to provide strong resistance to TSWV, TCSV, GRSV and CSNV (Rudolf *et al.* 2003).

For bunyaviruses (interferon-induced) MxA has been suggested to inhibit viral replication by sequestering viral RNPs to the perinuclear region (Frese *et al.*, 1996; Kanerva *et al*, 1996; Flohr *et al.*, 1999; Kochs *et al.*, 2002; Bridgen *et al.*, 2004). Tobacco plants transformed with a copy of the human MxA protein gene do not exhibit resistance to TSWV or a variety of other RNA viruses (Frese *et al.*, 2000), and suggest that expression of MxA alone is not sufficient to confer tospovirus resistance in plants.

Besides the use of engineered, transgenic virus resistance approaches, sources of natural resistance to tospoviruses are available. However, the number of resistance genes against tospoviruses and suitable for commercial breeding is limited. So far, two single dominant TSWV resistance genes have been well described. The first one has been identified in *L. peruvianum* for which high resistance levels to various TSWV isolates have been observed (Stevens *et al.*, 1992). This gene has been introgressed into the tomato variety 'Stevens' where it also showed resistance against tospoviruses GRSV and TCSV (Boiteux and Giordano, 1993). The resistance has been mapped on chromosome 9 to a single gene, *i.e.* the *Sw*5-b copy within the *Sw*5 gene cluster (Folkerstma *et al.*, 1999; Brommonschenkel *et al.*, 2000; Spassova *et al.*, 2001). *N. tabacum* cv. SR1 plants transformed with *Sw*5-b showed resistance against TSWV, which also demonstrated that *Sw*5-b is functional in other host plants (Spassova *et al.*, 2001). The *Sw*5-b gene codes for a protein of 1246 amino acids and is a member of the group of resistance genes that contain an amino terminal coiled-coil domain, a nucleotide-binding-site region called NB-ARC (due to the sequence similarity of this domain with apoptosis-related proteins in nematodes and mammals; van der Biezen and Jones, 1999), and leucine-rich repeats. For all single dominant resistant genes, the resistance is being triggered by a single protein from the pathogen referred to as avirulence determinant (*avr*). For *Sw*5-b, the TSWV NSm or glycoproteins are candidates for *avr*, since reassortant studies have shown that the viral gene which triggers the *Sw*5 resistance gene maps to the M RNA of TSWV (Hoffman *et al.*, 2001). Owing to the presence of multiple nucleotide sequence differences in the coding and non-coding regions of M RNA segments from several TSWV isolates (resistance-breaking and non-breaking isolates), resistance breaking nucleotide sequences have not been identified yet. TSWV strains overcoming *Sw*5 resistance have already been reported (Latham and Jones, 1998), indicating that this resistance appears not absolutely durable.

The second resistance gene, denoted *Tsw*, has been identified in *Capsicum chinense* 'PI' accessions and maps to the distal portion of chromosome 10 (Jahn *et al.*, 2000). The resistance is being displayed by a hypersensitive reaction (HR) to a broad range of TSWV isolates and is shown to be temperature sensitive, with high temperatures leading to reduced resistance performance (Roggero *et al.*, 1996, 1999; Rosello *et al.*, 1997; Moury *et al.*, 1998; Roggero, 2002). Like for *Sw5*, the resistance is governed by a single dominant gene (Black *et al.*, 1991; Boiteux, 1995) and also for this trait resistance-breaking variants have been identified (Roggero *et al.*, 2002; Aramburu and Marti, 2003). The avirulence gene has been mapped to the S RNA segment of the TSWV (Jahn *et al.* 2000), pointing towards either N or NSs as the *avr* determinant. Although two reports have appeared on the identification of the *avr* determinant, contradictory results leave this issue inconclusive (Margaria *et al.*, 2007; Lovato *et al.*, 2008).

Besides virus resistance, thrips resistance has been shown to provide significant protection to TSWV, even when a crop is susceptible to the virus. In a large, greenhouse-scale experiment TSWV spread has been restricted and delayed in plants that were resistant to thrips, which has been explained as a result from impeded thrips population development (Maris *et al.*, 2003a,b).

Owing to the emergence of resistance-breaking TSWV isolates and the possibility of persistent and heavy infestations with thrips, the use of virus- and thrips- resistant plants combined with other (sanitary and culture) practices is the most advisable strategy for disease management of TSWV.

Concluding remarks

During the last decade the insight into the molecular biology of tospoviruses has greatly advanced. Although a reverse genetics system for tospoviruses has hampered progress in certain areas, transient protein expression analyses in various cell systems have contributed to a further understanding on the roles of viral proteins at various stages in the infection cycle. The research on the interaction with host proteins, though, is only getting started and the issues of TSWV cell-to-cell movement and particle assembly, in specific the interaction of the structural proteins with the (plant) host endomembrane system, will present interesting targets for this. The mechanism of RNAi suppression by NSs is not yet understood either, nor if it interferes in gene regulation and/or other host processes, making this an additional interesting target to study the interplay with the host as well. It is expected that many of this work will not only contribute to the understanding of tospoviruses but also to more general (plant) virological and (cell) biological issues. One nice example for this is the trans-complementation and rescuing of movement defective TMV replicons with tospovirus NSm protein (Lewandowski and Adkins, 2005). For a long time TMV and TSWV, two very distinct plant viruses, have been assumed to exemplify two different cell-to-cell movement mechanisms, i.e. transport of naked viral RNA through modified plasmodesmata (TMV) versus tubule-assisted movement of infectious viral RNP complexes (TSWV) (Scholthof, 2005). The finding of trans-complementation of TMV with TSWV NSm has been very surprising but seems to point towards a more common cell-to-cell movement mechanism for plant viruses.

The advanced insight into the molecular biology of tospoviruses and availability of molecular tools have assisted in the development of a variety of transgenic (RNAi- and protein-mediated) resistance strategies. The appearance of resistance-gene breaking tospoviruses has slowly limited the durability of natural resistance genes, but by pyramiding these with various engineered forms of resistance will provide a broader and more durable resistance strategy against tospoviruses, as already successfully demonstrated by Gubba *et al.* (2002).

References

Amin, P.W., Reddy, D.V.R., and Ghanekar, A.M. (1981). Transmission of *Tomato spotted wilt virus*, causal agent of bud necrosis of peanut, by *Scirtothrips dorsalis* and *Frankliniella schultzei*. Plant Dis. *65*, 663–665.

Andersson, I., Bladh, L., Mousavi-Jazi, M., Magnusson, K.-E., Lundkvist, A., Haller, O., and Mirazimi, A. (2004a). Human MxA protein inhibits the replication of Crimean–Congo hemorrhagic fever virus. J. Virol. *78*, 4323–4329.

Andersson, I., Simon, M., Lundkvist, A., Nilsson, M., Holmstrom, A., Elgh, F., and Mirazimi, A. (2004b). Role of actin filaments in targeting of Crimean Congo hemorrhagic fever virus nucleocapsid protein to perinuclear regions of mammalian cells. J. Med. Virol. *72*, 83–93.

Aramburu, J., and Marti, M., (2003). The occurrence in north-east Spain of a variant of *Tomato spotted wilt virus* (TSWV) that breaks resistance in tomato (*Lycopersicon esculentum*) containing the Sw-5 gene. Plant Pathol. *52*, 407.

de Ávila, A.C., Huguenot, C., Resende, R., Kitajima, E.W., Goldbach, R.W., and Peters, D. (1990). Serological differentiation of twenty isolates of *Tomato spotted wilt virus*. J. Gen. Virol. *71*, 2801–2807

de Ávila, A.C., de Haan, P., Kitajima, E.W., Kormelink, R., Resende, R. de O., Goldbach, R., and Peters D. (1992). Characterization of a distinct isolate of *Tomato spotted wilt virus* (TSWV) from Iimpatiens sp. In The Netherlands. J. Phytopathol. *134*, 133–151.

de Ávila, A.C., de Haan, P., Kormelink, R., Resende, R. de O., Goldbach, R., and Peters D. (1993). Classification of tospoviruses based on phylogeny of nucleoprotein gene sequences. J. Gen. Virol. *74*, 153–159.

Bandla, M.D., Campbell, L.R., Ullman, D.E., and Sherwood, J.L. (1998). Interaction of tomato spotted wilt tospovirus (TSWV) glycoproteins with a thrips mid-gut protein, a potential cellular receptor for TSWV. Phytopathol. *88*, 98–104.

von Bargen, S., Salchert, K., Paape, M., Piechulla, B., and Kellmann, J.-W. (2001). Interactions between the *Tomato spotted wilt virus* movement protein and plant proteins showing homologies to myosin, kinesin and DnaJ-like chaperones. Plant Physiol. Biochem. *39*, 1083 – 1093.

Baulcombe, D.C. (1996). RNA as a target and an initiator of post-transcriptional gene silencing in transgenic plants. Plant Mol. Biol. *32*,79–88.

Bellocq, C., Ramaswamy, R., Patterson, J., and Kolakofsky, D. (1987). Translational requirement for La Crosse virus S-mRNA synthesis: *In vitro* studies. J. Virol. *61*, 87–95.

Bernstein, E., Caudy, A.A., Hammond, S.M., and Hannon, G.J. (2001). Role for a bidentate ribonuclease in the initiation step of RNA interference. Nature *409*, 363–366.

Best, R.J., and Katekar, G.F. (1964). Lipid and purified preparation of *Tomato spotted wilt virus*. Nature *203*, 671–6722.

Best, R.J., and Palk, B.A. (1964). Electron microscopy of strain E of *Tomato spotted wilt virus* and comments on its probable biosyhnthesis. Virol. *23*, 445–460.

van der Biezen, E.A., and Jones, J.D.G. (1999). The NB-ARC domain: a novel signalling motif shared by plant resistance gene products and regulators of cell death in animals. Curr. Biol. *8*, 226–227.

Black, L.L., Hobbs, H.A., and Gatti, J.M., Jr. (1991). *Tomato spotted wilt virus* resistance in *C. chinense* PI152225 and 159236. Plant Dis. *75*, 863.

Blakqori, G., Kochs, G., Haller, O., and Weber, F. (2003). Functional L polymerase of La Crosse virus allows *in vivo* reconstitution of recombinant nucleocapsids. J. Gen. Virol. *84*, 1207–1214.

Boiteux, L.S. (1995). Allelic relationships between genes for resistance to tomato spotted wilt tospovirus in *Capsicum. Chin.* Theor.Appl. Gen. *90*, 146–149.

Boiteux, L.S., and Giordano, L. de B., (1993). Genetic basis of resistance against two Tospovirus species in tomato (*Lycopersicon esculentum*). Euphytica *71*,151–154.

de Borbón, C.M., Gracia, O., and Piccolo, R. (2006). Relationships between tospovirus incidence and thrips populations on tomato in Mendoza, Argentina. J. Phytopathol. *154*, 93–99.

Bouloy, M., Plotch, S.J., and Krug, R.M. (1978). Globin messenger RNA are primers for the transcription of influenza viral RNA in-vitro. Proc. Nat. Acad. Sci. U.S.A. *75*, 4886–4890.

Brandizzi, F., Frangne, N., Marc-Martin, S., Hawes, C., Neuhaus, J.M and Paris, N. (2002a). The destination for single-pass membrane proteins is influenced markedly by the length of the hydrophobic domain. Plant Cell *14*, 1077–1092.

Brandizzi, F., Snapp, E.L., Roberts, A.G., Lippincott-Schwartz, J., Hawes, C. (2002b). Membrane protein transport between the endoplasmic reticulum and the Golgi in tobacco leaves is energy dependent but cytoskeleton independent: evidence from selective photobleaching. Plant Cell *14*, 1293–1309.

Bridgen, A., Dalrymple, D.A., Weber, F., and Elliott, R.M. (2004). Inhibition of Dugbe Nairovirus replication by human MxA protein. Virus Res. *99*, 47–50.

Brommonschenkel, S.H., Frary, A., and Tanksley, S.D. (2000). The broad-spectrum tospovirus resistance gene *Sw-5* of tomato is a homolog of the root-knot nematode resistance gene Mi. MPMI *13*, 1130–1138.

Bucher, E., Lohuis, D., van Poppel, P.M.J.A., Geerts-Dimitriadou, C., Goldbach, R., and Prins, M. (2006). Multiple virus resistance at a high frequency using a single transgene construct. J. Gen. Virol. *87*, 3697–3701.

Bucher, E., Sijen, T., de Haan, P., Goldbach, R., and Prins, M. (2003). Negative-strand tospoviruses and tenuiviruses carry a gene for a suppressor of gene silencing at analogous genomic positions. J. Virol. *77*, 1329–1336.

Chen, C.C., and Chiu, R.J. (1996). A tospovirus infecting peanut in Taiwan. Acta Hortic. *431*, 57–67.

Ciuffo, M., Tavella, L., Pacifico, D., Masenga, V. And Turina, M. (2008). A member of a new Tospovirus species isolated in Italy from wild buckwheat (*Polygonum convolvulus*). Arch. Virol. *153*, 2059–2068.

Ciuffo, M., Kurowski, C., Vivoda, E., Copes, B., Masenga, V., Falk, B.W., and Turina, M. (2009). A New *Tospovirus* sp. in Cucurbit Crops in Mexico. Plant Dis. *93*, 467–474.

Cortez, I., Livieratos, I.C., Derks, A., Peters, D., and Kormelink, R. (1998). Molecular and serological characterization of iris yellow spot virus, a new and distinct tospovirus species. Phytopathology *88*, 1276–1282.

Cortez, I., Saaijer, J., Wongjkaew, K.S., Pereira, A.M., Goldbach, R., Peters, D., and Kormelink, R. (2001). Identification and characterization of a novel tospovirus species using a new RT-PCR approach. Arch. Virol. *146*, 265–278.

Cortez, I., Aires, A., Pereira, A.-M., Goldbach, R., Peters, D., and Kormelink, R. (2002). Genetic organisation of *Iris yellow spot virus* M RNA: indications for functional homology between the G(c) glycoproteins of tospoviruses and animal-infecting bunyaviruses. Arch. Virol. *147*, 2313–2325.

DeAngelis, J.D., Sether, D.M., and Rossignol, P.A. (1994). Transmission of Impatiens necrotic spot virus in peppermint by western flower thrips (Thysanoptera: Thripidae). J. Econ. Entomol. *87*, 197–201.

Diaz-Pendon, J.A., Li, F., Li, W.-X., and Ding, S.W. (2007). Suppression of antiviral silencing by cucumber mosaic virus 2b protein in *Arabidopsis* is associated with drastically reduced accumulation of three classes of viral small interfering RNAs. Plant Cell *19*, 2053–2063.

Donaire, L., Wang, Y., Gonzalez-Ibeas, D., Mayer, K.F., Aranda, M.A., and Llave, C. (2009). Deep-sequencing of plant viral small RNAs reveals effective and widespread targeting of viral genomes. Virol. *392*, 203–214.

Dong, J.H., Cheng, X.F., Yin, Y.Y., Fang, Q., Ding, M., Li, T.T., Zhang, L.Z., Su, X.X., McBeath, J.H., and Zhang, Z.K. (2008). Characterization of tomato zonate spot virus, a new tospovirus in China. Arch. Virol. *153*, 855–864.

Duijsings, D., Kormelink, R., and Goldbach, R. (1999). Alfalfa mosaic virus RNAs serve as cap donors for *Tomato spotted wilt virus* transcription during coinfection of Nicotiana benthamiana. J. Virol. *73*, 5172–5175.

Duijsings, D., Kormelink, R., and Goldbach, R. (2001). *In vivo* analysis of the TSWV cap-snatching mechanism: ingle basse complementarity and primer length requirements. EMBO J. *20*, 2545–2552.

Dunn, E.F., Pritlove, D.C., Jin, H., and Elliott, R.M. (1995). Transcription of a recombinant Bunyavirus RNA template by transiently expressed Bunyavirus proteins. Virol. *211*, 133–143.

Fabian, M.R., and White, K.A. (2004). 5′–3′ RNA–RNA interaction facilitates cap- and poly(A) tail-independent translation of Tomato bushy stunt virus mRNA: a potential common mechanism for *Tombusviridae*. J. Biol. Chem. *279*, 28862–28872.

Filho, F.M. de Assis, Naidu, R.A., Deom, C.M., and Sherwood, J.L. (2002). Dynamics of *Tomato spotted wilt virus* replication in the alimentary canal of two thrips species. Phytopathol. *92*, 729–733.

Fire, A., Xu, S., Montgomery, M.K., Kostas, S.A., Driver, S.E., and Mello, C.C. (1998). Potent and specific genetic interference by double-stranded RNA in *Caenorhabditis elegans*. Nature *391*, 806–811.

Flick, R., and Petterson, R.F. (2001). Reverse genetics system for Uukuniemi virus (*Bunyaviridae*): RNA polymerase I-catalyzed expression of chimeric viral RNAs. J. Virol. *75*, 1643–1655.

Flick, R., Elgh, F., Hobom, G., and Pettersson, R.F. (2002). Mutational analysis of the Uukuniemi virus (*Bunyaviridae*) promoter reveals two regions of functional importance. J. Virol. *76*, 10849–10860.
Flick, R., Flick, K., Feldmann, H., and Elgh, F. (2003a). Reverse genetics for Crimean–Congo hemorrhagic fever virus. J. Virol. *77*, 5997–6006.
Flick, K., Hooper, J.W., Schmaljohn, C.S., Pettersson, R.F., Feldmann, H., and Flick, R. (2003b). Rescue of Hantaanvirus minigenomes. Virol. *306*, 219–224.
Flick, K., Katz, A., Overby, A., Feldmann, H., Pettersson, R.F., and Flick, R. (2004). Functional analysis of the noncoding regions of the Uukuniemi virus (*Bunyaviridae*) RNA segments. J. Virol. *78*, 11726–11738.
Flohr, F., Schneider-Schaulies, S., Haller, O., and Kochs, G. (1999). The central interactive region of human MxA GTPase is involved in GTPase activation and interaction with viral target structures. FEBS Lett. *364*, 24–28.
Folkertsma, R.T., Spassova, M.I., Prins, M., Stevens, M.R., Hille, J., and Goldbach, R.W. (1999). Construction of a bacterial artificial chromosome (BAC) library of *Lycopersicon esculentum* cv. Stevens and its application to physically map the *Sw-5* locus. Mol. Breed. *5*,197–207.
Francki, R.I.B., Fauquet, C.M., Knudson, D.L., and Brown, F. (1991). Classification and nomenclature of viruses: V Report of International Committee on Taxonomy of Viruses. Arch Virol. (*Suppl 2*),1–450.
Francki, R.I.B., and Grivell, C.J. (1970). An electron microscopy study of the distribution of *Tomato spotted wilt virus* in systemically infected *Datura stramonium* leaves. Virol. *42*, 969.
Francki, R.I.B., and Hatta, T. (1981). *Tomato spotted wilt virus.* In Plant Virus Infections and Comparative Diagnosis, Kurstak, E., ed. (Elsevier Biomedical Press, North Holland), pp. 491–512.
Francki, R.I.B., Milne, R.G., and Hatta, T. (1984). *Tomato spotted wilt virus* group. In Atlas of Plant Viruses, vol.1, Franki, R.I.B., Milne, R.G., and Hatta, T., eds (CRC Press, Boca Raton), pp.101–110.
Frese, M., Kochs, G., Feldmann, H., Hertkorn, C., and Haller, O. (1996). Inhibition of Bunyaviruses, Phleboviruses and Hantaviruses by human MxA protein. J. Virol. *70*, 915–923.
Frese, M., Prins, M., Ponten, A., Goldbach, R.W., Haller, O., and Zeltz, P. (2000). Constitutive expression of interferon-induced human MxA protein in transgenic tobacco plants does not confer resistance to a variety of RNA viruses. Transgen. Res. *9*, 429–438.
Fujisawa, I., Tanaka, K., and Ishii, M (1988). *Tomato spotted wilt virus* transmisssion by three species of thrips, *Thrips setosus, Thrips tabaci* and *Thrips palmi* (in Japanese). Ann. Phytopath. Soc. Japan *54*, 392.
Gallie, D.R. (1998). A tale of two termini: A functional interaction between termini of an mRNA is a prerequisite for efficient translation initiation. Gene Amsterdam *216*, 1–11.
Gallie, D.R., and Kobayashi, M. (1994). The role of the 3′-untranslated region of non-polyadenylated plant viral mRNAs in regulating translational efficiency. Gene Amsterdam *142*, 159–165.
Garcin, D., and Kolakofsky, D. (1990). A novel mechanism for the initiation of Tacaribe Arenavirus genome replication. J. Virol. *64*, 6196–6203.
Gardner, M.W.,Tompkins, C.M., and Whipple, O.C. (1935). Spotted wilt of truck crops and ornamental plants. Phytopathol. *25*, 17.
Gera, A., Cohen, J., Salamon, R., and Raccah, B. (1998a). Iris yellow spot tospovirus detected in onion (Allium cepa) in Israel. Plant Dis. *82*, 127.
Gera, A., Kritzman, A., Cohen, J., and Raccah, B. (1998b). Tospovirus infecting bulb crops in Israel. In Recent Progress in Tospovirus and Thrips Research, Peters, D., and Goldbach, R. eds. (Wageningen, The Netherlands), pp. 86–87.
Gielen, J.J.L., de Haan, P., Kool, A.J., Peters, D., van Grinsven, M.Q.J.M., and Goldbach, R.W. (1991). Engineered resistance to *Tomato spotted wilt virus,* a negative-strand RNA virus. Bio/Technol. *9*, 1363–1367.
Goldbach, R., and Kuo, G. (1996). Introduction: Proceedings of the international symposium on tospovirus and thrips of floral and vegetable crops. Act. Hort. *431*,21–26.
Goldbach, R.W., and Peters, D. (1994). Possible causes of the emergence of tospovirus diseases. Semin. Virol. *5*, 113–120.
Goto, K., Kobori, T., Kosaka, Y., Natsuaki, T., and Masuta, C. (2007). Characterisation of silencing suppressor 2b of Cucumber mosaic virus based on examination of its small RNA-binding abilities. Plant Cell Physiol. *48*, 1050–1060.
Granval de Milan, N., Piccolo, R., and Gracia, O. (1998). Potato (*Solanum tuberosum* L.) tuber transmission of tospoviruses (TSWV, GRSV and TCSV) in Mendoza province, Argentina. In Recent Progress in Tospovirus and Thrips Research, Peters, D., and Goldbach, R. eds. (Wageningen, The Netherlands), pp. 46–47.

Griep, R.A., Prins, M., van Twisk, C., Keller, H.J.H.G., Kerschbaumer, R.J., Kormelink, R., Goldbach, R.W., and Schots, A. (2000). Application of phage display in selecting *Tomato spotted wilt virus*-specific single-chain antibodies (scFvs) for sensitive diagnosis in ELISA. Phytopathol. *90*, 183–190.

Gubba, A., Gonsalves, C., Stevens, M.R., Tricoli, D.M., and Gonsalves, D. (2002). Combining transgenic and natural resistance to obtain broad resistance to tospovirus infection in tomato (*Lycopersicon esculentum mill*). Mol. Breed. *9*, 13–23

Guo, H.S., and Ding, S.W. (2002). A viral protein inhibits the long range signaling activity of the gene silencing signal. EMBO J. *21*, 398–407.

de Haan, P., Wagemakers, L., Peters, D., and Goldbach, R. (1989). Molecular cloning and terminal sequence determination of the S and M RNA species of *Tomato spotted wilt virus*. J. Gen. Virol. *70*, 3469–3474.

de Haan, P., Wagemakers, L., Peters, D., and Goldbach, R. (1990). The S RNA segment of *Tomato spotted wilt virus* has an ambisense character. J. Gen. Virol. *71*, 1001–1008.

de Haan, P., Kormelink, R., Resende, D., van Poelwijk, F., Peters, D., and Goldbach, R. (1991). *Tomato spotted wilt virus* L RNA encodes a putative RNA polymerase. J. Gen. Virol. *72*, 2207–2216.

de Haan, P., Gielen, J.J.L., Prins, M., Wijkamp, I.G., van Schepen, A., Peters, D., van Grinsven, M.Q.J.M., and Goldbach, R. (1992). Characterisation of RNA-mediated resistance to *Tomato spotted wilt virus* in transgenic tobacco plants. Bio/Technol. *10*, 1133–1137.

Hassani-Mehraban, A., Saaijer, J., Peters, D., Goldbach, R., and Kormelink, R. (2005). A new tomato-infecting tospovirus from Iran. Phytopathology *95*, 852–858.

Hassani-Mehraban, A., Saaijer, J., Peters, D., Goldbach, R., and Kormelink, R. (2008). Molecular and biological comparison of two Tomato yellow ring virus (TYRV) isolates: challenging the *Tospovirus* species concept. Arch. Virol. *152*, 85–96.

Hassani-Mehraban, A., Brenkman, A.B, van den Broek, N.J.F., Goldbach, R., and Kormelink, R. (2009). RNAi-mediated transgenic tospovirus resistance broken by intraspecies silencing suppressor protein complementation. MPMI *22*, 1250–1257.

Hassani-Mehraban, A., Botermans, M., Verhoeven, K., Meekes, E., Saaijer, J., Peters, D., Goldbach, R., and Kormelink, R. (2010). A distinct tospovirus species causing necrotic streak on *Alstroemeria* sp. in Columbia. Arch. Virol. *155*, 423–428.

Hassani-Mehraban, A., Cheewachaiwit, S., Relevante, C., Kormelink, R., and Peters, D. (2011). Serological and molecular characterization of *Tomato necrotic ring virus* (TNRV), a recently described tospovirus species infecting tomato and pepper in Thailand. Eur. J. Plant Pathol. doi: 10.1007/S10658-011-9771-9.

Heath, C.M., Windsor, M., and Wileman, T. (2001). Aggresomes resemble sites specialized for virus assembly. J. Cell Biol. *153*, 449–455.

Hemmes, H. (2007). Strategies underlying RNA silencing suppression by negative strand RNA viruses. PhD thesis, Wageningen University, The Netherlands.

Hemmes, H., Lakatos, L., Goldbach, R., Burgyan, J., and Prins, M. (2007). The NS3 protein of Rice hoja blanca tenuivirus suppresses RNA silencing in plant and insect hosts by efficiently binding both siRNAs and miRNAs. RNA *13*, 1079–1089.

Hoffmann, K., Qiu, W.P., and Moyer, J.W. (2001). Overcoming host- and pathogen-mediated resistance in tomato and tobacco maps to the M RNA of *Tomato spotted wilt virus*. MPMI *14*, 242–249.

Huiet, L., Feldstein, P.A., Tsai, J.H., and Falk, B.W. (1993). The maize stripe virus major noncapsid protein messenger RNA transcripts contain heterogeneous leader sequences at their 5′ termini. Virol. *197*, 808–812.

van den Hurk, J., Tas, P.W.L., and Peters, D. (1977). The RNA of *Tomato spotted wilt virus*. J. Gen. Virol. *36*, 81–91.

Iapalucci, S., Lopez, N., and Franze-Fernandez, M.T. (1991). The 3′ end termini of the *Tacaribe arenavirus* subgenomic RNAs. Virol. *182*, 269–278.

Ie, T.S. (1964). An electron microscope study of *Tomato spotted wilt virus* in the plant cell. Neth. J. Plant Path. *70*, 114–115.

Ie, T.S. (1971). Electron microscopy of developmental stages of *Tomato spotted wilt virus* in plant cells. Virology *43*, 468–479.

Ie, T.S. (1973). *Tomato spotted wilt virus* in the anthers of *Tropaeolum majus*. Neth. J. Plant Path. *79*, 387–391.

Ie., T.S. (1982). A sap-transmissible, defective form of *Tomato spotted wilt virus*. J. Gen. Virol. *59*, 387–391.

Jahn, M., Paran, I., Hoffmann, K., Radwansky,E., Livingstone, K., Grube, R., Aftergoot, E., Lapidot, M., and Moyer, J. (2000). Genetic mapping of the Tsw locus for resistance to the Tospovirus *Tomato spotted wilt virus* in *Capsicum* spp. and its relationship to the Sw-5 gene for resistance to the same pathogen in tomato. MPMI *13*, 673–682.

Jan, F.-J., Fagoaga, C., Pang, S.-Z., and Gonsalves, D. (2000). A minimum length of N gene sequence in transgenic plants is required for RNA-mediated tospovirus resistance. J. Gen. Virol. *81*, 235–242.
Jin, H., and Elliott, R.M. (1993). Characterization of Bunyamwera virus S RNA that is transcribed and replicated by the L protein expressed from recombinant vaccinia virus. J. Virol. *67*, 1396–1404.
Johansen, L.K., and Carrington, J.C. (2001). Silencing on the spot. Induction and suppression of RNA silencing in the Agrobacterium-mediated transient expression system. Plant Physiol. *126*, 930–938.
Johnston, J.A., Ward, C.L., and Kopito, R.R. (1998). Aggresomes: A cellular response to misfolded proteins. J. Cell Biol. *143*, 1883–1898.
Kameya-Iwaki, M., Honda, Y., Hanada, K., Tochihara, H., Yonaha, T., Hokama, K., and Yokoyama, T. (1984). Silver mottle disease of watermelon caused by tomato spotted witt virus. Plant Dis. *68*, 1006–1008.
Kanerva, M., Melen, K., Vaheri, A., and Julkunen, I. (1996). Inhibition of puumula and tula hantaviruses in Vero cells by MxA protein. Virol. *224*, 55–62.
Kato, K., and Hanada, K. (2000). Characterization of the S RNA segment of melon yellow spot virus. Jap. J. Phytopathol. *66*, 252–254.
Kellman, J.-F., Liebisch, P., Schmitz, K.-P., and Piechulla, B. (2001). Visual representation by atomif force microscopy (AFM) of tomato spotted wiltvirus ribonucleoproteins. Biol. Chem. *382*, 1559–1562.
Khvorova, A., Reynolds, A., and Jayasena, S.D. (2003). Functional siRNAs and miRNAs exhibit strand bias. Cell *115*, 209–216.
Kikkert, M., van Poelwijk, F., Storms, M., Bloksma, H., Karsies, W., Kormelink, R., and Goldbach, R. (1997). A protoplast system for studying *Tomato spotted wilt virus* infection. J. Gen. Virol. *78*, 1755–1763.
Kikkert, M., Meurs, C., van de Wetering, F., Kormelink, R., and Goldbach, R. (1998). Binding of *Tomato spotted wilt virus* to a 94 kDa thrips protein. Phytopathology *88*, 63–69.
Kikkert, M., van Lent, J., Storms, M., Bodegom, P., Kormelink, R., and Goldbach, R. (1999). *Tomato spotted wilt virus* particle morphogenesis in plant cells. J. Virol. *73*, 2288–2297.
Kikkert, M., Verschoor, A., Kormelink, R., Rottier, P., and Goldbach, R. (2001). *Tomato spotted wilt virus* glycoproteins exhibit trafficking and localization signals that are functional in mammalian cells. J. Virol. *75*, 1004–1012.
Kitajima, E.W. (1965). Electron microscopy of vira cabexa virus (Brazilian tomato spotted witl virus) within the host cell. Virology *26*, 89–99.
Kitajima, E.W., de Ávila, A.C., Resende, R. de O., Goldbach, R.W., and Peters, D. (1992). Comparative cytological and immunogold labelling studies on different isolates of *Tomato spotted wilt virus*. J. Submicrosc. Cytol. Pathol. *24*, 1–14.
van Knippenberg, I. (2005). Analysis of *Tomato spotted wilt virus* genome transcription. PhD thesis, Wageningen University, The Netherlands.
van Knippenberg, I., Goldbach, R., and Kormelink, R. (2002). Purified *Tomato spotted wilt virus* particles support both genome replication and transcription *in vitro*. Virology *303*, 278–286.
van Knippenberg, I., Lamine, M., Goldbach, R., and Kormelink, R. (2005a). *Tomato spotted wilt virus* transcriptase *in vitro* displays a preference for cap donors with multiple base complementarity to the viral template. Virology *335*, 122–130.
van Knippenberg, I., Goldbach, R., and Kormelink, R. (2005b). *Tomato spotted wilt virus* S-segment mRNAs have overlapping 3′-ends containing a predicted stem–loop structure and conserved sequence motif. Vir. Res. *110*,125–131.
Kochs, G., Janzen, C., Hohenberg, H., and Haller, O. (2002). Antivirally active MxA protein sequesters La Crosse virus nucleocapsid protein into perinuclear complexes. Proc. Natl. Acad. Sci. U.S.A. *99*, 3153–3158.
Kormelink, R., Kitajima, E.W., de Haan, P., Zuidema, D., Peters, D., and Goldbach, R. (1991). The nonstructural protein (NSs encoded by the ambisense S RNA of *Tomato spotted wilt virus* is associated with fibrous structures in infected plant cells. Virology *181*, 459–468.
Kormelink, R., De Haan, P., Meurs, C., Peters, D., and Goldbach, R. (1992a). The nucleotide sequence of the M RNA segment of *Tomato spotted wilt virus*: A bunyavirus with two ambisense RNA segments. J. Gen. Virol. *73*, 2795–2804.
Kormelink, R., de Haan, P., Peters, D., and Goldbach, R. (1992b). Viral RNA synthesis in *Tomato spotted wilt virus*-infected *Nicotiana rustica* plants. J. Gen. Virol. *73*, 687–693.
Kormelink, R., Van Poelwijk, F., Peters, D., and Goldbach, R. (1992c). Non-viral heterogeneous sequences at the 5′ ends of *Tomato spotted wilt virus* (TSWV) mRNAs. J. Gen. Virol. *73*, 2125–2128.
Kormelink, R., Storms, M., van Lent, J., Peters, D., and Goldbach, R. (1994). Expression and subcellular location of the NSm protein of *Tomato spotted wilt virus* (TSWV), a putative viral movement protein. Virology *200*, 56–65.

Lakatos, L., Szittya, G., Silhavy, D., and Burgyan, J. (2004). Molecular mechanism of RNA silencing suppression mediated by p19 protein of tombusviruses. EMBO J. *23*, 876–884.

Lakatos, L., Csorba, T., Pantaleo, V., Chapman, E.J., Carrington, J.C., Liu, Y.P., Dolja, V.V., Calvino, L.F., Lopez-Moya, J.J., and Burgyan, J. (2006). Small RNA binding is a common strategy to suppress RNA silencing by several viral suppressors. EMBO J. *25*, 2768–2780.

Lakshmi, K.V., Wightman, J.A., Reddy, D.V.R., Ranga Rao, G.V., Buiel, A.A.M., and Reddy, D.D.R. (1995). Transmission of peanut bud necrosis virus by *Thrips palmi* in India. In Thrips Biology and Management , Parker, B.L., and Lewis, T., eds. (Plenum Press, New York), pp.170–184.

Latham, L.J., and Jones, R.A.C. (1998). Selection of resistance breaking strains of tomato spotted wilt tospovirus. Ann. Appl. Biol. *133*, 385–402.

Law, M.D., and Moyer,J.W. (1990). A tomato spotted wilt-like virus with a serologically distinct N protein. J. Gen. Virol. *71*, 933–938.

Law, M.D., Speck, J., and Moyer, J.W. (1992). The M RNA of *Impatiens necrotic* spot Tospovirus (*Bunyaviridae*) has an ambisense genomic organization. Virology *188*, 732–741.

Lawson, R.H., Dienelt, M.M., and Hsu, H.T. (1996). Ultrastructural comparisons of defective, partially defective, and nondefective isolates of *Impatiens necrotic* spot virus. Phytopathol. *86*, 650–661.

Lee, Y., Ahn, C., Han, J., Choi, H., Kim, J., Yim, J., Lee, J., Provost, P., Radmark, O., Kim, S., and Kim, V.N. (2003). The nuclear RNase III Drosha initiates microRNA processing. Nature *425*, 415–419.

Leonard, S., Viel, C., Beauchemin, C., Daigneault, N., Fortin, M.G., and Laliberte, J.F. (2004). Interaction of VPg-pro of Turnip mosaic virus with the translation initiation factor 4E and the poly(A)-binding protein *in planta*. J. Gen. Virol. *85*, 1055–1063.

Lewandowski, D.J., and Adkins, S. (2005). The tubule-forming NSm protein from *Tomato spotted wilt virus* complements cell-to-cell and long-distance movement of Tobacco mosaic virus hybrids. Virol. *342*, 26–37.

Li, W., Lewandowski, D.J., Hilf, M.E., and Adkins, S. (2009). Identification of domains of the *Tomato spotted wilt virus* NSm protein involved in tubule formation, movement and symptomatology. Virol. *390*, 110–121.

Lin, Y.-H., Chen, T.-C., Hsu, H.-T., Liu, F.-L., Chu, F.-H., Chen, C.-C., Lin, Y.-Z., and Yeh, S.-D. (2005). Serological comparison and molecular characterization for verification of Calla lily chlorotic spot virus as a new tospovirus species belonging to *Watermelon silver mottle virus* serogroup. Phytopathology 95,1482–1488.

Llave, C., Kasschau, K.D., Rector, M.A., and Carrington, J.C. (2002). Endogenous and silencing-associated small RNAs in plants. Plant Cell *14*, 1605–1619.

Lopez, N., Muller, R., Prehaud, C., and Bouloy, M. (1995). The L protein of Rift valley fever virus can rescue viral ribonucleoproteins and transcribe synthetic genome-like RNA molecules. J. Virol. *69*, 3972–3979.

Louro, D. (1996). Detection and identification of *Tomato spotted wilt virus* and impatiens necrotic spot virus in Portugal. Act. Hort. *431*, 99–105.

Lovatoa, F.A., Inoue-Nagatab, A.K., Nagata, T., de Ávila, A.C., Rodrigues Pereira, L.A., and Resende, R. de (2008). The N protein of *Tomato spotted wilt virus* (TSWV) is associated with the induction of programmed cell death (PCD) in *Capsicum chinense* plants, a hypersensitive host to TSWV infection. Vir. Res. *137*, 245–25.

McMichael, L.A., Persley, D.M., and Thomas, J.E. (2002). A new tospovirus serogroup IV species infecting capsicum and tomato in Queensland, Australia. Austral. Plant Pathol. *31*, 231–239.

Mallory, A.C., Reinhart, B.J., Bartel, D., Vance, V.B., and Bowman, L.H. (2002). A viral suppressor of RNA silencing differentially regulates the accumulation of short interfering RNAs and micro-RNAs in tobacco. Proc. Nat. Acad. Sci. U.S.A. *99*, 15228–15233.

Marchoux, G., Gebreselassie, K., and Villevieille, M. (1991). Detection of *Tomato spotted wilt virus* and transmission by *Frankliniella occidentalis* in France. Plant Pathol. *40*, 347–380.

Margaria, P., Ciuffo, M., Pacifico, D. And Turina, M. (2007). Evidence that the nonstructural protein of *Tomato spotted wilt virus* is the avirulence determinant in the interaction with resistant pepper carrying the *Tsw* gene. MPMI *20*, 547–558.

Maris, P.C., Joosten, N.N., Goldbach, R.W., and Peters, D. (2003a). Restricted Spread of *Tomato spotted wilt virus* in Thrips-resistant pepper. Phytopathology 93, 1223–1227.

Maris, P.C., Joosten, N.N., Peters, D., and Goldbach, R.W. (2003b). Thrips resistance in pepper and its consequences for the acquisition and inoculation of *Tomato spotted wilt virus* by the western flower thrips Phytopathol. 93, 96–101.

Martin, M.M. (1964). Purification and electron microscope studies of tomato spotted witl virus (TSWV) from tomato roots. Virol. *22*, 646–649.

Matsuda, D., and Dreher, T.W. (2004). The tRNA-like structure of Turnip yellow mosaic virus RNA is a 3'-translational enhancer. Virol. *321*, 36–46.

Medeiros, R.B., Resende, R. de O., and de Avila, A.C. (2004). The plant virus *Tomato Spotted Wilt Tospovirus* activates the immune system of its main insect vector, *Frankliniella occidentalis*. J. Virol. *78*, 4976–4982.

Medeiros, R.B., Ullman, D.E., Sherwood, J.L., and German, T.L. (2000). Immunoprecipitation of a 50-kDa protein: a candidate receptor component for tomato spotted wilt tospovirus (*Bunyaviridae*) in its main vector, *Frankliniella occidentalis*. Vir. Res. *67*, 109–118.

Meulewaeter, F., van Lipzig, R., Gultayev, A.P., Pleij, C.W., van Damme, D., Cornelissen, M., and van Eldik, G. (2004). Conservation of RNA structures enables TNV and BYDV 5' and 3' elements to cooperate synergistically in cap-independent translation. Nucl. Acids. Res. *32*, 1721–1730

Milne, R.G. (1970). An electron microscopic study of *Tomato spotted wilt virus* in sections of infected cells and in negative stain preparations. J. Gen. Virol. *6*, 267–276.

Mohamed, N.A. (1981). Isolation and characterisation of subviral structures from *Tomato spotted wilt virus*. J. Gen. Virol. *53*, 197–206.

Mohamed, N.A., Randles, J.W., and Francki, R.I.B. (1973). Protein composition of *Tomato spotted wilt virus*. Virology *56*, 12–21.

Molnár, A., Csorba, T., Lakatos, L., Várallyay, E., Lacomme, C., and Burgyán, J. (2005). Plant virus-derived small interfering RNAs originate predominantly from highly structured single-stranded viral RNAs. J. Virol. *79*, 7812–7818.

Moreau, P., Brandizzi, F., Hanton, S., Chatre, L., Melser, S., Hawes, C., and Satiat-Jeunemaitre, B. (2007) The plant ER–Golgi interface: a highly structured and dynamic membrane complex. J. Exp. Bot. *58*, 49–64.

Moury B., Selassie K.G., Marchoux G., Daubèze A-M., and Palloix A. (1998). High temperature effects on hypersensitive resistance to Tomato spotted wilt tospovirus (TSWV) in pepper (Capsicum chinense Jacq.). Eur J Plant Pathol *104*, 489–498.

Nagata, T., Storms, M.M.H., Goldbach, R., and Peters, D. (1997). Multiplication of *Tomato spotted wilt virus* in primary cell cultures derived from two thrips species. Vir. Res. *49*, 59–66.

Nagata, T., and de Ávila, A.C. (2000). Transmission of chrysanthemum stem necrosis virus, a recently discovered Tospovirus, by two thrips species. J. Phytopathol. Berlin *148*, 123–125.

Nagata, T., Inoue-Nagata, A.K., van Lent, J., Goldbach, R., and Peters, D. (2002). Factors determining vector competence and specificity for transmission of *Tomato spotted wilt virus*. J. Gen. Virol. *83*, 663–671.

Nakahara S., and Monteiro R.C. (1999). *Frankliniella zucchini* (Thysanoptera: Thripidae), a new species and vector of tospovirus in Brazil. Proc. Entomol. Soc. Washington *101*, 290–294.

Neeleman, L., Olsthoorn, R.C.L, Linthorst, H.J.M., and Bol, J.F. (2001). Translation of a nonpolyadenylated viral RNA is enhanced by binding of viral coat protein or polyadenylation of the RNA. Proc. Natl. Acad. Sci. U.S.A. *98*, 14286–14291.

Ohnishi, J., Katsuzaki, H., Tsuda, S., Sakurai, T., Akutsu, K., and Murai, T. (2006). *Frankliniella cephalica*, a new vector for *Tomato spotted wilt virus*. Plant Dis. *90*, 685.

Okuda, M., Kato, K., Hanada, K., and Iwanami, I. (2006). Nucleotide sequence of melon yellow spot virus M RNA segment and characterization of non-viral sequences in subgenomic RNA. Arch. Virol. *151*, 1–11.

Paape, M., Solovyev, A.G., Erokhina, T.N., Minina, E.A., Schepetilnikov, M.V., Lesemann, D.-E., Schiemann, J., Morozov, S.U., and Kellmann, J.W. (2006). At-4/1, an interactor of the *Tomato spotted wilt virus* movement protein, belongs to a new family of plant proteins capable of directed intra- and intercellular trafficking. MPMI *19*, 874–883.

Palmer, J.M., Reddy, D.V.R., Wightman, J.A., and Ranga Rao, G.V. (1990). New information on the thrips vectors of *Tomato spotted wilt virus* in groundnut crops in India. Int. Arachis Newsl. *7*, 24–25.

Pang, S.-Z., Jan, F.-J., and Gonsalves, D. (1997). Nontarget DNA sequences reduce the transgene length necessary for RNA-mediated tospovirus resistance in transgenic plants. Proc. Natl. Acad. Sci. U.S.A. *94*, 8261–8266.

Peters, D. (1998). An updates list of plant species susceptible to tospoviruses. In Recent Progress in Tospovirus and Thrips Research, Peters, D., and Goldbach, R., eds. (Wageningen, The Netherlands), pp.107–110.

Pittman, H.A. (1927). Spotted wilt of tomatoes. Preliminary note concerning the transmission of the 'spotted wilt' of tomatoes by an insect vector (Thrips tabaci Lind.). Counc. Sci. Ind. Res. Bull. *1*, 74–77.

Plotch, S., Bouloy, M., and Krug, R.M. (1979). Transfer of 5-prime terminal cap of globin messenger rna to influenza viral complementary RNA during transcription in-vitro. Proc. Nat. Acad. Sci. U.S.A. *76*, 1618–1622.

Plotch, S.J., Bouloy, M., Ulmanen, I., and Krug, R.M. (1981). A unique cap 7 methyl guanosine 5′ tri phosphoryl 5′-2-O methyl nucleoside dependent Influenza virion endo nuclease cleaves capped RNA to generate the primers that initiate viral RNA transcription. Cell *23*, 847–858.
van Poelwijk, F., Boye, K., Oosterling, R., Peters, D., and Goldbach, R. (1993). Detection of the L protein of *Tomato spotted wilt virus*. Virology *197*, 468–470.
van Poelwijk, F., Kolkman, J., and Goldbach, R. (1996). Sequence analysis of the 5′ ends of *Tomato spotted wilt virus* N mRNAs. Arch. Virology *141*, 177–184.
Prehaud, C., Lopez, N., Blok, M.J., Obry, V., and Bouloy, M. (1997). Analysis of the 3′ terminal sequence recognized by the Rift valley fever virus transcription complex in its ambisense S segment. Virol. *227*, 189–197.
Premachandra, W.T.S.D., Borgemeister, C., Maiss, E., Knierim, D., and Poehling, H.-M. (2005). *Ceratothripoides claratris*, a new vector of a *Capsicum chlorosis* virus isolate infecting tomato in Thailand. Phytopathology *95*, 659–663.
Prins, M., and Goldbach, R. (1998). The emerging problem of tospovirus infection and nonconventional methods of control. Trends Microb. *6*, 31–35.
Prins, M., de Haan, P., Luyten, R., van Veller, M., van Grinsven, M.Q.J.M., and Goldbach, R. (1995). Broad resistance to tospoviruses in transgenic tobacco plants expressing three tospoviral nucleoprotein gene sequences. MPMI *8*, 85–91.
Prins, M., Resende, de O., Anker, Corine, van Schepen, de Haan, P., and Goldbach, R. (1996). Engineered RNA-mediated resistance to *Tomato spotted wilt virus* is sequence specific. MPMI *9*,416–418.
Prins, M., Kikkert, M., Ismayadi, C., de Graauw, W., de Haan, P., and Goldbach, R. (1997a). Characterisation of RNA-mediated resistance to *Tomato spotted wilt virus* in transgenic tobacco plants expressing NS_M gene sequences. Plant Mol. Biol. 33, 235–243.
Prins, M., Storms, M.M.H., Kormelink, R., de Haan, P., and Goldbach, R. (1997b). Transgenic tobacco plants expressing the putative movement protein of Tomato spotted wilt tospovirus exhibit aberrations in growth and appearance. Trans. Res. *6*, 245–251.
Prins, M., Lohuis, D., Schot, A., and Goldbach, R. (2005). Phage display-selected single-chain antibodies confer high levels of resistance against *Tomato spotted wilt virus*. J. Gen. Virol. *86*, 2107–2113.
Raju, R., Raju, L., Hacker, D., Garcin, D., Compans, R., and Kolakofsky, D. (1990). Nontemplated bases at the 5′ ends of Tacaribe virus messenger RNA. Virol. *174*, 53–59.
Ravkov, E.V., and Compans, R.W. (2001). Hantavirus nucleocapsid protein is expressed as a membrane-associated protein in the perinuclear region. J. Virol. *75*, 1808–1815.
Ravkov, E. V., Nichol, S.T., Peters, C.J., and Compans, R.W. (1998). Role of actin microfilaments in Black Creek Canal virus morphogenesis. J. Virol. *72*, 2865–2870.
Reddy, D.V.R., Sudarshana, M.R., Ratna, A.S., Reddy, A.S., Amin, P.W., Kumar, I.K., and Murphy, A.K. (1991). The occurrence of yellow spot virus, a member of tomato spotted wilt group, on peanut (Arachis hypogaea L.) in India. In Virus–Thrips–plant interactions of TSWV, Hsu, H.-T., and Lawson, R.H., eds., (Proc. USDA Workshop, Beltsville, USA), pp. 77–78.
Reddy, D.V.R., Ratna, A.S., Sudarshana, M.R., Poul, F., and Kumar, I.K. (1992). Serological relationships and purification of bud necrosis virus, a tospovirus occurring in peanut (Arachis hypogaea L.) in India. Ann. Appl. Biol. *120*, 279–286.
Reichelt, M., Stertz, S., Krijnse-Locker, J., Haller, O., and Kochs, G. (2004). Missorting of LaCrosse virus nucleocapsid protein by the interferon-induced MxA GTPase involves smooth ER membranes. Traffic *5*, 772–784.
Resence, R. De O., de Haan, P., de Avila, A., Kitajima, E.W., Kormelink, R., Goldbach, R., and Peters, D. (1991). Generation of envelope and defective interfering RNA mutants of *Tomato spotted wilt virus* by mechanical passage. J. Gen. Virol. 72, 2375–2383.
Resende, R. de O., Posser, L., Nagata, I., Bezerra, I.C., Lima, M.I., Kitajima, E.W., and de Ávila, A.C. (1996). New tospoviruses found in Brazil. Act. Hort. *431*, 78–79.
Ribeiro, D. (2007). Towards understanding TSWV particle assembly: Analysis of the interacellular behaviour of the viral structural proteins. PhD thesis. Wageningen University, The Netherlands.
Ribeiro, D., Foresti, O., Denecke, J., Wellink, J., Goldbach, R., and Kormelink, R. (2008). *Tomato spotted wilt virus* glycoproteins induce the formation of ER and Golgi-derived pleomorphic membrane structures in protoplasts. J. Gen. Virol. *89*, 1811–1818.
Ribeiro, D., Borst, J.-W., Goldbach, R., and Kormelink, R. (2009a). *Tomato spotted wilt virus* nucleocapsid protein interacts with both viral glycoproteins Gn and Gc *in planta*. Virol. 383, 121–130.
Ribeiro, D., Goldbach, R., and Kormelink, R. (2009b). Trafficking of *Tomato spotted wilt virus* glycoproteins: requirements for ER-arrest and exit to the Golgi. Traffic *10*, 664–672.

Rinne, P.L.H., Boogaard, R.van den, Mensink, M., Kopperud, C., Kormelink, R., Goldbach, R., and Schoot, C. van der (2005). Tobacco plants respond to the constitutive expression of the tospovirus movement protein NSm with a heat-reversible sealing of plasmodesmata that impairs development. The Plant J. *43*, 688–707.

Roggero, P. (2002). Infection of tospoviruses in pepper and control by resistant plants. Biológico, Sao Paulo *64*, 179–181.

Roggero, P., Lisa, V., Nervo, G., and Pennazio, S. (1996). Continuous high temperature can break the hypersensitivity of *Capsicum chinense* 'PI152225' to tomato spotted wilt tospovirus (TSWV). Phytopathol. Medit. 35, 117–120.

Roggero, P., Pennazio, S., Masenga, V., and Tavella, L. (1999). Resistance to tospoviruses in pepper. In Proceedings of the 7th International Symposium on Thysanoptera, Reggio di Calabria, Italy, 2001.

Roggero, P., Masenga, V., and Tavella, L. (2002). Field isolates of *Tomato spotted wilt virus* overcoming resistance in pepper and their spread to other hosts in Italy. Plant Dis *86*, 950–954.

Rosello, S., Diez, M.J., and Nuez, F. (1997). Utilization of *Capsicum* sp. resistance to TSWV in pepper breeding. Capsicum Eggplant Newsletters *16*, 87–90.

Rudolph, C., Schreier, P.H., and Uhrig, J.F. (2003). Peptide-mediated broad spectrum plant resistance to tospoviruses. Proc. Natl. Acad. Sci. U.S.A. *100*, 4429–4434.

Sakimura, K. (1963). Frankliniella fusca, an additional vector for the *Tomato spotted wilt virus*, with notes on *Thrips tabacii*, another vector. Phytopathology 53, 412–415.

Samuel, G., Bald, J.G., and Pittman, H.A. (1930). Investigations of 'spotted wilt' of tomatoes. Scien. Ind. Res. Bull. *44*,1–64.

Satyanarayana, T., Mitchell, S.E., Reddy, D.V.R., Brown, S., Kresovich, S., Jarret, R., Naidu, R. A., and Demski, J.W. (1996a). Peanut bud necrosis tospovirus S RNA: Complete nucleotide sequence, genome organization and homology to other tospoviruses. Arch. Virol. *141*, 85–98.

Satyanarayana, T., Mitchell, S.E., Reddy, D.V.R., Kresovich, S., Jarret, R., Naidu, R.A., Gowda, S., and Demski, J.W. (1996b). The complete nucleotide sequence and genome organization of the M RNA segment of peanut bud necrosis tospovirus and comparison with other tospoviruses. J. Gen. Virol. *77*, 2347–2352.

Satyanarayana, T., Gowda, S., Lakshminarayana Reddy, K., Mitchell, S.E., Dawson, W.O., and Reddy, D.V.R. (1998). Peanut yellow spot virus is a member of a new serogroup of Tospovirus genus based on small (S) RNA sequence and organisation. Arch. Virol. *143*, 353–364

Schnettler, E., Hemmes, H., Huismann, R., Goldbach, R., Prins, M., and Kormelink, R. (2010)· Diverging affinity of tospovirus RNA silencing suppressor proteins, NSs, for various RNA duplex molecules. J. Virol. *84*, 11542–11554.

Scholthof, H.B. (2005). Plant virus transport: motions of functional equivalence. Trends Plant Sci. *10*, 376–382.

Schwarz, D.S., Hutvagner, G., Du, T., Xu, Z., Aronin, N., and Zamore, P.D. (2003). Asymmetry in the assembly of the RNAi enzyme complex. Cell *115*, 199–208.

Simons, J.F., and Pettersson, R.F. (1991). Host-derived 5′ ends and overlapping complementary 3′ ends of the two messenger RNAs transcribed from the ambisense S segment of Uukuniemi virus. J. Virol. *65*, 4741–4748.

Sin, S.H., McNulty, B.C., Kennedy, G.G., and Moyer, J.W. (2005). Viral genetic determinants for thrips transmission of *Tomato spotted wilt virus*. Proc. Natl. Acad. Sci. USA *102*, 5168–5173.

Snippe, M., Goldbach, R., and Kormelink, R. (2005a). *Tomato spotted wilt virus* particle assembly and the prospects of fluorescence microscopy to study protein–protein interactions involved. Adv. Virus Res. *65*, 63–120.

Snippe, M., Borst, J.-W., Goldbach, R., and Kormelink, R. (2005b). The use of fluorescence microscopy to visualise homotypic interactions of *Tomato spotted wilt virus* nucleocapsid protein in living cells. J. Virol. Meth. *125*, 15–22.

Snippe, M., Smeenk, L., Goldbach, R., and Kormelink, R. (2007a). The cytoplasmic domain of *Tomato spotted wilt virus* Gn glycoprotein is required for Golgi localization and interaction with Gc. Virology *363*, 272–279.

Snippe, M., Borst, J.-W., Goldbach, R., and Kormelink, R. (2007b). *Tomato spotted wilt virus* Gc and N proteins interact *in vivo*. Virology 357, 115–123.

Soellick, T.R., Uhrig, J.F., Bucher, G.L., Kellmann, J.W., and Schreier, P.H. (2000). The movement protein NSm of Tomato spotted wilt tospovirus (TSWV): RNA binding, interaction with the TSWV N protein, and identification of interacting plant proteins. Proc. Natl. Acad. Sci. U.S.A. 97, 2373–2378.

Spassova, M.I., Prins, T.W., Folkertsma, R.T., Klein-Lankhorst, R.M., Hille, J., Goldbach, R.W., and Prins, M. (2001). The tomato gene *Sw-5* is a member of the coiled coil, nucleotide binding, leucine-rich repeat class of plant resistance genes and confers resistance to TSWV in tobacco. Mol. Breed. *7*, 151–161.

Stevens, M.R., Scott, S.J., and Gergerich, R.C. (1992). Inheritance of a gene for resistance to *Tomato spotted wilt virus* (TSWV) from Lycopersicon peruvianum Mill. Euphytica *59*, 9–17.

Storms, M.M.H., Kormelink, R., Peters, D., van Lent, J.W.M., and Goldbach, R.W. (1995). The nonstructural NSm protein of *Tomato spotted wilt virus* induces tubular structures in plant and insect cells. Virology *214*, 485–493.

Storms, M.M.H., van der schoot, C., Prins, M., Kormelink, R., van Lent, J.W.M., and Goldbach, R.W. (1998). A comparison of two methods of microinjection for assessing altered plasmodesmal gating in tissues expressing viral movement proteins. Plant J. *13*, 131–140.

Szittya, G., Molnar, A., Silhavy, D., Hornyik, C., and Burgyan, J. (2002). Short defective interfering RNAs of tombusviruses are not targeted but trigger post-transcriptional gene silencing against their helper virus. Plant Cell *14*, 359–372.

Takeda, A., Sugiyama, K., Nagano, H., Mori, M., Kaido, M., Mise, K., Tsuda, S., and Okuno, T. (2002). Identification of a novel RNA silencing suppressor, NSs protein of *Tomato spotted wilt virus*. FEBS Lett. *532*, 75–79.

Takeuchi, S., Okuda, M., Hanada, K., Kawada, Y., and Kameya-Iwaki, M. (2001). Spotted wilt disease of cucumber (*Cucumis sativus*) caused by melon yellow spot virus. Jap. J. Phytopathol. *67*, 46–51.

Tas, P.W.L., Boerjan, M.L., and Peters, D. (1977). The structural proteins of *Tomato spotted wilt virus*. J. Gen. Virol. *36*, 267–279.

Tsuda, S., Fujisawa, I., Ohnishi, J., Hosokawa, D., and Tomaru, K. (1996). Localization of Tomato spotted wilt tospovirus in larvae and pupae of the insect vector *Thrips setosus*. Phytopathology *86*, 1199–1203.

Ullman, D.E., Westcot, D.M., Hunter, W.B., and Mau, R.F.L. (1989). Internal anatomy and morphology of *Frankliniella occidentalis* (Pergande) (Thysanoptera: Thripidae) with special reference to interactions between thrips and *Tomato spotted wilt virus*. Int. J. Insect Morphol. Embryol. *18*, 289–310.

Ullman, D.E., Cho, J.J., Mau, R.F.L., Westcoc, D.M., and Custer, D.M. (1992). A mid-gut barrier to *Tomato spotted wilt virus* acquisition by adult Western flower thrips. Phytopathology *82*, 1333–1342.

Ullman, D.E., German, T.L., Sherwood, J.L., Westcot, D.M., and Cantone, F.A. (1993). Tospovirus replication in insect vector cells: Immunocytochemical evidence that the nonstructural protein encoded by the S RNA of Tomato spotted wilt tospovirus is present in thrips vector cells. Phytopathology *83*, 456–463.

Ullman, D.E., Westcot, D.M., Chenault, K.D., Sherwod, J.L., German, T.L., Bandla, M.D., Cantone, F.A., and Duer, H.L. (1995). Compartmentalization, intracellular transport, and autophagy of tomato spotted wilt tospovirus proteins in infected thrips cells. Phytopathology *85*, 644–654.

Vaira, A. M., Roggero, P., Luisoni, E., Milne, R. G., and Lisa, V. (1993). Characterization of two tospoviruses in Italy: *Tomato spotted wilt virus* and Impatiens necrotic spot virus. Plant Pathol. *42*, 530–542.

Van Kammen, A., van Henstra, S., and Ie, T.S. (1966). Morphology of *Tomato spotted wilt virus*. Virol. *30*, 574–577.

Verkley, F. N., and Peters, D. (1983). Characterization of a defective form of *Tomato spotted wilt virus*. J. Gen. Virol. *64*, 677–682.

Verkleij, F. N., de Vries, P., and Peters, D. (1982). Evidence that *Tomato spotted wilt virus* RNA is a positive strand. J. Gen. Virol. *58*, 329–338.

Vijayalakshmi, K. (1994). Transmission and ecoology of Thrips palmi Karny, the vector of peanut bud necrosis virus. PhD thesis, Andrha Pradesh Agricultural Universityy, Hyderabad.

Ward, T.H., and Brandizzi, F. (2004). Dynamics of proteins in Golgi membranes: comparisons between mammalian and plant cells highlighted by photobleaching techniques. Cell. Mol. Life Sci. *61*, 172–185.

Webb, S., Tsai, J., and Mitchell, F. (1998). Bionomics of Frankliniella bispinosa and its transmission of *Tomato spotted wilt virus*. In Recent Progress in Tospovirus and Thrips Research, Peters, D., and Goldbach, R., eds. Recent Progress in Tospovirus and Thrips Research, (Wageningen, The Netherlands), p. 67.

van de Wetering, F., Goldbach, R., and Peters, D. (1996). Tomato spotted wilt tospovirus ingestion by first instar larvae of *Frankliniella occidentalis* is a prerequisite for transmission. Phytopathology *86*, 900–905.

Whitfield, A.E., Kumar, N.K.K., Rotenberg, D., Ullman, D.E., Wyman, E.A., Zietlow, C., Willis, D.K., and German, T.L. (2008). A soluble form of the *Tomato spotted wilt virus* (TSWV) glycoprotein Gn (Gn-S) Inhibits transmission of TSWV by *Frankliniella occidentalis*. Phytopathology *98*,45–50.

Whitfield, A.E., Ullman, D.E., and German, T.L. (2004). Expression and characterization of a soluble form of *Tomato spotted wilt virus* glycoprotein Gn. J. Virol. *78*, 13197–13206.

Whitfield, A.E., Ullman, D.E., and German, T.L. (2005). Tospovirus–thrips interactions. Ann. Rev. Phytopathol. *43*, 459–489.

Wijkamp, I. (1995). Virus–vector relationships in the transmission of tospoviruses. PhD thesis, Wageningen University, The Netherlands.

Wijkamp, I., and Peters, D. (1993). Determination of the median latent period of two tospoviruses in *Frankliniella occidentalis*, using a novel leaf disk assay. Phytopathology *83*, 986–991.

Wijkamp, I., van Lent, J., Kormelink, R., Goldbach, R., and Peters, D. (1993). Multiplication of *Tomato spotted wilt virus* in its insect vector, *Frankliniella occidentalis*. Phytopathol. *83*, 986–991.

Wijkamp, I., Almarza, N., Goldbach, R., and Peters, D. (1995). Distinct levels of specificity in thrips transmission of tospoviruses. Phytopathology *85*, 1069–1074.

Yeh, S.D., and Chang, T. (1995). Nucleotide sequence of the N gene of watermelon silver mottle virus, a proposed new member of the genus Tospovirus. Phytopathology *85*, 58–64.

Yeh, S.D., Lin, Y.-C., Cheng, Y.-H., Jih, C.-L., Chen, M.-J., and Chen, C.-C. (1992). Identification of tomato spotted wilt-like virus on watermelon in Taiwan. Plant Dis. *76*, 835–840.

Zhang, M-Y., Zimmermann, S., Fischer, R., and Schillberg, S. (2008). Generation and evaluation of movement protein-specific single-chain antibodies for delaying symptoms of *Tomato spotted wilt virus* infection in tobacco. Plant Pathol. 57, 854–860.

Zhang, X., Yuan, Y.R., Pei, Y., Lin, S.S., Tuschl, T., Patel, D.J., and Chua, N.H. (2006). Cucumber mosaic virus-encoded 2b suppressor inhibits Arabidopsis Argonaute1 cleavage activity to counter plant defense. Gen. Dev. *20*, 3255–3268.

Bunyaviridae Infections and their Diagnostics: *Hantavirus*, *Nairovirus*, *Orthobunyavirus* and *Phlebovirus* Genera

7

Antti Vaheri and Olli Vapalahti

Abstract

The large family of *Bunyaviridae*, comprising more than 300 viruses, includes four genera which have important human and animal pathogens, many of which are of global public health importance. Hantaviruses are carried by rodents and insectivores, nairoviruses by ticks, orthobunyaviruses by mosquitoes, and phleboviruses by phlebotomous flies, mosquitoes or ticks in humans. In humans hanta- and nairoviruses and some phleboviruses cause haemorrhagic fevers, hantaviruses also renal and cardiopulmonary syndromes, and orthobunya- and phlebo-viruses affect often the central nervous system. Several members of *Bunyaviridae* such as Crimean–Congo haemorrhagic fever and Rift Valley fever viruses are important pathogens to domestic animals (cattle, sheep). In the following we review the current methods of choice to detect bunyaviruses in the carrier animals (rodents, insectivores, arthropods) and to diagnose bunyavirus infections in humans and animals, mainly serology to assay specific IgM and IgG antibodies and reverse-transcription polymerase chain reaction (RT-PCR) to detect viral RNA. The viruses within each genus show antigenic cross-reactions, reflecting their phylogeny, and therefore for virus typing neutralization tests or sequencing are often required.

Introduction

The large family of *Bunyaviridae* contains five genera (Nichol *et al.*, 2005; Schmaljohn and Nichol, 2007). Members of the four genera listed above infect vertebrates and contain important human pathogens. Members of the *Tospovirus* genus infect plants and are dealt with separately (see Chapter 6 by Kormelink). Hantaviruses are carried by rodents and insectivores, but thus far only rodent-borne viruses have been associated with human infections. Transmission of hantaviruses to man is thought to occur mainly by inhalation of aerosolized excreta (faeces, urine, saliva) from infected rodents. Nairo-, orthobunya- and phleboviruses are transmitted to their vertebrate hosts by arthropods (ticks, mosquitoes, phlebotomous flies), although occasional transmission from viraemic animals (e.g. to abattoir workers) or patients (to health care personnel) may occur in the case of Rift Valley Fever (RVF) and Crimean–Congo haemorrhagic fever (CCHF) viruses.

Genus *Hantavirus*

The three gene segments of hantaviruses encode from the large (L) gene segment the RNA-dependent RNA polymerase (L protein), from the medium (M) segment two glycoproteins, Gn and Gc, and from the small (S) segment the nucleocapsid (N) protein (for reviews see Plyusnin and Morzunov, 2001; Vaheri *et al.*, 2008) and some hantaviruses also a small-sized ~10 kDa NSs protein which can function to evade interferon signalling (Jääskeläinen *et al.*, 2007). Antibodies have been generated experimentally to L and NSs proteins, but only Gn, Gc and N proteins have been reported to induce an immune response in natural infections (for review see Vaheri *et al.*, 2008).

Infections by hantaviruses in rodents are almost asymptomatic – increased aggressive behaviour in male rats by Seoul virus (Glass *et al.*, 1988) and impaired breeding and winter survival of bank voles by Puumala virus (Kallio *et al.*, 2006b, 2007) – and persistent, while in humans the infections are transient, and some, but not all, hantaviruses cause disease. Several hantaviruses are human pathogens, some with 10–35% mortality, and cause two diseases: haemorrhagic fever with renal syndrome (HFRS) in Eurasia, and hantavirus cardiopulmonary syndrome (HCPS) in the Americas (Mills and Childs, 2001; Vapalahti *et al.*, 2003; Heyman *et al.*, 2009). The principal HFRS pathogens (with mortalities) are in Asia Hantaan (3–7%, carried by *Apodemus agrarius mantchuricus*, striped field mouse, eastern subspecies), Seoul (1–2%; *Rattus rattus* and *R. norvegicus*, black and brown rat) and Amur-Soochong (10%, carried by *A. peninsulae*, Korean field mouse) and in Europe Puumala (0.1%, *Myodes glareolus*, bank vole), Dobrava-Belgrade (10%, *A. flavicollis*, yellow-necked mouse) and Saaremaa (mild disease, A. *agrarius agrarius*, western subspecies). Seoul virus is cosmopolitan and has been detected in several continents in harbour rats and in the past in laboratory rats and recently also in wild rats in Europe (France and Belgium). The major HCPS pathogens are in Sin Nombre, New York and Andes viruses (30–40% mortality). American rodents carrying HCPS-causing hantaviruses are all Sigmodontinae and Neotominae rodents which do not exist in Eurasia. In addition, Prospect Hill virus (*Microtus pennsylvanicus*, meadow vole) in North America and Tula virus (carried by *M. arvalis*, European common vole and its sister species *M. levis*, Sibling vole) in Europe have been associated with human infection but not firmly with human disease. Hantaviruses do not infect naturally any other species except their natural hosts, rodents and insectivores, and humans. Experimental infections of, for example, hamsters with Andes virus (Hooper *et al.*, 2001) and macaques with Puumala virus (Klingström *et al.*, 2002; Sironen *et al.*, 2008), result in symptoms and signs mimicking HCPS and HFRS, respectively. Hantaviruses, being enveloped and containing in the virions an RNA-dependent RNA polymerase, are unexpectedly stable, >10 days at room temperature (Kallio *et al.*, 2006a).

Detection of hantavirus infection in rodents and insectivores, in our experience, can follow the following sequence:

1. Detection of antibodies from blood samples (conveniently collected on paper strips in field conditions) or heart extracts; a special 5-min immunochromatographic test is available for use in field conditions
2. immunoblotting of tissues of seropositive individuals;
3. reverse transcriptase-polymerase chain reaction (RT-PCR) of the S segment of antigen-positive individuals; and

4 sequencing and virus isolation by serial passaging of Vero cells with the inoculate if needed.

These procedures have been recently discussed extensively (Vaheri *et al.*, 2008) and will therefore not be treated here in more detail.

Acute onset of fever and headache, followed by thrombocytopenia, and in the case of HFRS by proteinuria, should raise the suspicion of hantavirus infection (Schmaljohn and Hjelle, 1997; Kanerva *et al.*, 1998; Peters *et al.*, 1999; Peters and Khan, 2002; Vapalahti *et al.*, 2003). It should be noted that HFRS is commonly associated with abnormal cardiac and/or pulmonary findings (Linderholm *et al.*, 1997; Mäkelä *et al.*, 2009) and renal findings are not rare in HCPS (Schmaljohn and Hjelle, 1997). Notably, severe PUUV-HFRS is strongly associated with HLA-B8, DR3, DQ2 alleles (Mustonen *et al.*, 1996) and mild PUUV-HFRS with HLA-B27 (Mustonen *et al.*, 1998). About 5% of hospitalized PUUV-HFRS patients require transient dialysis treatment. The mortality in PUUV-HFRS is very low (~0.1–0.2%), mainly due to fluid imbalance after shock, haemorrhages and necrosis in the pituitary gland and encephalitis. Long-term consequences, hypertension following Seoul virus infection (Glass *et al.*, 1993) were first detected already in the 1990s in Baltimore. After a 5-year follow-up, of PUUV-HFRS include increased systolic blood pressure and proteinuria in about 20% of the patients (Mäkelä *et al.*, 2000). After ten years of follow-up the effect had largely but not totally disappeared (Miettinen *et al.*, 2006). Some patients surviving the viral attack to the pituitary gland (Settergren *et al.*, 1992; Hautala *et al.*, 2002; Pekic *et al.* 2005) require hormone replacement therapy.

Diagnostics of HFRS (for a recent review see Vaheri *et al.*, 2008) is mainly based on detection of IgM antibodies although viral RNA may be usually, but not invariably found in blood, urine and saliva – and sometimes in the cerebrospinal fluid – during the first week of illness (Plyusnin *et al.*, 1997, 1999; Evander *et al.*, 2007; Mähönen *et al.*, 2007; Pettersson *et al.*, 2008). High viral RNA load in blood predicts poor outcome both in HCPS and in HFRS. In HCPS high titre of neutralizing antibodies is associated with a favourable outcome (Ye *et al.*, 2004). Notably, RNA has been found before IgM antibodies and symptoms in HCPS.

Antibodies are usually, but not always, detectable in the first sample taken. At the latest, hantavirus antibodies have been negative up to 5 days after onset of illness (Kallio-Kokko *et al.*, 1998), but these cases are only a few per cent of, for example, all cases of Puumala virus infection. For serology, most laboratories use enzyme immunoassay based on recombinant N protein produced by the baculovirus or *Drosophila* system and several commercial EIA tests are available (Vapalahti *et al.*, 1996; Elgh *et al.*, 1997; Sjölander *et al.*, 1997). In countries where multiple HFRS-causing hantaviruses co-circulate it is advisable to use both Puumala (PUUV) and a mouse/rat-derived virus (Dobrava, Saaremaa, Hantaan, Seoul, Amur/Soochong). Many laboratories use also indirect immunofluorescence tests on, for example, Puumala-infected acetone-fixed Vero E6 cells in which the acute-phase sera give a typical granular staining pattern (reaction with N protein); later convalescent sera give a more diffuse staining (reaction with glycoproteins; the anti-glycoprotein IgG response seems to develop more slowly than the anti-N response which is usually present already in the earliest samples) (Vapalahti *et al.*, 2001; Vaheri *et al.*, 2008). IgM antibodies may be also detected by point-of-care (POC) immunochromatographic 5-min tests that are commercially available today for several hantaviruses (Hujakka *et al.*, 2001a,b, 2003). More recently, an optical reader has become available for POC tests. Hantaviruses show extensive serological cross-reactivity, especially within each of the four virus subgroups (Muridae-, Arvicolinae-, Neotominae-, Sigmodontinae-borne), but for accurate typing, neutralization tests or RNA sequencing are needed. Use of vesicular stomatitis virus pseudotypes bearing hantavirus envelope proteins (Araki *et al.*, 2001), EIA with truncated recombinant N proteins (Ogino *et al.*, 2003) and co-expression of selected segments/neutralizing epitopes of Gn and/or Gc from separate constructs have been described for serotyping (Jenison *et al.*, 1994) but these procedures have not been widely applied. Antibodies to Dobrava-Belgrade and Saaremaa viruses cross-react strongly but it is important to distinguish them since the former is associated with a severe and the latter with mild HFRS (Vapalahti *et al.*, 2003; Vaheri *et al.*, 2008). There is a high incidence of HFRS in Far East Asia (parts of China and Korea) and in many areas of Europe: central and northern Finland, northern Sweden, parts of European Russia (Republics of Bashkortostan, Tatarstan and Udmurtia, and Samara region), Ardennes region (Belgium, France) and south-western Germany (Heyman *et al.*, 2009).

Specific diagnosis of HCPS may be achieved either serologically or by RT-PCR from blood (serum, plasma) (Hjelle *et al.*, 1994; Terajima *et al.*, 1999; Xiao *et al.*, 2006) but the latter way is not reliable and sensitive enough for use as the single diagnostic test. The major reasons for this are, like in the case of HFRS, that often when HCPS is suspected, viraemia is already fading and secondly during transportation of the sample viral RNA gets degraded. For serology, detection of IgM and/or IgG antibodies both viral lysates or recombinant antigens have been successfully used either in EIA or strip immunoblot assay (SIA) (Jenison *et al.*, 1994; Ksiazek *et al.*, 1995; Hjelle *et al.*, 1997; Padula *et al.*, 2000; Figueiredo *et al.* 2009). In IgM EIA use of the μ-capture format eliminates the problems from IgM rheumatoid factor, an autoantibody to the Fc fragment of IgG which develops commonly in a variety of infections (Salonen *et al.*, 1980).

Detection of hantaviral antigen by immunohistochemistry (IHC), of viral RNA by *in situ* hybridization or virus isolation has worked sometimes from kidney biopsies (not used for diagnostics of HFRS since early 1990s), blood, urine, saliva or cerebrospinal samples or autopsy samples, for example kidney and hypophysis in HFRS (Poljak and Avsic-Zupanc, 1994; Hautala *et al.*, 2002). IHC analysis of HCPS autopsy samples (Zaki *et al.*, 1995) showed widespread presence of Sin Nombre hantaviral antigens in endothelial cells of the microvasculature, particularly in the lung. Hantaviral antigens were also observed within follicular dendritic cells, macrophages, and lymphocytes.

Genus *Nairovirus:* Crimean–Congo haemorrhagic fever virus

The two-partite name of the virus is derived from the fact that the illness was first recognized in 1944 in the Crimean peninsula and virus isolated in 1967 from Congo. CCHFV has indeed an extensive geographic distribution of medically significant tick-borne viruses occurring from Western China across Southern Asia to Southern and Eastern Europe and South Africa (for recent reviews see Whitehouse, 2004; Kallio-Kokko *et al.*, 2005; Ergonul, 2006). The incidence of sporadic cases and outbreaks of CCHF has been increasing in the endemic regions especially in Turkey (Ergonul, 2006) but also in Iran, Pakistan and Balkans (Chikikar *et al.* 2008; Papa *et al.* 2008; Christova *et al.*, 2009). The vector of CCHFV is Hyalomma spp. ticks, although other ixodid tick species may occasionally carry the virus. The virus may be transmitted to humans also from cattle organs/excreta in, for example, slaughter-houses or occasionally in hospitals from CCHF patients. Sheep, goats and cattle develop high titres of virus in blood, but tend not to fall ill. Of all the birds studied so far, ostriches can also become infected by CCHFV (Swanepoel *et al.*, 1998). The incubation time in human infection is generally short (2–9 days) leading often after a pre-haemorrhagic period (sudden fever, dizziness, neck pain, aching muscles, soreness in the eyes and sensitivity to light) to severe haemorrhages and a case fatality rate as high as 5–30%. Nairobi sheep disease virus (NSDV) causes a highly pathogenic disease, haemorrhagic gastroenteritis, in sheep and goats in Africa. Dugbe virus infects cattle in West Africa and is a mild human pathogen (Burt *et al.*, 1996) and is genetically a closely related nairovirus (Honig *et al.*, 2004). Ganjam virus is an Asian variant of NSDV occurring in Sri Lanka and India. NSD and Ganjam viruses can cause a mild, influenza-like disease in humans (Davies, 1997; Marczinke and Nichol, 2002; Joshi *et al.*, 2005; Kahn *et al.*, 2006).

Like other members of the *Bunyaviridae* family CCHFV, as well as NSDV, a haemorrhagic disease affecting sheep and goats in Africa, have a tripartite genome very similar to that in Genus *Hantavirus*. Early diagnosis of CCHF is critical to prevent nosocomial infections and in view of the therapeutic potential of ribavirin; several observations indicate that ribavirin has clear antiviral activity against CCHFV (Ergonul, 2008). Diagnostics of infections in humans or livestock (see Ergonul, 2006) is based today on (i) serological tests such enzyme immunoassay using recombinant CCHFV nucleocapsid protein or indirect immunofluorescence on virus-infected or N-expressing transfected cells and (ii) detection and quantification of CCHFV RNA by real-time reverse transcription PCR. Patients who die of CCHF do not usually develop a measurable antibody response and severe or lethal outcome is associated with high viral loads (Cevic *et al.*, 2007). IgM and IgG antibodies are detectable by EIA and immunofluorescence assays from about 7 days after the onset of illness. Specific IgM antibodies decline to undetectable levels by four months post-infection but IgG remains for years. Antigen-capture EIA has also been shown to be useful, especially to detect the most severe cases that would require antiviral therapy. Any attempts at isolating and culturing the virus should be performed in maximum biocontainment laboratory (i.e. BSL-4) either in newborn mice or more conveniently in cell cultures. For detection of CCHFV in ticks either RT-PCR or virus isolation may be used. Since CCHV is a highly pathogenic BSL-4 level agent and can be transmitted from person-to-person, it has been considered a potential agent of bioterrorism and/or biowarfare.

Genus *Orthobunyavirus*

Orthobunyavirus infections are very common in many parts of the world, with seroprevalences over 50% in adult or elderly populations, for example in Finland (Brummer-Korvenkontio *et al.*, 1973; Putkuri *et al.*, 2007) or the Czech Republic (Hubalek, 2008). However, actual diagnosis of acute illnesses in most countries is rare owing to limited use of laboratory diagnostics and variable research interest and clinical alert, but can be typically be made by serology (at least in California encephalitis group virus infections, usually at the time of neurological illness and entering the health care, the patients have antibodies). Methods for detection of viral RNA in clinical samples have been also developed, and the virus has been isolated in many – sometimes anecdotal – cases from human samples, mostly blood or CSF human diseases. Diseases associated with orthobunyaviruses include febrile diseases and central nervous system infection by California encephalitis serogroup (CE-group). In particular, La Crosse virus (LACV) is a major cause of paediatric encephalitis in

North America and infects up to 300,000 persons each year of which 70–130 result in severe disease of the central nervous system, but also the Tahyna-like viruses and the Jamestown Canyon/Inkoo viruses have been associated with CNS symptoms. Febrile diseases, also CNS infections, and stillbirth and congenital abnormalities in domestic animals have been associated with viruses (e.g. Batai and Batai-like viruses) (for review see Hubalek *et al.*, 2008) in Bunyamwera serogroup and (e.g. Akabame and Aino viruses) in Simbu serogroup and large haemorrhagic fever outbreaks in Africa with Ngari virus, a recombinant between Batai and Bunyamwera-like viruses (Gerrard *et al.*, 2004; Briese *et al.*, 2006). Oropouche virus in the Aino serogroup is probably the most important orthobunyavirus in South America with occasional outbreaks of febrile illness including headache, chills, dizziness, photophobia, myalgia, nausea, and vomiting fever with hundreds of thousands of cases (Vasconcelos *et al.*, 2009).

Viral isolations from humans include California encephalitis serogroup include CE-group Tahyna virus in Europe (from febrile children in Czech Republic (Bardos *et al.*, 1975); CE-group La Crosse virus (post mortem brains of encephalitis patients) and Bunyamwera serogroup Cache Valley virus (CVV) in at least two cases (blood and CSF) in USA/ North America; Bunyamwera, Germiston, Ilesha, and Shokwe viruses from Africa; Xingu and Oropouche viruses in South America; Wyeomyia virus and a CVV-like virus in Central America (Karabatsos, 1985; Calisher and Karabatsos,1988; Gonzales and Georges, 1988; Mangiafico, 1988; Sexton *et al.*, 1997). However, more recently, along with virus isolation, RT-PCR has been demonstrated to work in principle for Ngari virus haemorrhagic fever outbreaks in Africa (Briese *et al.*, 2006) and Oropouche outbreaks in Brazil (Vasconcelos *et al.*, 2009). Apart from these, many other orthobunyaviruses have been specifically implicated in human disease through serology and cross-neutralization assays, for example recent demonstration of acute Tahyna virus infection humans in China (Lu *et al.*, 2009).

The serological response is strongest in the acute phase towards the N protein and recombinant N protein antigens have been shown to react with acute human cases of, for example, Oropouche and Inkoo infections. However, it seems that the N response (at least in the case of Inkoo virus) declines and even disappears in some cases, whereas the glycoprotein (Gn, Gc) responses become stronger and prevail (Putkuri *et al.*, 2007).

While the nucleocapsid protein response evidently is cross-reactive *at least* within each 18 orthobunyavirus serogroup, and EIA, IFA or classically complement fixation tests can be used to measure the serogroup-specific response, the more species/type-specific glycoprotein response can be used to distinguish between infecting orthobunyaviruses by e.g. neutralization (provided that the relevant viruses are included in the panel for comparison). Also it should be taken into account that reassortants where M (glycoproteins) and S (nucleocapsid protein) have different origins, may occur. Several neutralization tests, as well as IFA and EIA formats have been used for diagnosing orthobunyavirus cases.

The RT-PCR assays have been widely used for detection and typing of orthobunyaviruses in mosquitoes, and typically neutralization assays have been used for animal studies to show geographical spread of various orthobunyaviruses (Brummer-Korvenkontio *et al.*, 1973; for review see Hubalek, 2008).

Genus *Phlebovirus*

The most important phleboviruses relevant for discussing diagnostics are (1) Rift Valley fever virus (RVFV) and the two sandfly fever virus groups; (2) Toscana virus (TOSV) along with the more related viruses Naples and Massilia; and (3) Sandfly fever Sicilian virus (SFSV). Both genetically and serologically, at least three separate antigens and RT-PCR primers/probes are needed to cover the diagnostics of each of these two diseases and three virus groups. While most infections are probably mild or unspecific febrile illnesses, RVFV occurs in outbreaks where small proportion of patients develop sever ocular, central nervous system or (perhaps in less than 1% of cases) haemorrhagic manifestations, and the sandfly fever viruses cause meningitis and encephalitis.

Phleboviruses can be readily isolated in e.g. Vero, LLC-MK2, and BHK-21 cells, or in suckling mice (Kallio-Kokko *et al.*, 2005), and several RT-PCR protocols have been developed for sensitive RVFV RNA detection (Garcia *et al.*, 2001; Drosten *et al.*, 2002; Näslund *et al.*, 2008; Weidmann *et al.*, 2008), although only recently have some studies evaluated the usefulness of these protocols in clinical settings. As for RVFV, some protocols are based on regular real-time TaqMan assay, some intended for less equipped laboratories, based on loop-mediated isothermal amplification LAMP (Peyrefitte *et al.*, 2008; Leroux *et al.*, 2009), or tested for mobile TaqMan devices (Wiedman *et al.*, 2009). The RT-PCR approaches have been used, apart for human, also for animal and mosquito samples (Ibrahim, 1997). In human RVFV infections, it seems that the virus can be found approximately 5 days after onset of illness (not usually found any more during the second week of illness), and at the time of disappearance of viral RNA, the IgM upsurges, followed by IgG. Furthermore, it has been clearly shown that the higher viral load and low antibody levels are associated with severe and lethal outcome of infection (Bird *et al.*, 2007; Njenga *et al.*, 2009). Therefore, monitoring the RNA values in RVFV infection has also prognostic value and it has been shown that a cut-off level for viral RNA can be set to predict fatality of the disease with 50% probability of death for e.g. RNA levels above (and 4.5% below) a certain threshold (Njenga *et al.*, 2009). Detection of antigenaemia has only 50% sensitivity as compared with RT-PCR. Animals (sheep, cattle) also seem to usually harbour RVFV viraemia only for a few days but can be detected positive. Antibodies are detectable in experimentally infected sheep 4–6 days after infection by neutralization and 6–7 days after infection by ELISA (Paweska *et al.*, 2003). Recently, a specific sandwich ELISA assay to detect RVFV antigen in animal and human tissues, even mosquitoes, was developed and might be useful in less equipped laboratories in endemic areas (Jansen van Wuren and Paweska, 2009). Would a vaccine based on engineered virus lacking the virulence factor NSs become widely used, methods differentiating vaccinated animals from infected would be available (McElroy *et al.*, 2009). Another important 'diagnostic method' in a larger scale is satellite prediction, which has been proven to predict the rainfall-driven RVFV outbreaks in East Africa several months in advance (Linthicum *et al.*, 1999; Anyamba *et al.*, 2009).

The sandfly fever viruses can be isolated or detected from acute cases by RT-PCR (along with enteroviruses; they are they second cause of illness in Italy in the summer months) from cerebrospinal fluid. However, early on from the onset of illness, also IgM and IgG antibodies can be found (Dionisio *et al.*, 2003). Several real-time PCR-methods for detection of Toscana virus or sandfly fever causing viruses are available (Pérez-Ruiz *et al.*, 2007, Weidmann *et al*, 2008).

Serological assays are applicable for diagnostics, and are often the method of choice, and the only choice after the first week of illness. The nucleocapsid protein is the most prominent antigen in early serological response (Magurano *et al.*, 1999, Paweska *et al.*, 2007), and N has linear epitopes detectable in immunoblotting, whereas the glycoprotein epitopes are usually visible only in assays with intact three-dimensional protein structure (e.g. RIPA). The immunodominant role of phlebovirus nucleocapsid antigen, for which also recombinant assays have been developed (Soldateschi *et al.*, 1999), is similar to hanta- and orthobunyaviruses where the anti-glycoprotein response is weaker especially in the beginning of the serological response (Kallio-Kokko *et al.*, 2001; Putkuri *et al.*, 2007). Serological tests have been based on immunofluorescence (IFAT), EIA, or neutralization; also tests (neutralization) employing virus-like particles with reporter genes have been developed. It should be emphasized that the serological cross reactions between TOSV and SFSV groups are weak and antigens from both groups are needed in diagnosing sandfly fever in the Mediterranean region.

Concluding remarks

WHO World Health Report (published on 23 August 2007) predicted new infectious diseases to become a major threat to global public health in this century and noted: 'In our increasingly interconnected world, new diseases are emerging at an unprecedented rate, often with the ability to cross borders rapidly and spread. Since 1967, at least 39 new pathogens have been identified, including HIV, Ebola haemorrhagic fever, Marburg fever and SARS'. Most of these new, emerging and re-emerging infectious diseases are viral zoonoses and the *Bunyaviridae* presented in this chapter contain several, some still poorly studied, virus groups which could have emerging outbreak potential, especially considering the capability of genetic shift, as demonstrated recently by the Ngari haemorrhagic fever outbreak in Africa (Briese *et al.*, 2006). In many parts of the world these infections are not recognized by the medical or veterinary community, for example HFRS in several East- European countries and CCHF in Asia and Africa. This is also because diagnostics is not actively used. This means that there will be a growing demand for user-friendly economical diagnostic tests. We need rapid serological tests to detect the infections in humans, domestic animals and in wildlife. The hantaviral 5-min immunochromatographic strip tests are a good example. Such a point-of-care test would be welcome also for the diagnostics of CCHF. We need both rapid screening tests such as high-throughput sequencing and microarrays (Drosten *et al.*, 2002; Lipkin, 2008; Haagmans *et al.*, 2009; MacLean *et al.*, 2009) to detect simultaneously multiple viral RNA and DNA genomes and to detect multiple virus-specific IgM/IgG antibodies. Specific screening tests are needed also because the same symptoms are (e.g. haemorrhagic fevers, encephalitides) are caused by a plethora of viruses, because new antivirals are being developed, and because the spectrum of viral zoonoses widens (Jones *et al.*, 2008). Furthermore, while many new tests are developed, only few international quality control rounds are available to confirm proper performance of the tests. The regular distribution of such quality control tests involving geographically representative strain panels would be a most welcome tool to be able to compare – and develop – testing systems in various research and diagnostic laboratories.

References

Anyamba, A., Chretien, J.P., Small, J., Tucker, C.J., Formenty, P.B., Richardson, J.H., Britch, S.C., Schnabel, D.C., Erickson, R.L., and Linthicum, K.J. (2009). Prediction of a Rift Valley fever outbreak, Proc. Natl. Acad. Sci. U.S.A. *106*, 955–959.

Araki, K., Yoshimatsu, K., Ogino, M., Ebihara, H., Lundkvist, A., Kariwa, H., Takashima, I., and Arikawa, J. (2001). Truncated hantavirus nucleocapsid proteins for serotyping Hantaan, Seoul, and Dobrava hantavirus infections. J. Clin. Microbiol. *39*, 2397–2404.

Bardos, V., Medek, M., Kania, V., and Hubalek, Z. (1975). Isolation of Tahyna virus from the blood of sick children. Acta Virol. *19*, 447.

Bird, B.H., Bawiec, D.A., Ksiazek, T.G., Shoemaker T.R., and Nichol S.T. (2007). Highly sensitive and broadly reactive quantitative reverse transcription-PCR assay for high-throughput detection of Rift Valley fever virus, J. Clin. Microbiol. *45*, 3506–3513.

Briese, T., Bird, B., Kapoor, V., Nichol, S.T., and Lipkin, W.I. (2006)., Batai and Ngari viruses: M segment reassortment and association with severe febrile disease outbreaks in East Africa, J. Virol. *80*, 5627–5630.

Brummer-Korvenkontio, M. (1973). Arboviruses in Finland. V. Serological survey of antibodies against Inkoo virus (California group) in human, cow, reindeer, and wildlife sera. Am. J. Trop. Med. Hyg. *22*, 654–661.

Burt, F.J., Spencer, D.C., Leman, P.A., Patterson, B., and Swanepoel, R. (1996). Investigation of tick-borne viruses as pathogens of humans in South Africa and evidence of Dugbe virus infection in a patient with prolonged thrombocytopenia. Epidemiol. Infect. *116*, 353–361.

Calisher, C.H., and Karabatsos, N. (1988). Arbovirus serogroups: definition and geographic distribution. In The Arboviruses: Epidemiology and Ecology, vol. I. Monath, T.P., ed. (CRC Press, Boca Raton, FL), pp. 19–57.

Calisher, C.H., Francy, D.B., Smith, G.C., Muth, D.J., Lazuick, J.S., Karabatsos, N., Jakob, W.L., and McLean, R.G. (1986). Distribution of Bunyamwera serogroup viruses in North America, 1956–1984. Am. J. Trop. Med. Hyg. *35*, 429–443.

Cevik, M.A., Erbay, A., Bodur, H., Eren, S.S., Akinci, E., Onguru, P., and Kubar, A. (2007). Viral load as a predictor outcome in Crimean–Congo hemorrhagic fever. Clin. Infect. Dis. *45*, e96–100.

Chikikar, S., Goya, M.M., Shirzadi, M.R., Ghiasi, S.M., Mirahmadi, R., Haeri, A., Moradi, M., Afzali, N., Rahpeyma, M., Zeinali, M., *et al.* (2008). Surveillance and loboratory detection system of Crimean–Congo hemorrhagic fever in Iran. Transbound. Emerg. Dis. *55*, 200–204.

Christova, I., Di Caro, A., Papa, A., Castelli, C., Andonova, L., Kalvatchev, N., Papadimitriou, E., Carletti, F., Mohareb, E., Cabobianchi, M.R., *et al.* (2009). Crimean–Congo hemorrhagic fever, southwestern Bulgaria. Emerg. Infect. Dis. *15*, 983–985.

Davies, F.G. (1997). Nairobi sheep disease. Parassitologia *39*, 95–98

Dionisio, D., Esperti, F., Vivarelli, A., and Valassina, M. (2003). Epidemiological, clinical and laboratory aspects of sandfly fever. Curr. Opin. Infect. Dis. *16*, 383–388.

Drosten, C., Göttig, S., Schilling, S., Asper, M., Panning, M., Schmitz, H., and Günther, S. (2002). Rapid detection and quantification of RNA of Ebola and Marburg viruses, Lassa virus, Crimean–Congo hemorrhagic fever virus, Rift Valley fever virus, dengue virus, and yellow fever virus by real-time reverse transcription-PCR. J. Clin. Microbiol. *40*, 2323–2330.

Dunn, E.F., Pritlove, D.C., and Elliott, R.M. (1994). The S RNA genome segments of Batai, Cache Valley, Guaroa, Kairi, Lumbo, Main Drain and Northway bunyaviruses: sequence determination and analysis. J. Gen. Virol. *75*, 597–608.

Elgh, F., Lundkvist, Å., Alexeyev, O.A., Stenlund, H., Avsic-Zupanc, T., Hjelle, B., Lee, H.W., Smith, K.J., Vainionpää, R., and Wiger, D. (1997). Serological diagnosis of hantavirus infections by an enzymelinked immunosorbent assay based on detection of immunoglobulin G andMresponses to recombinant nucleocapsid proteins of five viral serotypes. J. Clin. Microbiol. 35, 1122–1130.

Ergönül, Ö. (2006). Crimean–Congo haemorrhagic fever. Lancet Infect. Dis. *6*, 203–214.

Ergonul, O. (2008). Treatment of Crimean–Congo hemorrhagic fever. Antivir. Res. *78*, 125–131.

Evander, M., Eriksson, I., Pettersson, L., Juto, P., Ahlm, C., Olsson, G.E., Bucht, G., and Allard, A. (2007). Puumala hantavirus viremia diagnosed by real-time reverse transcriptase PCR using samples from patients with hemorrhagic fever and renal syndrome. J. Clin. Microbiol. *45*, 2491–2497.

Figueiredo, L.T., Moreli, M.L., Borges, A.A., de Figueiredo, G.G., Badra, S.J., Bisordi, I., Suzuki, A., Capria, S., and Padula, P. (2009). Evaluation of an enzyme-linked immunosorbent assay based on Araraquara virus recombinant nucleocapsid protein. Am. J. Trop. Med. Hyg. *81*, 273–276.

Garcia, S., Crance, J.M., Billecocq, A., Peinnequin, A., Jouan, A., Bouloy, M., and Garin, D. (2001). Quantitative real-time PCR detection of Rift Valley fever virus and its application to evaluation of antiviral compounds. J. Clin. Microbiol. *39*, 4456–4461.

Gerrard, S.R., Li, L., Barrett, A.D., and Nichol, S.T. (2004). Ngari virus is a Bunyamwera virus reassortant that can be associated with large outbreaks of hemorrhagic fever in Africa. J. Virol. *78*, 8922–8926.

Glass, G.E., Childs, J.E., Korch, G.W., and LeDuc, J.W. (1988). Association of intraspecific wounding with hantaviral infection in wild rats (*Rattus norvegicus*). Epidemiol. Infect. *101*, 459–472.

Glass, G.E., Watson, A.J., LeDuc, J.W., Kelen, G.D., Quinn, T.C., and Childs, J.E. (1993). Infection with a ratborne hantavirus in US residents is consistently associated with hypertensive renal disease. J. Infect. Dis. *167*, 614–620.

Gonzalez, J.P., and Georges, A-J. (1988). Other bunyaviral fevers: Bunyamwera, Ilesha, Germiston, Bwamba, and Tataguine. In The Arboviruses: Epidemiology and Ecology, vol. II, Monath, T.P., ed. (CRC Press ,Boca Raton, FL), pp. 87–98.

Haagmans, B.L., Andeweg, A.C., and Osterhaus, A.D.M.E. (2009). The application of genomics to emerging zoonotic viral diseases. PLoS Pathog. *5*, 1–5.

Hautala, T., Sironen, T., Vapalahti, O., Paakko, E., Sarkioja, T., Salmela, P.I., Vaheri, A., Plyusnin, A., and Kauma, H. (2002). Hypophyseal hemorrhage and panhypopituitarism during Puumala virus Infection: magnetic resonance imaging and detection of viral antigen in the hypophysis. Clin. Infect. Dis. *35*, 96–101.

Heyman, P., Vaheri, A., Lundkvist Å., and Avsic-Zupanc, T. (2009). Hantavirus infections in Europe: from virus carriers to a major public-health problem. Expert. Rev. Anti. Infect. Ther. *7*, 205–217.

Hjelle, B., Spiropoulou, C.F., Torrez-Martinez, N., Morzunov, S., Peters, C.J., and Nichol, S.T. (1994). Detection of Muerto Canyon virus RNA in peripheral blood mononuclear cells from patient with hantavirus pulmonary syndrome. J. Infect. Dis. *170*, 1013–1017.

Hjelle, B, Jenison, S, Torrez-Martinez, N, Herring, B, Quan, S, Polito, A, Pichuantes, S, Yamada, T, Morris, C, Elgh, F, Lee, HW, Artsob, H., and Dinello, R. (1997). Rapid and specific detection of Sin Nombre virus antibodies in patients with hantavirus pulmonary syndrome by a strip immunoblot assay suitable for field diagnosis. J. Clin. Microbiol. *35*, 600–608.

Honig, J.E., Osborne, J.C., and Nichol, S.T. (2004). The high genetic variation of viruses of the genus Nairovirus reflects the diversity of their predominant tick hosts. Virology *318*, 10–16.

Hooper, J.W., Larsen, T., Custer, D.M., and Schmaljohn, C.S. (2001). A lethal disease model for hantavirus pulmonary syndrome. Virology *289*, 6–14.

Hubálek, Z. (2008). Mosquito-borne viruses in Europe. Parasitol. Res. *103 (Suppl 1)*, S29–S43.

Hujakka, H., Koistinen, V., Eerikäinen, P., Kuronen, I., Mononen, I., Parviainen, M., Lundkvist, Å., Vaheri, A., Närvänen, A., and Vapalahti, O. (2001a). New immunochromatographic rapid test for diagnosis of acute Puumala virus infection. J. Clin. Microbiol. *39*, 2146–2150.

Hujakka, H., Koistinen, V., Eerikäinen, P., Kuronen, I., Laatikainen, A., Kauppinen, J., Vaheri, A., Vapalahti, O., and Närvänen, A. (2001b). Comparison of a new immunochromatographic rapid test with a commercial EIA for the detection of Puumala virus specific IgM antibodies. J. Clin. Virol. *23*, 79–85.

Hujakka, H., Koistinen, V., Eerikäinen, P., Kuronen, I., Mononen, I., Parviainen, M., Lundkvist, Å., Vaheri, A., Vapalahti, O., and Närvänen, A. (2003). Diagnostic rapid tests for acute hantavirus infections: specific tests for Hantaan, Dobrava and Puumala viruses versus a hantavirus combination test. J. Virol. Methods *108*, 117–122.

Ibrahim, M.S., Turell, M.J., Knauert, F.K., and Lofts, R.S. (1997). Detection of Rift Valley fever virus in mosquitoes by RT-PCR. Mol. Cell Probes *11*, 49–53.

Jääskeläinen, K.M., Kaukinen, P., Minskaya, E.S., Plyusnina, A., Vapalahti, O., Elliott, R.M., Weber, F., Vaheri, A., and Plyusnin, A. (2007). Tula and Puumala hantavirus NSs ORFs are functional and the products inhibit activation of the interferon-beta promoter. J. Med. Virol. *79*, 1527–1536.

Jansen van Vuren, P., and Paweska, J.T.J. (2009). Laboratory safe detection of nucleocapsid protein of Rift Valley fever virus in human and animal specimens by a sandwich ELISA. J. Virol Methods *157*,15–24.

Jenison, S., Yamada, T., Morris, C., Anderson, B., Torrez-Martinez, N., Keller, N., and Hjelle, B. (1994). Characterization of human antibody responses to four corners hantavirus infections among patients with hantavirus pulmonary syndrome. J. Virol. *68*, 3000–3006.

Jones, K.E., Patel, N.G., Levy, M.A., Storeygard, A., Balk, D., Gittleman, J.L., and Daszak, P. (2008). Global trends in emerging infectious diseases. Nature *451*, 990–994.

Joshi, M.V., Geevarghese, G., Joshi, G.D., Ghodke, Y.S., Mourya, D.T., and Mishra, A.C. (2005). Isolation of Ganjam virus from ticks collected off domestic animals around Pune, Maharashtra, India. J. Med. Entomol. *42*, 204–206.

Kahn, C.M., and Line S. (2006). The Merck Veterinary Manual [online]. Nairobi Sheep Disease (Merck and Co, Whitehouse Station, NJ). Available at: http://www.merckvetmanual.com/mvm/index.jsp?cfile=htm/bc/55800.htm.

Kallio, E.R., Klingstrom, J., Gustafsson, E., Manni, T., Vaheri, A., Henttonen, H., Vapalahti, O., and Lundkvist, A. (2006a). Prolonged survival of Puumala hantavirus outside the host: evidence for indirect transmission via the environment. J. Gen. Virol. *87*, 2127–2134.

Kallio, E.R., Poikonen, A., Vaheri, A., Vapalahti, O., Henttonen, H., Koskela, E., and Mappes, T. (2006b). Maternal antibodies postpone hantavirus infection and enhance individual breeding success. Proc. Biol. Sci. *273*, 2771–2776.

Kallio, E.R., Voutilainen, L., Vapalahti, O., Vaheri, A., Henttonen, H., Koskela, E., and Mappes, T. (2007). Endemic hantavirus infection impairs the winter survival of its rodent host. Ecology *88*, 1911–1916.

Kallio-Kokko, H., Vapalahti, O., Lundkvist, A., and Vaheri, A. (1998). Evaluation of Puumala virus IgG and IgM enzyme immunoassays based on recombinant baculoviruses-expressed nucleocapsid protein for early nephropathia epidemica diagnosis. Clin. Diagn. Virol. *10*, 83–90.

Kallio-Kokko, H., Leveelahti, R., Brummer-Korvenkontio, M., Lundkvist, A., Vaheri, A., and Vapalahti, O. (2001). Human immune response to Puumala virus glycoproteins and nucleocapsid protein expressed in mammalian cells. J. Med. Virol. *65*, 605–613.

Kallio-Kokko, H., Uzcategui, N., Vapalahti, O., and Vaheri A. (2005). Viral zoonoses in Europe. FEMS Microbiol. Rev. *29*, 1051–1077.

Kanerva, M., Mustonen, J., and Vaheri, A. (1998). Pathogenesis of Puumala and other hantavirus infections. Rev. Med. Virol. *8*, 67–86.

Karabatsos, N., ed. (1985). International catalogue of arboviruses, including certain other viruses of vertebrates. (American Society of Tropical Medicine and Hygiene, San Antonio, TX).

Klingström, J., Plyusnin, A., Vaheri, A., and Lundkvist, A. (2002). Wild-type Puumala hantavirus infection induces cytokines, C-reactive protein, creatinine, and nitric oxide in cynomolgus macaques. J. Virol. *76*, 444–449.

Ksiazek, T.G., Peters, C.J., Rollin, P.E., Zaki, S., Nichol, S., Spiropoulou, C., Morzunov, S., Feldmann, H., Sanchez, A., Khan, A.S., *et al.* (1995). Identification of a new North American hantavirus that causes acute pulmonary insufficiency. Am. J. Trop. Med. Hyg. *52*, 117–123.

Kuno, G., Mitchell, C.J., Chang, G.J., and Smith, G.C. (1996). Detecting bunyaviruses of the Bunyamwera and California serogroups by a PCR technique. J. Clin. Microbiol. *34*,1184–1188.

Le Roux, C.A., Kubo, T., Grobbelaar, A.A., van Vuren, P.J., Weyer, J., Nel L.H., Swanepoel, R., Morita, K., and Paweska, J.T. (2009). Development and evaluation of a real-time reverse transcription-loop-mediated isothermal amplification assay for rapid detection of Rift Valley fever virus in clinical specimens. J. Clin. Microbiol. *47*, 645–651.

Linderholm, M., Sandstrom, T., Rinnstrom, O., Groth, S., Blomberg, A., and Tarnvik, A. (1997). Impaired pulmonary function in patients with hemorrhagic fever with renal syndrome. Clin. Infect. Dis. *25*, 1084–1089.

Linthicum, K.J., Anyamba, A., Tucker, C.J., Kelley, P.W., Myers, M.F., and Peters, C.J. (1999). Climate and satellite indicators to forecast Rift Valley fever epidemics in Kenya. Science *285*, 397–400.

Lipkin, E.I. (2008). Pathogen discovery. PLoS Pathog. *4*, 1–3.

Lu, Z., Lu, X-J., Fu, S-H., Zhang, S., Li, Z-X., Yao, X-H., Feng, Y-P., Lambert, A.J., Ni, D-X., Wang, F-T., *et al.* (2009). Tahyna Virus and human infection, China. Emerg. Infect. Dis. *15*, 306–309.

McElroy, A.K., Albariño, C.G, and Nichol, S.T. (2009). Development of a RVFV ELISA that can distinguish infected from vaccinated animals. Virol. J. *13*,125.

MacLean, D., Jones, J.D.G., and Studholme, D.J. (2009). Application of 'next-generation' sequencing technologies to microbial genetics. Nature Rev. Microbiol. 7, 287–296.

Magurano, F., and Nicoletti, L. (1999). Humoral response in Toscana virus acute neurologic disease investigated by viral-protein-specific immunoassays. Clin. Diagn. Lab. Immunol. *6*, 55–60.

Mähönen, S.M., Sironen, T., Vapalahti, O., Pääkkö, E., Hautala, N., Ilonen, J., Glumoff, V., Vainio, O., Kauma, H., Vaheri, A., Plyusnin, A., and Hautala, T. (2007). Puumala virus RNA in cerebrospinal fluid in a patient with uncomplicated nephropathia epidemica. J. Clin. Virol. *40*, 248–251.

Mäkelä, S., Ala-Houhala, I., Mustonen, J., Koivisto, A.M., Kouri, T., Turjanmaa, V., Vapalahti, O., Vaheri, A., and Pasternack, A. (2000). Renal function and blood pressure five years after Puumala virus-induced nephropathy. Kidney Int. *58*, 1711–1718.

Mäkelä, S., Kokkonen, L., Ala-Houhala, I., Groundstroem, K., Harmoinen, A., Huhtala, H., Hurme, M., Paakkala, A., Pörsti, I., Virtanen, V., Vaheri, A., and Mustonen, J. (2009). More than half of the patients with acute Puumala hantavirus infection have abnormal cardiac findings. Scand. J. Infect. Dis. *41*, 57–62.

Mangiafico, J.A., Sanchez, J.L., Figueiredo, L.T., LeDuc, J.W., and Peters, C.J. (1988). Isolation of a newly recognized Bunyamwera serogroup virus from a febrile human in Panama. Am. J. Trop. Med. Hyg. *39*, 593–596.

Marczinke, B.I., and Nichol, S.T. (2002). Nairobi sheep disease virus, an important tick-borne pathogen of sheep and goats in Africa, is also present in Asia. Virology *303*, 146–151.

Miettinen, M.H., Mäkelä, S.M., Ala-Houhala, I.O., Huhtala, H.S., Koobi, T., Vaheri, A.I., Pasternack, A.I., Pörsti, I.H., and Mustonen, J.T. (2006). Ten-year prognosis of Puumala hantavirus-induced acute interstitial nephritis. Kidney Int. *69*, 2043–2048.

Mills, J.N., and Childs, J.E. (2001). Rodent-borne hemorrhagic fever viruses. In Infectious Diseases of Wild Mammals, Williams, E.S., and Barker, I.K., eds. (Iowa State University, Ames, IA), pp. 254–270.

Mustonen, J., Partanen, J., Kanerva, M., Pietilä, K., Vapalahti, O., Pasternack, A., and Vaheri, N. (1996). Genetic susceptibility to severe course of nephropathia epidemica caused by Puumala hantavirus. Kidney Int. *49*, 217–221.

Mustonen, J., Partanen, J., Kanerva, M., Pietilä, K., Vapalahti, O., Pasternack, A., and Vaheri, A. (1998). Association of HLA B27 with benign clinical course of nephropathia epidemica caused by Puumala hantavirus. Scand. J. Immunol. *47*, 277–279.

Näslund, J., Lagerqvist, N., Lundkvist, A., Evander, M., Ahlm, C., and Bucht, G. (2008). Kinetics of Rift Valley fever virus in experimentally infected mice using quantitative real-time RT-PCR. J. Virol. Methods *151*, 277–282.

Nichol, S.T., Beaty, B.J., Elliott, R.M., Goldbach, R., Plyusnin, A., Schmaljohn, C.S., and Tesh, R.B. (2005). Family *Bunyaviridae*. In Virus Taxonomy: Eighth Report of the International Committee on Taxonomy of Viruses, Fauquet, C.M., Mayo, M.A., Maniloff, J., Desselberger, U., and Ball, L.A., eds (Elsevier Academic Press, San Diego), pp. 695–716.

Njenga, M.K., Paweska, J., Wanjala, R., Rao, C.Y., Weiner, M., Omballa, V., Luman, E.T., Mutonga, D., Sharif, S., Panning, M., *et al.* (2009). Using a field quantitative real-time PCR test to rapidly identify highly viraemic Rift Valley fever cases. J. Clin. Microbiol. *47*, 1166–1171.

Ogino, M., Ebihara, H., Lee, B.H., Araki, K., Lundkvist, Å., Kawaoka, Y., Yoshimatsu, K, and Arikawa, J. (2003). Use of vesicular stomatitis virus pseudotypes bearing Hantaan or Seoul virus envelope proteins in a rapid and safe neutralisation test. Clin. Diagn. Lab. Immunol. *10*, 154–160.

Padula, P.J., Rossi, C.M., Della Valle, M.O., Martínez, P.V., Colavecchia, S.B., Edelstein, A., Miguel, S.D., Rabinovich, R.D., and Segura, E.L. (2000). Development and evaluation of a solid-phase enzyme immunoassay based on Andes hantavirus recombinant nucleoprotein. J. Med. Microbiol. *49*, 149–155.

Papa, A., Bino, S., Papadimitriou, E., Velo, E., Dhimolea, M., and Antoniadis, A. (2008). Suspected Crimean Congo haemorrhagic fever cases in Albania. Scand. J. Infect. Dis. 3, 1–3.

Paweska, J.T., Smith, S.J., Wright, I.M., Williams, R., Cohen, A.S., Van Dijk, A.A., Grobbelaar, A.A., Croft, J.E., Swanepoel, R., and Gerdes, G.H. (2003). Indirect enzyme-linked immunosorbent assay for the detection of antibody against Rift Valley fever virus in domestic and wild ruminant sera. Onderstepoort J. Vet. Res. *70*, 49–64.

Paweska, J.T., Jansen van Vuren, P., and Swanepoel, R. (2007). Validation of an indirect ELISA based on a recombinant nucleocapsid protein of Rift Valley fever virus for the detection of IgG antibody in humans. J. Virol. Methods.*146*, 119–124.

Pekic, S., Cvijovic, G., Stojanovic, M., Kendereski, A., Micic, D., and Popovic, V. (2005). Hypopituitarism as a late complication of hemorrhagic fever. Endocrine *26*, 79–82.

Pérez-Ruiz, M., Collao, X., Navarro-Marí, J.M., and Tenorio, A. (2007). Reverse transcription, real-time PCR assay for detection of Toscana virus. J. Clin. Virol. *39*, 276–281.

Peters, C.J., and Khan, A.S. (2002). Hantavirus pulmonary syndrome: the new American hemorrhagic fever. Clin. Infect. Dis. *34*, 1224–1231.

Peters, C.J., Simpson, G.L., and Levy, H. (1999). Spectrum of hantavirus infection: hemorrhagic fever with renal syndrome and hantavirus pulmonary syndrome. Annu. Rev. Med. *50*, 531–545.

Pettersson, L., Klingstrom, J., Hardestam, J., Lundkvist, A., Ahlm, C., and Evander, M. (2008). Hantavirus RNA in saliva from patients with hemorrhagic fever with renal syndrome. Emerg. Infect. Dis. *14*, 406–411.

Peyrefitte, C.N., Boubis, L., Coudrier, D., Bouloy, M., Grandadam, M., Tolou, H.J., and Plumet, S. (2008). Real-time reverse-transcription loop-mediated isothermal amplification for rapid detection of rift valley Fever virus, J. Clin. Microbiol. *46*, 3653–3659.

Plyusnin, A., and Morzunov, S.P. (2001). Virus evolution and genetic diversity of hantaviruses and their rodent hosts. Curr. Top. Microbiol. *256*, 47–75.

Plyusnin, A., Horling, J., Kanerva, M., Mustonen, J., Cheng, Y., Partanen, J., Vapalahti, O., Kukkonen, S.K.J., Niemimaa, J., Henttonen, H., *et al.* (1997). Puumala hantavirus genome in patients with nephropathia epidemica: correlation of PCR positivity with HLA haplotype and link to viral sequences in local rodents. J. Clin. Microbiol. *35*, 1090–1096.

Plyusnin, A., Mustonen, J., Asikainen, K., Plyusnina, A., Niemimaa, J., Henttonen, H., and Vaheri, A. (1999). Analysis of Puumala hantavirus genome in patients with Nephropathia epidemica and rodent carriers from the sites of infection. J. Med. Virol. *59*, 397–405.

Poljak, M., and Avsic-Zupanc, T. (1994). Immunohistochemical detection of Hantaan virus antigen in renal tissue from patient with hemorrhagic fever with renal syndrome. Nephron *67*, 252.

Putkuri, N., Vaheri, A., and Vapalahti, O. (2007). Prevalence and protein specificity of human antibodies to Inkoo virus infection, Clin. Vacc. Immunol. *14*, 1555–1562.

Rai, M.A., Khanani, M.R., Warraich, H.J., Hayat, A., and Ali, S.H. (2008). Crimean–Congo hemorrhagic fever in Pakistan. J. Med. Virol. *80*, 1004–1006.

Salonen, E.-M., Vaheri, A., Suni, J., and Wager, O. (1980). Rheumatoid factor in acute viral infections: Interference with IgM, IgG and IgA antibody determinations in enzyme-immunoassay. J. Infect. Dis. *142*, 250–255.

Schmaljohn, C.S., and Hjelle, B. (1997). Hantaviruses: A global disease problem. Emerg. Infect. Dis. 3, 95–104.

Schmaljohn C.S., and Nichol S.T. (2007). *Bunyaviridae*. In Fields Virology, Fields, B.N., Knipe, D.M., and Howley, P.M. (Lippincott-Raven, Philadelphia, PA), pp. 1741–1789.

Settergren, B., Boman, J., Linderholm, M., Wistrom, J., Hagg, E., and Arvidsson, P.A. (1992). A case of nephropathia epidemica associated with panhypopituitarism and nephrotic syndrome. Nephron *61*, 234–235.

Sexton, D.J., Rollin, P.E., Breitschwerdt, E.B., Corey, G.R., Myers, S.A., Dumais, M.R. Bowen, M.D., Goldsmith, C.S., Zaki, S.R., and Nichol, S.T. (1997). Life-threatening Cache Valley virus infection. N. Engl. J. Med. *336*, 547–549.

Sironen, T., Klingstrom, J., Vaheri, A., Andersson, L.C., Lundkvist, A., and Plyusnin, A. (2008). Pathology of Puumala hantavirus infection in macaques. PLoS ONE 3, e3035.

Sjölander, K.B., Elgh, F., Kallio-Kokko, H., Vapalahti, O., Hägglund, M., Palmcrantz, V., Juto, P., Vaheri, A., Niklasson, B., and Lundkvist, Å. (1997). Evaluation of serological methods for diagnosis of Puumala hantavirus infection (nephropathia epidemica). J. Clin. Microbiol. *35*, 3264–3268.

Soldateschi, D., Dal Maso, G.M., Valassina, M., Santini, L., Bianchi, S., and Cusi, M. G. (1999). Laboratory diagnosis of Toscana virus infection by enzyme immunoassay with recombinant viral nucleoprotein, J. Clin. Microbiol. *37*, 649–652.

Swanepoel, R., Leman, P.A., Burt, F.J., Jardine, J., Verwoerd, D.J., Capua, I., Brückner, G.K., and Burger, W.P. (1998). Experimental infection of ostriches with Crimean–Congo haemorrhagic fever virus. Epidemiol. Infect. *121*, 427–432.

Terajima, M., Hendershot, J.D., 3rd, Kariwa, H., Koster, F.T., Hjelle, B., Goade, D., DeFronzo, M.C., and Ennis, F.A. (1999). High levels of viremia in patients with the hantavirus pulmonary syndrome. J. Infect. Dis. *180*, 2030–2034.

Vaheri, A., Vapalahti, O., and Plyusnin, A. (2008). How to diagnose hantavirus infections and detect them in rodents and insectivores. Rev. Med. Virol. *18*, 277–288.

Vapalahti, O., Lundkvist, Å., Kallio-Kokko, H., Paukku, K., Julkunen, I., Lankinen, H., and Vaheri, A. (1996). Antigenic properties and diagnostic potential of Puumala virus nucleocapsid protein expressed in insect cells. J. Clin. Microbiol. *34*, 119–125.

Vapalahti, O., Lundkvist, Å., and Vaheri, A. (2001). Human immune response, host genetics, and severity of disease. In Hantaviruses, Schmaljohn, C.S., and Nichol, S.T., eds. Curr. Top. Microbiol. Immunol. *256*,153–169.

Vapalahti, O., Mustonen, J., Lundkvist, A., Henttonen, H., Plyusnin, A., and Vaheri, A. (2003). Hantavirus infections in Europe. Lancet Infect. Dis. 3, 653–661.

Vasconcelos, H.B., Azevedo, R.S.S., Casseb, S.M., Nunes-Neto, J.P., Chiang, J.O., Cantuária, P.C., Segura, M.N.O., Martins, L.C., Monteiro, H.A.O., Rodrigues, S.G., *et al.* (2009). Oropouche fever epidemic in Northern Brazil: Epidemiology and molecular characterization of isolates. J. Clin. Virol. *44*, 129–133.

Walters, L.L., Tirrell, S.J., and Shope, R.E. (1999). Seroepidemiology of California and Bunyamwera serogroup (*Bunyaviridae*) virus infections in native populations of Alaska. Am. J. Trop. Med. Hyg. *60*, 806–821

Weidmann, M., Sanchez-Seco, M.P., Sall A.A., Ly, P.O., Thiongane, Y., Lô, M.M., Schley, H., and Hufert, F.T. (2008). Rapid detection of important human pathogenic Phleboviruses. J. Clin. Virol. *41*, 138–142.

Whitehouse, C. A. (2004). Crimean–Congo hemorrhagic fever, Antivir. Res. *64,* 145–160.
Xiao, R., Yang, S., Koster, F., Ye, C., Stidley, C., and Hjelle, B. (2006). Sin Nombre viral RNA load in patients with hantavirus cardiopulmonary syndrome. J. Infect. Dis. *194,* 1403–1409.
Ye, C., Prescott, J., Nofchissey, R., Goade, D., and Hjelle, B. (2004). Neutralizing antibodies and Sin Nombre virus RNA after recovery from hantavirus cardiopulmonary syndrome. Emerg. Infect. Dis. *10,* 478–482.
Yu, X.J., Liang, M.F., Zhang, S.Y., Liu, Y., Li, J.D., Sun, Y.L., Zhang, L., Zhang, Q.F., Popov, V.L., Li, C., *et al.* (2011). Fever with thrombocytopenia associated with a novel bunyavirus in China. N. Engl. J. Med. *364,* 1523–1532 (epub ahead of print, 16 March 2011).
Zaki, S.R., Green, P.W., Coffield, L.M., Goldsmith, C.S., Nolte, K.B., Foucar, K., Feddersen, R.M., Zumwalt, R.E., Miller, G.L., Khan, A.S. *et al.* (1995). Hantavirus pulmonary syndrome. Pathogenesis of an emerging infectious disease. Am. J. Pathol. *146,* 552–579.
Zelená, H., Januska, J., and Raszka, J. (2008). Micromodification of virus-neutralisation assay with vital staining in 96-well plate and its use in diagnostics of Tahyna virus infections. Epidemiol. Mikrobiol. Imunol. *57,* 106–110.
Zeller, H.G., Diallo, M., Angel, G., Traoré-Lamizana, M., Thonnon, J., Digoutte, J.P., and Fontenille, D. (1996). [Ngari virus (*Bunyaviridae*: Bunyavirus). First isolation from humans in Senegal, new mosquito vectors, its epidemiology] [in French]. Bull. Soc. Pathol. Exot. *89,* 12–16.

Concluding Remarks

8

Alexander Plyusnin and Richard M. Elliott

Abstract

The preceding chapters in this book have revealed the breadth and diversity of viruses classified within the family *Bunyaviridae*. Although unified by common characteristics such as tripartite single stranded RNA genome, virion composed of four structural proteins and cytoplasmic site of replication, bunyaviruses show major differences in biological behaviour and replication strategy. The level of understanding of viruses in the five genera is equally diverse. The availability of reverse genetics systems for orthobunyaviruses and phleboviruses, that permit generation of viruses containing defined genetic alterations, has opened up new insights into aspects of replication, genome plasticity and interactions with the host innate immune response. On the other hand, extensive nucleotide sequence analyses of hantaviruses have provided unparalleled knowledge of their evolution and relationships. However, it is clear that much remains to be discovered about this most interesting of virus families that will keep (bunya)-virologists occupied for many years to come. Below we briefly comment on the most important venues for future studies.

Wide-scale sequence analyses

We view comprehensive wide-scale sequence determination of bunyavirus genomes as fundamental to understanding their relationships, with practical consequences such as developing robust classification and taxonomic schemes that will aid diagnostics. As mentioned in Chapter 1, more than 170 orthobunyaviruses have been described that are currently classified as about 50 species, but this demarcation is highly fluid. Only by obtaining complete nucleotide sequence information on these isolates can a more informed classification be explored. Progress in this direction would depend on improving conventional reverse transcriptase polymerase chain reaction (RT-PCR)-sequencing techniques (Lambert and Lanciotti, 2009), particularly as many of these viruses grow poorly in cell culture. This type of project suits multinational, multipartner collaboration, as exemplified by the VIZIER project, which as part of its activities successfully determined the sequences, among others, of all tick-borne flaviviruses, all *Aedes*-borne flaviviruses, all *Culex*-borne flaviviruses, and all flaviviruses with no known vector (i.e. > 40 viruses) (Coutard and Canard, 2010). With regard to the *Bunyaviridae*, encouraging examples are given here by direct recovery of four complete Puumala hantavirus genomes from a fatal case of nephropathia epidemica and rodents at the site of infection (Plyusnina *et al.*, 2011) and 33 complete genomes of RVFV cell culture isolates (Bird *et al.*, 2007). The introduction of new high throughput sequencing technologies into routine bunyavirus research can be foreseen in the near future. Existing, quite impressive, collections of cell culture isolates of orthobunyaviruses and tospoviruses and also virus-containing samples (insect, mammal and plant tissues) for other members of the family should aid in this quest.

Reverse genetic systems for hantaviruses, nairoviruses and tospoviruses

Although nowadays it is the 'must' in molecular virology of RNA viruses (Kawaoka, 2004; de la Torre, 2008; Jackson *et al.*, 2011), major breakthroughs are still eagerly awaited in order that reverse genetic systems that enable recovery of infectious virus from cloned cDNA be established for hanta-, nairo- and tospo-viruses. The reason(s) for slow or no progress in this field remains largely unknown, but surely it is not for lack of effort. The problem seems rather rooted in limited data on structure–functional organization of bunyaviral RNA-dependent RNA polymerase (the L protein, which is particularly large in tospo- and especially in nairoviruses) and also basics of transcription/replication mechanisms employed by these three bunyavirus genera. The generation of cDNA that encodes a *functional* sequence is difficult, as is developing assays that can be used to test the functionality of proteins expressed from such cDNAs. Encouraging progress has been made with a minigenome system for nairoviruses (Bergeron *et al.*, 2010), but virus rescue so far remains elusive.

Bunyavirus receptors

So far, not a single primary bunyavirus receptor has been unequivocally determined. For hantaviruses, beta-1 and beta-3 integrins were shown to be involved (Gavrilovskaya *et al.*, 1999) and also cell surface proteins of 70 kDa and 30 kDa in size have been reported as 'suspects' (Kim *et al.*, 2002; Mou *et al.*, 2006), as well as decay-accelerating factor (DAF/CD55) (Krautkramer and Zeier, 2008). Essentially no information is available for viruses in other genera. A recent paper by Lozach *et al.* (2010) reported that attachment of Uukuniemi phlebovirus to mammalian cells was specific (although no ligand was identified) and penetration occurred by acid-activated membrane fusion from late endosomal compartments. Clearly much further work is needed in this area.

Role of arthropod vector

With the exception of genus *Hantavirus*, all other bunyaviruses are transmitted by arthropods: mosquitoes, other flies, ticks, or thrips. This means that the molecular machinery of these viruses has to work efficiently in both mammalian (or plant) and arthropod (or thrips) cells. However, the outcomes of these infections are different, usually acute and cytolytic in mammalian cells but non-cytolytic leading to persistence in vector cells. Yet, almost exclusively bunyaviruses have been studied using mammals (plants) or corresponding cell cultures. In arthropod cells the virus will have to overcome host innate immunity, or at least cope with it sufficiently to allow viral replication while at the same time not killing the cell. Understanding of arthropod innate immunity with respect to viruses is a relatively young field, but one that is growing, aided enormously by the availability of vector genome sequences. While RNAi is currently regarded as the major anti-viral defence mechanism in arthropod cells, it would not be unsurprising if other defence pathways are revealed in the next few years. The other major research avenue is dissection of the molecular basis for vector competence, i.e. the ability of a particular arthropod to transmit a particular bunyavirus. Again we see great strides being made in this area over the coming years with our ability to manipulate viral genes on the one hand and knowledge of vector genomes on the other. These studies have the power to inform on novel control measures for the arthropod-transmitted bunyavirus diseases.

Persistent infections

One of the largest mysteries of virology rests within this particular field. It is a well-established dogma that a virus would never kill its natural host, or at least would not do so before it is given an opportunity to attack new victims in which it would survive till the next act of transmission. One way to achieve this goal is to establish a persistent infection in which the virus is surviving without causing serious damage to the host and constantly (or from time to time) spreading. In recent years it became more and more apparent that, in a course of long co-evolution and co-speciation with their respective natural hosts, viruses developed efficient molecular strategies helping them to evade clearance by host defence mechanisms such as, for example, innate immunity response (Garcia-Sastre, 2004). Some bunyaviruses (e.g. hantaviruses) can survive in animals in the presence of significant titres of specific antibodies, including neutralizing ones; the mechanisms behind this phenomenon, i.e. those that involve regulatory T cells, await further investigation.

References

Bergeron, E., Albarino, C.G., Khristova, M.L., and Nichol, S.T. (2010). Crimean–Congo Hemorrhagic fever virus-encoded ovarian tumor protease activity is dispensable for virus RNA polymerase function. J. Virol. *8,* 216–226.

Bird, B.H., Khristova, M.L., Rollin, P.E., Ksiazek, T.G., and Nichol, S.T. (2007). Complete genome analysis od 33 ecologically and biologically diverse Rift Valley Fever Virus strains reveals widespread virus movement and low genetic diversity due to recent common ancestry. J. Virol. *81,* 2805–2816.

Coutard, B., and Canard, B. (2010). The VIZIER project: Overview, expectations, and achievements. Antiviral Res. *87,* 85–94.

Garcia-Sastre, A. (2004). Identification and characterization of viral antagonists of type I interferon in negative-strand RNA viruses. In Biology of Negative Strand RNA Viruses: the Power of Reverse Genetics, Kawaoka Y., ed. (Springer-Verlag, Berlin-Heidelberg–New York), pp. 249–280.

Gavrilovskaya, I.N., Brown, E.J., Ginsberg, M.H., and Mackow, E.R. (1999). Cellular entry of hantaviruses which cause hemorrhagic fever with renal syndrome is mediated by beta3 integrins. J. Virol. *73,* 3951–3959.

Jackson, D., Elderfield, R.A., and Barclay, W.S. (2011). Molecular studies of influenza B virus in the reverse genetics era. J. Gen. Virol. *92,* 1–17.

Kawaoka, Y., ed. (2004). Biology of Negative Strand RNA Viruses: the Power of Reverse Genetics. (Springer-Verlag, Berlin-Heidelberg–New York).

Kim, T.Y., Choi Y., Cheong, H.S., and Choe, J. (2002). Identification of a cell surface 30 kDa protein as a candidate receptor for hantaan virus. J. Gen. Virol. *83,* 767–773.

Krautkramer, E., and Zeier, M. (2008). Hantavirus causing hemorrhagic fever with renal syndrome enters from the apical surface and requires decay-accelerating factor (DAF/CD55). J. Virol. *82,* 4257–4264.

Lambert, A.J., and Lanciotti, R.S. (2009). Consensus amplification and novel multiplex sequencing method for S segment species identification of 47 viruses of the *Orthobunyavirus, Phlebovirus,* and *Nairovirus* genera of the family *Bunyaviridae*. J. Clin. Microbiol. *47,* 2398–2404.

Lozach, P.-Y., Mancini, R., Bitto, D., Meier, R., Oestereich, L., Överby, A.K., Pettersson, R.F., and Helenius, A. (2010). Entry of Bunyaviruses into mammalian cells. Cell Host Microbe 7, 488–499.
Mou, D.L., Wang, Y.P., Huang, C.X., Li, G.Y., Pan, L., Yang, W.S., and Bai, X.F. (2006). Cellular entry of hantaan virus A9 strain: Specific interactions with beta3 integrins and a novel 70 kDa protein. Biochem. Biophys. Res. Commun. *339*, 611–617.
Plyusnina, A., Razzauti, M., Sironen, T., Henttonen, H., and Plyusnin, A. (2011). Analysis of complete Puumala hantavirus genomes recovered from a fatal case of nephropathia epidemica and bank voles (*Myodes glareolus*) from the site of infection. Ms submitted.
de la Torre, J.C. (2008). Reverse genetics approaches to combat pathogenic arenaviruses. Antiviral Res. *80*, 239–250.

Index